Advanced Research on Glucosinolates in Food Products

Advanced Research on Glucosinolates in Food Products

Editors

Franziska S. Hanschen
Sascha Rohn

MDPI • Basel • Beijing • Wuhan • Barcelona • Belgrade • Manchester • Tokyo • Cluj • Tianjin

Editors
Franziska S. Hanschen
Technische Universität Berlin
Germany

Sascha Rohn
Technische Universität Berlin
Germany

Editorial Office
MDPI
St. Alban-Anlage 66
4052 Basel, Switzerland

This is a reprint of articles from the Special Issue published online in the open access journal *Foods* (ISSN 2304-8158) (available at: https://www.mdpi.com/journal/foods/special_issues/Glucosinolates_Food_Products).

For citation purposes, cite each article independently as indicated on the article page online and as indicated below:

LastName, A.A.; LastName, B.B.; LastName, C.C. Article Title. *Journal Name* **Year**, *Volume Number*, Page Range.

ISBN 978-3-0365-2977-6 (Hbk)
ISBN 978-3-0365-2976-9 (PDF)

Contents

About the Editors

Franziska S. Hanschen, born in 1985, is a research group leader at the Leibniz Institute for Vegetable and Ornamental Crops (IGZ) in Grossbeeren, Germany. After her studies on food chemistry, she obtained her PhD in 2012 at the Technische Universität Berlin, Germany, then went to IGZ to continue her research on glucosinolate breakdown pathways as a Post-Doc. Here, she focuses on the enzymatic well as chemical degradation pathways of glucosinolates and unravels the underlying mechanisms. Since 2018, she leads the Leibniz Junior Research Group of Optimization of Glucosinolate Degradation Pathways (OPTIGLUP). Her research was published in 50 articles so far (h-index is 21 and more than 1,100 citations). In 2020, she did her habilitation at the University of Hamburg in the subject of food chemistry and was awarded the "Werner-Baltes-Preis des Jungen Wissenschaftlers" by the German Food Chemical Society (LChG), section of the German Chemical Society (GDCh). Since 2019, she is deputy chairwoman of the Northeastern branch of the LChG.

Sascha Rohn, born 1973, is a full professor for Food Chemistry at the Technische Universität Berlin (Germany). He graduated from the University of Frankfurt/Main, Germany, with the first and second state examination in Food Chemistry, 1999. In 2002, he obtained his Ph.D. in Food Chemistry from the Institute of Nutritional Science, University of Potsdam, Germany, working on interactions of polyphenols with food proteins. After two years as a postdoc, he left Potsdam towards Berlin, where he did a habilitation at the Institute of Food Technology and Food Chemistry of the Technische Universität Berlin. From October 2009 to October 2020, he was a full professor at the Hamburg School of Food Science, Institute of Food Chemistry, University of Hamburg, Germany. His group is dealing with the analysis of bioactive food compounds. Especially, they are characterizing the reactivity and stability of bioactive compounds. The aim is to identify degradation products that serve as quality parameters, as process markers during food/feed processing, or as biomarkers in nutritional physiology. Results of their work have been presented in more than 200 publications so far (Scopus h-index is 50 and more than 8,000 citations). More than 30 well-known scientific journals ask Prof. Rohn regularly to review scientific manuscripts. From 2006 to 2012, he was chairman of the Northeastern branch of the German Food Chemical Society (LChG). From 2014 to 2018, he was a member of the steering committee of the German Nutrition Society (DGE). In April 2015, he became the director of the Institute for Food and Environmental Research (ILU e.V.) in Bad Belzig, Germany. As a non-profit organization, this institute conducts applied, technologically-oriented research and development for the food industry with regard to preserve the environment. The organization pursues objectives that are exclusively and directly exploitable for everyday use.

foods

MDPI

Editorial

Advanced Research on Glucosinolates in Food Products

Franziska S. Hanschen [1,*] and Sascha Rohn [2]

[1] Leibniz Institute of Vegetable and Ornamental Crops (IGZ) e.V., Theodor-Echtermeyer-Weg 1, D-14979 Grossbeeren, Germany
[2] Institute of Food Technology and Food Chemistry, Technische Universität Berlin, Gustav-Meyer-Allee 25, D-13355 Berlin, Germany; rohn@tu-berlin.de
* Correspondence: Hanschen@igzev.de

Glucosinolate-containing foods, such as vegetables from the plant order Brassicales and its derivative products, are valued for their health-beneficial properties. The latter are linked to glucosinolate hydrolysis products, such as isothiocyanates [1].

The Special Issue "Advanced Research on Glucosinolates in Food Products" aimed to collect the latest research on the impact of the whole food supply chain, including production, as well as domestic food preparation, on glucosinolates and the formation and chemistry of their breakdown products in vegetables and further foods. In this context, the consequences for human health are important, too. This Special Issue includes eleven research articles that cover research on the effect of pre-harvest factors on glucosinolates, their hydrolyzing enzymes, and the formation of volatile hydrolysis products [2–5]. Further topics include the linkage between glucosinolates and sensory aspects [2,6,7], and the effects of food preparation and follow-up reactivity [7–10]. Finally, two articles focus on the bioavailability and functional effects of isothiocyanates for human health [1,11].

Given that they are evolutionary plant defense compounds, glucosinolates and their hydrolysis in Brassicaceae vegetables are strongly affected by plant genotype and ecophysiological parameters. One of the studies described by Oloyede et al. revealed that cabbage grown in the field at lower temperatures exhibited higher myrosinase activity compared to greenhouse-grown plants, but the effect on glucosinolates and their hydrolysis products differed between cabbage morphotypes [5]. In another study, Chowdhury et al. showed that the growth conditions in a plant factory that are optimal for the plant growth of kale can differ from the optimal conditions for higher glucosinolate contents, as glucosinolate levels decrease as temperatures and humidity levels increase, though they are positively affected by increased CO_2 levels [3]. In contrast, Bell et al. revealed a positive correlation between the presence of glucosinolates and the average growth temperature of rocket (*Diplotaxis tenuifolia* and *Eruca sativa*), demonstrating this with multiple leaf cuts, showing how the process affected the perception of pepperiness, bitterness, and hotness, which in turn reduced consumer acceptance [2]. These findings pose a challenge for the aim to increase health-promoting isothiocyanate levels, as this could reduce consumer acceptance and thereby the consumption of these products. Growth conditions also seem to affect glucosinolate hydrolysis outcomes, as in a study with commercial white and red cabbages purchased over a 3-month period. Red cabbages especially showed higher isothiocyanate release in late summer compared to cabbages purchased in late autumn, with the latter also being more intense producers of nitriles and epithionitriles [4].

The processing of these vegetables during food preparation can considerably affect glucosinolates, their product formation, and also the flavor of the products. Thermal treatments inactivate myrosinase, thereby stopping enzymatic glucosinolate hydrolysis, resulting in diminished isothiocyanate formation [10]. Stir-frying maintains myrosinase activity and, consequently, reduced glucosinolate contents were observed in stir-fried foods. In contrast, steaming retains glucosinolate levels the best and often favors a forced isothiocyanate formation [10]. However, a high content of glucosinolates in boiled vegetables can

Citation: Hanschen, F.S.; Rohn, S. Advanced Research on Glucosinolates in Food Products. *Foods* **2021**, *10*, 3148. https://doi.org/10.3390/foods10123148

Received: 8 December 2021
Accepted: 15 December 2021
Published: 20 December 2021

affect consumer acceptance as well, because intact glucosinolates were positively correlated with a bitter taste [7]. In contrast, isothiocyanates are very reactive compounds and undergo a transformation with nucleophiles, thereby reducing their contents in a food matrix and forming reaction products. Krell et al. investigated the fate of glucosinolates during the baking of bread enriched with nasturtium (*Tropaeolum majus* L.). Next to the thermal degradation of glucosinolates to nitriles, Krell et al. showed that isothiocyanates react with the lysine residues of the wheat proteins in the bread crumb [9]. Another example of follow-up reactivity is the isothiocyanate 4-methylthio-3-butenyl isothiocyanate being known as a precursor to the yellow pigment 2-[3-(2-thioxopyrrolidin-3-ylidene)methyl]-tryptophan, present in the traditional Japanese salted radish root 'takuan-zuke'. Kobayashi et al. studied the formation mechanism and found that the yellowing of takuan-zuke is accelerated at pH values below 5 and that the color of air-dried takuan-zuke was deeper than that of the salt-pressed product. Moreover, the findings in the study described by Kobayashi et al. led to the assumption that a so far unknown pigment is responsible for the color, as 2-[3-(2-thioxopyrrolidin-3-ylidene)methyl]-tryptophan levels were too low to account for the color alone [8].

When eating Brassicales plants or products thereof, glucosinolates, and especially their isothiocyanates, are valued for their positive effects on human health. However, conversion to isothiocyanates and bioavailability determines the effects of these compounds tremendously. Shekarri and Dekker described a physiologically based model to simulate the bioavailability and absorption kinetics of sulforaphane (4-(methylsulfinyl)butyl isothiocyanate, which is found in broccoli and also in red cabbage). This model represents a preliminary step in enabling the prediction of the biological effect of isothiocyanates and can be used in the growing field of personalized nutrition [11]. Novel insights into the immunomodulating effects of watercress (*Nasturtium officinale*) isothiocyanates on exercise-induced inflammation are presented in a human intervention study described by Schulze et al. [1] After the consumption of fresh watercress (a source of 2-phenylethyl isothiocyanate), a mild pro-inflammatory reaction was observed, though the immune response was more pronounced for both pro-inflammatory and anti-inflammatory markers after an exercise unit compared to a control meal. However, during the recovery phase, watercress consumption led to a stronger anti-inflammatory downregulation of the pro-inflammatory cytokines IL-6 and TNF-α. Thus, the study came to the conclusion that fresh watercress causes a stronger pro-inflammatory response and anti-inflammatory counter-regulation during and after exercise [1].

Due to the inclusion of well-designed studies and well-written articles, this Special Issue offers a valuable contribution to the ambition of producing vegetables rich in valuable glucosinolates as a source of health-promoting isothiocyanates. However, the research also reveals that this ambition might negatively affect consumer liking and acceptance, and therefore this needs to be considered in the future as well. New insights into the effect of food processing on glucosinolates and the formation and fate of their breakdown in follow-up products are presented. These processes are still far away from being fully understood. Evaluating the role of glucosinolates in supporting human health still remains challenging.

Author Contributions: Conceptualization, F.S.H. and S.R.; writing—original draft preparation, F.S.H. and S.R.; writing—review and editing, F.S.H. and S.R.; funding acquisition, F.S.H. All authors have read and agreed to the published version of the manuscript.

Funding: Franziska S. Hanschen is funded by the Leibniz-Association (Leibniz-Junior Research Group OPTIGLUP; J16/2017).

Conflicts of Interest: The authors declare no conflict of interest.

References

1. Schulze, H.; Hornbacher, J.; Wasserfurth, P.; Reichel, T.; Günther, T.; Krings, U.; Krüger, K.; Hahn, A.; Papenbrock, J.; Schuchardt, J.P. Immunomodulating Effect of the Consumption of Watercress (*Nasturtium officinale*) on Exercise-Induced Inflammation in Humans. *Foods* **2021**, *10*, 1774. [CrossRef] [PubMed]
2. Bell, L.; Lignou, S.; Wagstaff, C. High Glucosinolate Content in Rocket Leaves (*Diplotaxis tenuifolia* and *Eruca sativa*) after Multiple Harvests Is Associated with Increased Bitterness, Pungency, and Reduced Consumer Liking. *Foods* **2020**, *9*, 1799. [CrossRef] [PubMed]
3. Chowdhury, M.; Kiraga, S.; Islam, M.N.; Ali, M.; Reza, M.N.; Lee, W.-H.; Chung, S.-O. Effects of Temperature, Relative Humidity, and Carbon Dioxide Concentration on Growth and Glucosinolate Content of Kale Grown in a Plant Factory. *Foods* **2021**, *10*, 1524. [CrossRef] [PubMed]
4. Wermter, N.S.; Rohn, S.; Hanschen, F.S. Seasonal Variation of Glucosinolate Hydrolysis Products in Commercial White and Red Cabbages (*Brassica oleracea* var. *capitata*). *Foods* **2020**, *9*, 1682. [CrossRef] [PubMed]
5. Oloyede, O.O.; Wagstaff, C.; Methven, L. Influence of Cabbage (*Brassica oleracea*) Accession and Growing Conditions on Myrosinase Activity, Glucosinolates and Their Hydrolysis Products. *Foods* **2021**, *10*, 2903. [CrossRef]
6. Bell, L.; Kitsopanou, E.; Oloyede, O.O.; Lignou, S. Important Odorants of Four Brassicaceae Species, and Discrepancies between Glucosinolate Profiles and Observed Hydrolysis Products. *Foods* **2021**, *10*, 1055. [CrossRef] [PubMed]
7. Nor, N.D.M.; Lignou, S.; Bell, L.; Houston-Price, C.; Harvey, K.; Methven, L. The Relationship between Glucosinolates and the Sensory Characteristics of Steamed-Pureed Turnip (*Brassica Rapa* subsp. *Rapa* L.). *Foods* **2020**, *9*, 1719. [CrossRef] [PubMed]
8. Kobayashi, T.; Kumakura, K.; Takahashi, A.; Matsuoka, H. Low pH Enhances the Glucosinolate-Mediated Yellowing of Takuanzuke under Low Salt Conditions. *Foods* **2020**, *9*, 1524. [CrossRef] [PubMed]
9. Krell, M.; Cvancar, L.; Poloczek, M.; Hanschen, F.S.; Rohn, S. Determination of Isothiocyanate-Protein Conjugates in a Vegetable-Enriched Bread. *Foods* **2021**, *10*, 1300. [CrossRef] [PubMed]
10. Oloyede, O.O.; Wagstaff, C.; Methven, L. The Impact of Domestic Cooking Methods on Myrosinase Stability, Glucosinolates and Their Hydrolysis Products in Different Cabbage (*Brassica oleracea*) Accessions. *Foods* **2021**, *10*, 2908. [CrossRef]
11. Shekarri, Q.; Dekker, M. A Physiological-Based Model for Simulating the Bioavailability and Kinetics of Sulforaphane from Broccoli Products. *Foods* **2021**, *10*, 2761. [CrossRef] [PubMed]

foods

MDPI

Article

Effects of Temperature, Relative Humidity, and Carbon Dioxide Concentration on Growth and Glucosinolate Content of Kale Grown in a Plant Factory

Milon Chowdhury [1,2], Shafik Kiraga [2], Md Nafiul Islam [1,2], Mohammod Ali [1], Md Nasim Reza [1,2], Wang-Hee Lee [1,2] and Sun-Ok Chung [1,2,*]

[1] Department of Agricultural Machinery Engineering, Graduate School, Chungnam National University, Daejeon 34134, Korea; chowdhurym90@cnu.ac.kr (M.C.); nafiulislam@cnu.ac.kr (M.N.I.); sdali77@o.cnu.ac.kr (M.A.); reza5575@cnu.ac.kr (M.N.R.); wanghee@cnu.ac.kr (W.-H.L.)

[2] Department of Smart Agricultural Systems, Graduate School, Chungnam National University, Daejeon 34134, Korea; kiragashafik@o.cnu.ac.kr

* Correspondence: sochung@cnu.ac.kr; Tel.: +82-42-821-6712

Citation: Chowdhury, M.; Kiraga, S.; Islam, M.N.; Ali, M.; Reza, M.N.; Lee, W.-H.; Chung, S.-O. Effects of Temperature, Relative Humidity, and Carbon Dioxide Concentration on Growth and Glucosinolate Content of Kale Grown in a Plant Factory. *Foods* **2021**, *10*, 1524. https://doi.org/10.3390/foods10071524

Academic Editors: Franziska S. Hanschen and Sascha Rohn

Received: 16 May 2021
Accepted: 29 June 2021
Published: 1 July 2021

Abstract: The growth of plants and their glucosinolate content largely depend on the cultivation environment; however, there are limited reports on the optimization of ambient environmental factors for kale grown in plant factories. This study was conducted to investigate the effects of temperature, relative humidity, and the carbon dioxide (CO_2) concentration on kale growth and glucosinolate content in different growth stages of cultivation in a plant factory. Kale was grown under different temperatures (14, 17, 20, 23, and 26 °C), relative humidities (45, 55, 65, 75, and 85%), and CO_2 concentrations (400, 700, 1000, 1300, and 1600 ppm) in a plant factory. Two and four weeks after transplantation, leaf samples were collected to evaluate the physical growth and glucosinolate contents. The statistical significance of the treatment effects was determined by two-way analysis of variance, and Duncan's multiple range test was used to compare the means. A correlation matrix was constructed to show possible linear trends among the dependent variables. The observed optimal temperature, relative humidity, and CO_2 range for growth (20–23 °C, 85%, and 700–1000 ppm) and total glucosinolate content (14–17 °C, 55–75%, and 1300–1600 ppm) were different. Furthermore, the glucosinolate content in kale decreased with the increase of temperature and relative humidity levels, and increased with the increase of CO_2 concentration. Most of the physical growth variables showed strong positive correlations with each other but negative correlations with glucosinolate components. The findings of this study could be used by growers to maintain optimum environmental conditions for the better growth and production of glucosinolate-rich kale leaves in protected cultivation facilities.

Keywords: *Brassica*; plant growth; glucosinolates; protected horticulture; environmental conditions

1. Introduction

Kale (*Brassica oleracea* var. *alboglabra Bailey*) is a salad species that is one of the most versatile and commercially valuable vegetables due to its short growth period, various uses, and desirable metabolic and nutritional profiles [1–3]. This crisp and hearty vegetable is often consumed raw in salads and smoothies but can also be consumed in steamed, sautéed, or cooked states. Kale originates from China and has since gained particular attention in other countries due to its constituent cancer-preventive and human-health-promoting phytochemicals (i.e., glucosinolates, carotenoids, phenols, and vitamins) [4–6]. Glucosinolates are amino-acid-derived, active secondary metabolites that mainly contain sulfur- and nitrogen-related compounds (i.e., β-D-thioglucose, tryptophan, phenylalanine, sulfonated oxime moiety). They can be classified into aliphatic, aromatic, and indole groups [7], where each group consists of several chemical constituents. Progoitrin, sinigrin, glucoraphanin, and gluconapin are the major constituents of the aliphatic group. Similarly, 4-hydroxyglucobrassicin, glucobrassicin, 4-methoxyglucobrassicin and neoglucobrassicin, and gluconasturtiin are the major indole and aromatic glucosinolate constituents, respectively [1]. Glucosinolates are composed of relatively few amino acids and chain-elongated homologs through an independent metabolic pathway (Figure 1) and are available in all parts of almost all varieties of plants of the Brassicales order; however, the content is higher in the reproductive tissues (i.e., flowers and seeds) than in vegetative tissues [8]. The breakdown products of glucosinolates have a significant amount of anticarcinogenic activity for decreasing the risk of developing lung, stomach, colon, and rectum cancers; helping to maintain low blood pressure and reducing the risk of developing type 2 diabetes [6,9,10].

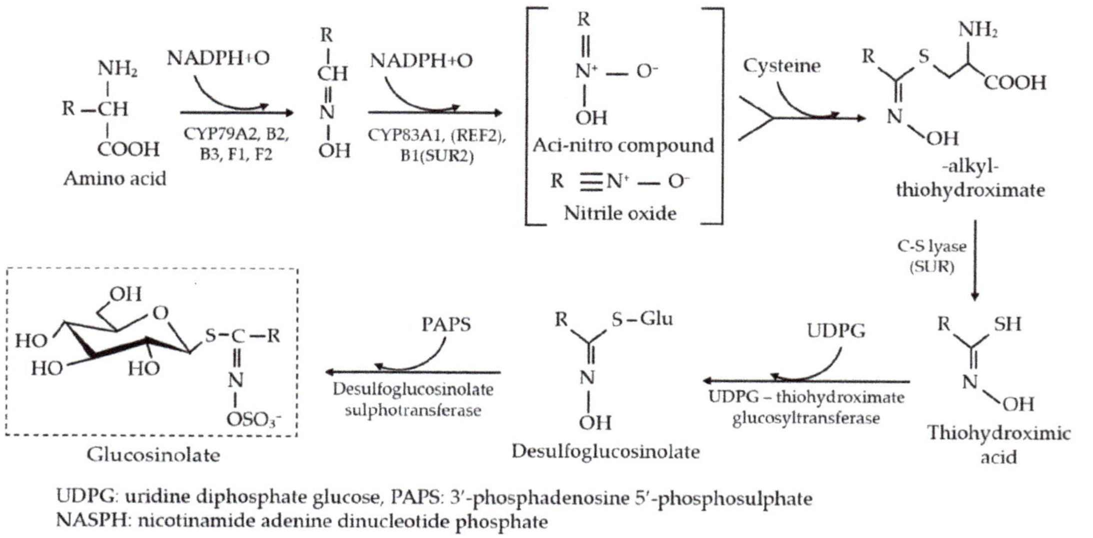

Figure 1. Synthesis of glucosinolates in Brassicaceous plants [11].

Kale growth and the formation of glucosinolates depend on crop genetic factors, tissue type, crop health, agronomic factors (i.e., water supply and fertigation), cultivation facilities (i.e., plant factory, greenhouse, and open field), and environmental factors such as temperature, relative humidity, carbon dioxide (CO_2), light type, intensity, photoperiod, and cultivation methods [12–14]. The physical development stage is also a major determinant of the glucosinolates composition in kale [15]. Although kale can be easily cultivated in open fields using traditional methods, the quality and quantity of the growth and glucosinolate content cannot be ensured, as they are extremely sensitive to climatic and field conditions [16]. In recent years, farmers have produced kale in protected cultivation facilities, such as plant factories and greenhouses, due to the possibility of adjusting the growth environment and achieving fast and sustainable growth rates, functional component-rich and high-quality yield, lower rates of disease and pest infestation, and lower labor costs in addition to the possibility of year-round production with minimum influence from geological and climatic conditions [17–19]. Moreover, hydroponic cultivation systems with ion-specific (ISE-sensor-based) nutrient management could enhance the growth and nutritional profile of kale by 15% to 60% [20–25]. However, major environmental factors (i.e., temperature, relative humidity, and CO_2) have to be specifically optimized according to crop to ensure sustainable kale growth and glucosinolate formation.

The physical growth of kale can be easily determined by measuring its physical properties such as plant length, width, weight, number of leaves, and stem diameter, whereas the glucosinolate content needs to be identified by laboratory analysis. The deposition of glucosinolates in growing plants and their distribution to plant organs are significantly affected by environmental factors [26], with temperature being one of the key factors. Several studies have been conducted to determine the process and effects of temperature on seed germination, physical development, flower formation, and yield [27–30]. However, physiological processes and their integration are sped up under higher temperatures with both positive and negative effects. For example, high temperatures promote faster growth and greater fruit production of plants, especially in cereal crops, but they also remove functional components from leaves through high transpiration rates [31]. Generally, elevated temperatures affect the structural components of chloroplasts significantly, causing effects such as variation in thylakoids, granum stacking, and swelling with photosystem II reduction, resulting in disruption to the cellular cytoplasm, cell breakdown and, ultimately, cell death. In addition, rising temperatures interrupt protein mechanisms, RNA synthesis, enzymatic interactions, and cell function. As a result, these imbalances and abnormal cell functioning affect the growth and accumulation of glucosinolate synthesis [32,33].

The relative humidity of the ambient environment also directly affects plant growth by resisting water and nutrient consumption. During transpiration, the relative humidity level becomes saturated. As a result, plants halt transpiration and nutrient uptake from the soil or growing media at high relative humidity levels where there is a lack of air circulation, resulting in gradual rotting in cases of long-term humidity saturation [34–36]. The maintenance of optimum relative humidity is essential for better growth and glucosinolate accumulation. Several researchers have reported that the photosynthesis rate is proportional to the relative humidity level as a higher range of relative humidity lowers water stress in the leaves and increases stomatal conductance. Although higher relative humidity increases the nutrient concentration, the nutrient solution supply and plant transpiration rate need to be monitored carefully [34,37].

The CO_2 concentration influences the photosynthetic rate, metabolism, and physiological and chemical defense of plants [13,38]. A lack of CO_2 would not only result in a lower biomass but the plants would also be of inferior quality and strength. As an essential substrate of the photosynthesis process, CO_2 is directly absorbed by plants. CO_2 also influences the transpiration process of plants. A meta-analysis was conducted, and it was reported that elevated CO_2 could reduce transpiration by up to 22% in different plant species [39]. CO_2 also preserves the essential nutrient components along with water by reducing the transpiration rate [40,41]. La et al. [38] investigated the effects of CO_2 elevation at different nitrogen levels on the growth and glucosinolate content of Chinese kale and reported that all physical growth variables significantly increased with the elevation of CO_2 at each nitrogen level; however, total glucosinolate content was only increased under low nitrogen level and elevated CO_2 concentration.

The temperature, relative humidity, and CO_2 concentration are the basic environmental factors that affect kale growth and, especially, glucosinolate formation. As they are interrelated, these factors should not be studied in isolation. The proper combination of these factors needs to be specifically confirmed for each crop to ensure optimal growth, a favorable nutritional profile, and identification of the ideal harvesting time. To date, very few studies have investigated the effects of these environmental factors on kale, especially when grown in plant factories using hydroponic cultivation methods. Therefore, the objective of this study was to investigate the effects of temperature, relative humidity, and CO_2 on the growth and glucosinolate content at different stages of kale growth based on cultivation in a plant factory.

2. Materials and Methods

2.1. Plant Factory and Seedling Preparation

Plant factories are fully-closed crop cultivation systems that are fitted with artificial lights and used to grow high-value vegetables and medicinal plants throughout the year by utilizing artificially controlled ambient environmental factors [17,19]. In this study, five small chambers were prepared, as shown in Figure 2, to implement five different treatment conditions with varied temperature, relative humidity, and CO_2 concentrations. The targeted environmental factors (i.e., temperature, relative humidity, and CO_2) and other environmental factors (i.e., light sources, light intensity, photoperiod, and nutrient solution (EC and pH)) were maintained according to the experimental plan (Table 1). A wireless sensor network (XBee-Pro, Digi, Hopkins, MN, USA) was used to monitor the ambient environmental parameters and control the relevant actuators, as detailed by Chung et al. [42]. Three plant beds were placed vertically in each cultivation chamber, and a nutrient solution tank was kept at the bottom (floor). Each plant bed had 24 planting positions and 6 mist spray nozzles for spraying the nutrient solution onto plant roots as a fine mist for a duration of 2 min at 13-min intervals. Commercial nutrient solutions A and B (Daeyu Co., Ltd., Seoul, Korea) were used, and the target nutrient level was monitored and managed once a day using EC and pH sensors.

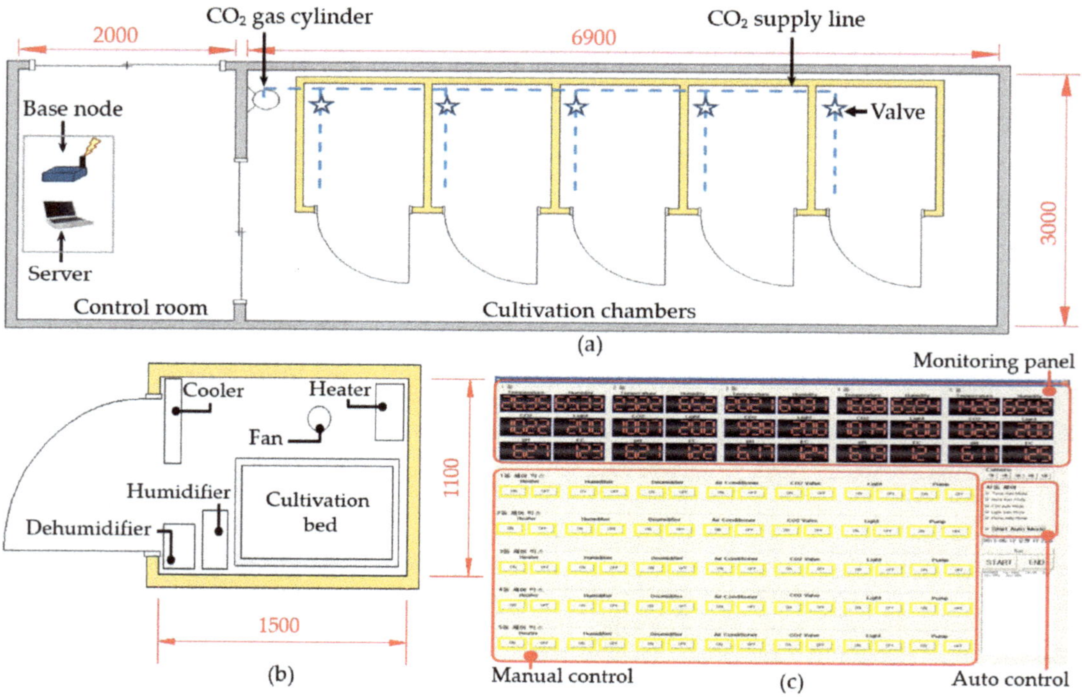

Figure 2. (**a**) Layout of the plant factory (control room and cultivation chambers); (**b**) fabricated individual chamber; and (**c**) ambient environment monitoring and control system. All dimensions are presented in millimeters (mm).

A commercial kale variety with green smooth leaves and a hard and fibered stem was cultivated in the experiments using a recycle-type aeroponic nutrient management system. Kale seeds were sown in a hydroponic germination sponge, covered with wet paper (until germination), kept in the plant factory under a controlled environment for germination, and grown until transplantation (Figure 3). The maintained temperature, relative humidity, CO_2 concentration, light type, and photoperiod were 25 ± 3 °C, 65 ± 5%, 1000 ± 100 ppm, fluorescent, and 16/8 (day/night hours), respectively. Nutrient-rich water was provided into the root zone, and the EC and pH of the nutrient solution were 1.2 ± 1.00 (dS m^{-1}) and 6.5 ± 0.5, respectively. After three weeks of germination, healthy seedlings with true leaves were transplanted into the plant bed with the sponge.

Table 1. Different treatments of temperature, relative humidity, and CO_2 during the kale cultivation in the plant factory.

Environmental Variables	Targeted Levels			Monitored Levels	Used Sensor
	Experiment 1 (Temp.)	Experiment 2 (Humi.)	Experiment 3 (CO_2)		
Temperature (°C)	14 ± 1 17 ± 1 20 ± 1 23 ± 1 26 ± 1	20 ± 1	20 ± 1	14.58 ± 0.74 17.34 ± 1.80 20.25 ± 0.69 23.26 ± 0.52 25.97 ± 1.64	ETH-01DV, ECONARAE, Seoul, Korea
Relative humidity (%)	65 ± 5	45 ± 5 55 ± 5 65 ± 5 75 ± 5 85 ± 5	65 ± 5	44.78 ± 5.23 56.06 ± 4.35 67.66 ± 4.67 76.85 ± 4.49 82.66 ± 5.65	ETH-01DV, ECONARAE, Seoul, Korea
CO_2 (ppm)	1000 ± 100	1000 ± 100	400 ± 100 700 ± 100 1000 ± 100 1300 ± 100 1600 ± 100	475.62 ± 106.30 723.29 ± 140.60 980.75 ± 125.36 1318.34 ± 125.11 1672.30 ± 93.21	SH-300-DS, SOHA TECH CO. Ltd., Seoul, Korea
Light source (LED color ratio)	R:B = 11:7			-	-
Light intensity (μmol m^{-2} s^{-1})	160			160 ± 25	GY-30, ROHM Co. Ltd., Kyoto, Japan
Photoperiod (day/night hrs)	16/8			-	MaxiRex 5QT, Legrand Korea Co., Ltd., Seoul, Korea
pH	6.50 ± 0.5			6.55 ± 0.52	PH-BTA, Vernier, OR, USA
EC (dS m^{-1})	1.2 ± 1.00			1.28 ± 0.29	CON-BTA, Vernier, OR, USA

Figure 3. Preparation of kale seedlings for transplantation: (**a**) kale seeds were sown and covered; (**b**) germinated seeds; and (**c**) two-week-old seedlings under controlled environment conditions.

2.2. *Experimental and Analytical Procedures*

2.2.1. Experimental Design

Different separate experiments were conducted to investigate the influences of temperature, relative humidity, and CO_2 on kale growth and glucosinolate content. Five treatments with various environmental factors were applied in each experiment. For example, temperatures of 14, 17, 20, 23, and 26 °C were varied while other factors were kept constant. Similarly, five relative humidity levels and CO_2 concentrations were implemented in experiments 2 and 3 to evaluate the effects of relative humidity and CO_2, respectively. The targeted and monitored levels of temperature, relative humidity, and CO_2 along with other growth factors are summarized in Table 1. The light source and ratio, intensity, photoperiod, and pH and EC levels were selected following the findings of Zhang et al. [43], Lefsrud et al. [44], Naznin et al. [45], and Jones [24], respectively.

2.2.2. Sample Collection and Data Acquisition

Two and four weeks after transplantation (Figure 4), sample collection was performed in two steps. First, mature and healthy plants were visually selected and collected from the plant beds for physical growth evaluation. Three plants from each bed and nine plants from three beds (replicates) of each cultivation chamber were collected randomly among 72 plants (24 plants/bed × 3 plant beds). To analyze the glucosinolate content, three normal-sized, mature, healthy leaves were harvested from each plant bed (one leaf from each collected plant), and a total of nine leaves were collected from three plant beds (as a replication) from each cultivation chamber. The measured values for each growth parameter and the glucosinolate content were averaged to represent one data point. As a result, nine data points for growth parameters and one data point for the glucosinolate content were recorded from each plant bed. In total, 270 data points were collected for growth evaluation (9 data points/bed × 2 sampling times × 3 replications × 5 treatments) and 30 data points were collected to assess the glucosinolate content (1 data point/bed × 2 sampling times × 3 replications × 5 treatments) for each experiment. The physical growth variables, namely, the plant height, width, weight, number of leaves, stem diameter, chlorophyll level, leaf length, width, and weight were measured, and the leaves were transferred to the chemical laboratory immediately (to minimize the degradation) for glucosinolate analysis using a commercial high-performance liquid chromatography (HPLC) machine (model: 1200 series, Agilent Technologies, Santa Clara, CA, USA). The chlorophyll concentration was also measured using a commercial device (model: SPAD 502DL, Spectrum Technology Inc., Aurora, IL, USA).

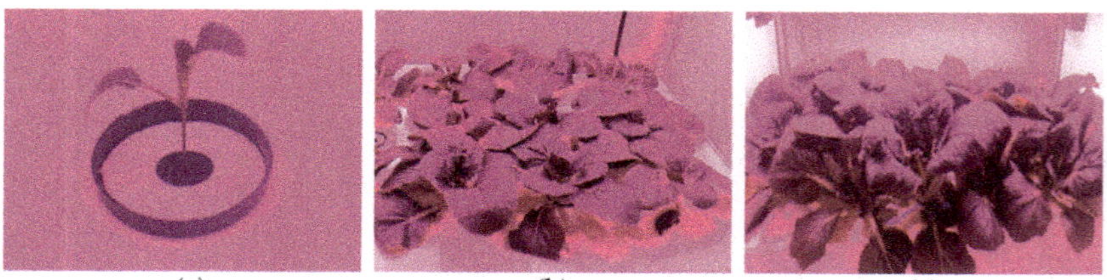

(a) (b) (c)

Figure 4. Growth condition of kale after different periods of transplantation: (**a**) transplantation day; (**b**) 2 weeks after transplantation; and (**c**) 4 weeks after transplantation.

2.2.3. Estimation of Glucosinolate Content

The glucosinolates of the freshly harvested kale leaves were extracted and analyzed as described by Doheny-Adams et al. [46]. The whole process was conducted according to the ISO 9167:2019 [47], and the process is divided into four major steps: (a) tissue disruption, (b) extraction in methanol, (c) purification and desulfation, and (d) separation and identification of glucosinolates by HPLC analysis (1200 series, Agilent Technologies, Santa Clara, CA, USA). The collected leaf samples were stored in an airtight box, taken to the chemical laboratory immediately to freeze in liquid nitrogen, and stored at −80 °C for 48 h to reduce the activity of myrosinase. For freeze-drying, samples were lightly wrapped with aluminum foil and transported on dry ice to load into the freeze drier (Lyotrap, LTE scientific Ltd., Oldham, UK) within 30 s. The freeze-dried leaf samples were ground to make a homogenized fine powder using a grinder (EK2311, Salter, Tonbridge, UK). Then, 100 mg of the freeze-dried samples was preheated for 3 min at 75 °C and 4.5 mL of preheated 70% methanol at 75 °C was added. The sample was incubated for 10 min at 75 °C (with manual shaking every 2 min) and then centrifuged by a rotor at 4000 rpm (B 3.11, Jouan, Nantes, France) for 10 min. In the purification step, 25 mg of sulfatase and 1ml of 40% ethanol were mixed and centrifuged for 1 min at 8000 rpm. The supernatant was shifted to a new Eppendorf tube and 1 mL of pure ethanol was injected for precipitating the sulfatase before the second centrifugation. Finally, the sulfatase pellet was air dried after separating from the supernatant and diluted in 2 mL of water. For desulfation, 0.5 cc of Sephadex slurry was used to prepare the columns and 2 mL of imizadole formate (6 M) was added on each for activation. The columns were cleaned twice with 1 mL of water each time. The columns were washed again using 1 mL of 20 mM sodium acetate, and 75 µL of purified sulfatase (0.05–0.3 U/mL) was injected. After that, columns were incubated for 24 h at 28 °C before desulfoglucosinolates were eluted with two 1 mL volumes of water. After 24 h of incubation, elution of desulfoglucosinolates was performed thrice using 1.5 mL of distilled water and filtered through 0.45-µm polytetrafluoroethylene (PTFE) syringe filters (Millipore, Bedford, MA, USA) into an HPLC vial. A reverse phase C18 column (150 × 3.0 mm, 3 µm, Inertsil ODS-3, GL Sciences, Tokyo, Japan) was used, which was equilibrated for 30 min using ultrapure water (solvent A) and 100% acetonitrile (solvent B) with detection at 227 nm. The flow rate was 0.4 mL min^{-1}, and separation was performed according to the default program. As an external standard, sinigrin (0.1 mg/mL; Sigma, St. Louis, MO, USA) was utilized. The identification and quantification of individual glucosinolate components was performed by comparing the sinigrin retention time and using their HPLC areas and response factor, respectively. In this study, the obtained retention time for progoitrin, sinigrin, glucobrassicin, 4-methoxyglucobrassicin, and neoglucobrassicin were 5.97, 7.13, 21.93, 24.68, and 30.37 min, respectively.

2.2.4. Statistical Analysis

All the presented physical growth parameters and glucosinolate content values are the means of independent measurements for different treatments of each environmental factor. The significance of differences between mean values was determined by two-way analysis of variance (ANOVA). Data were analyzed considering 95% confidence levels and two-sided confidence intervals. Duncan's multiple range test was used to simultaneously compare means (SAS Institute Inc, Campus drive Cary, NC, USA). A correlation matrix recording correlation coefficients was created to show the inter-relationships between variables.

3. Results

3.1. ANOVA of the Environmental Factors

The effects of ambient environmental factors (temperature, relative humidity, and CO_2) on plant physical growth variables and total glucosinolate content were analyzed using two-way ANOVA analysis. Five different treatment conditions for each environmental factor and two sampling times were considered when conducting the ANOVA analysis for each growth variable and the glucosinolate content. The results of the two-way ANOVA analysis for the plant height, width, weight, and total glucosinolate content are shown in Table 2 out of nine physical variables and five identified glucosinolate components. The F-values of the treatments and sampling times were higher than the F crit values, except for some growth and glucosinolate variables under the CO_2 treatments, which confirms the adequacy of the hypothesis. This ANOVA analysis indicates that the treatments and sampling times had significant impacts ($p < 0.05$) on the growth and glucosinolate content (except for some CO_2 treatments). However, some P-values under the CO_2 treatments were higher than 0.05, which also indicates that those growth or glucosinolate variables were not notably affected by the CO_2 treatments. The overall results show that a single unit change of each environmental factor will affect the plant growth and glucosinolate content.

Table 2. Two-way ANOVA test showing the individual effects of the treatments (Tr) and sampling times (ST) on growth variables and total glucosinolate content of kale.

SV	Plant Height			Plant Width			Plant Weight			Total Glucosinolates		
	Tr	ST	Err	Tr	ST	Err	Tr	ST	Err	Tr	ST	Err
Temperature effect												
SS	4.1×10^4	2.3×10^5	6.6×10^3	1.4×10^4	4.3×10^4	3.2×10^3	206.9	1.03×10^3	298.2	8.7×10^3	3.7×10^3	3.2×10^3
df	4	1	20	4	1	20	4	1	20	4	1	20
MS	1.0×10^4	2.3×10^5	334.1	3.5×10^3	4.3×10^4	162.1	51.71	1.3×10^3	14.91	2.1×10^3	3.7×10^3	162.6
F-value	30.89	714.63		21.71	269.3		3.46	69.01		13.45	22.99	
p-value	<0.001	<0.001		<0.001	<0.001		<0.05	<0.001		<0.001	<0.001	
F crit	2.87	4.35		2.87	4.35		2.87	4.35		2.87	4.35	
Relative humidity effect												
SS	5.1×10^3	2.03×10^4	4708	1.8×10^4	1.1×10^5	1.3×10^4	6.53	192.53	7.33	0.34	1.16	1.93
df	4	1	20	4	1	4	4	1	20	4	1	20
MS	1.3×10^3	2.03×10^4	235.4	4607.4	1.1×10^5	632.4	1.63	192.53	0.37	0.08	1.16	0.09
F-value	5.49	86.37		7.29	189.27		4.45	525.09		0.88	12.01	
p-value	<0.05	<0.001		<0.001	<0.001		<0.05	<0.001		0.49	<0.05	
F crit	2.867	4.35		2.87	4.35		2.87	4.35		2.87	0.41	
CO_2 effect												
SS	652.8	3020	1187.3	311.67	4.4×10^4	1.4×10^4	0.252	898.7	4.08	64.46	3.18	55.65
df	4	1	20	4	1	20	4	1	20	4	1	20
MS	163.2	3020	59.37	77.91	4.4×10^4	748.5	0.06	898.7	0.204	16.11	3.18	2.78
F-value	2.75	50.8		0.10	59.31		0.31	4398.3		5.79	1.14	
p-value	0.05	<0.001		0.97	<0.001		0.86	<0.001		<0.05	0.29	
F crit	2.87	4.35		2.87	4.35		2.87	4.35		2.87	4.35	

SV: source of variation, SS: sum of square, df: degree of freedom, MS: mean square, F crit: critical value in the F distribution, Tr: treatment, ST: sampling times, Err: error, E: exponential.

3.2. Correlation of the Glucosinolates Components

Table 3 shows the magnitude, direction, and linear pairwise relationship between the identified glucosinolate variables under the considered ambient environmental factors (temperature, relative humidity, and CO_2). Among the five identified glucosinolate variables under the temperature experiments, sinigrin and glucobrassicin were strongly and positively correlated among them and identified glucosinolate variables under the relative humidity experiments, and they were strongly and negatively correlated with the variables identified in the CO_2 experiments. The correlations were statistically significant at a 0.1% level (except for some variables). Although, progoitrin had a significant positive correlation with each of the five identified glucosinolate variables under the CO_2 experiments, no significant correlations were observed with other variables. Except for some strong correlations, 4-methoxyglucobrassicin, and neoglucobrassicin were also not significantly correlated with other identified glucosinolate variables. Strong negative correlations with a 0.1% significance level were observed for most of the identified glucosinolate variables under the relative humidity and CO_2 experiments. However, identified glucosinolate variables under the CO_2 experiments were strongly and positively correlated. They were statistically significant at a 0.1% level (except for the C_Sin). The multicollinearity issue can also be predicted from the correlation matrix. A highly correlated value (>0.7) hinders the evaluation of the true effects of the predictor variables. According to Table 3, some of the glucosinolate variables had notable evidence of strong correlation. For example, C_Pro showed positive correlations of 0.90, 0.90, and 0.99 with C_Glu, C_4-met, and C_Neo, respectively, and C_4-met showed negative correlations of −0.91 and −0.93, with T_Sin and T_Glu, respectively. Variance inflation factor (VIF) was also investigated and the values varied from 1 to 3 for most of the variables, indicating that the variables were slightly explained by other independent variables. However, the VIF values of T_Sin (4.72) and T_Glu (6.46) were relatively high [48].

3.3. Evaluation of Temperature Effects

A statistical analysis was conducted to evaluate the effects of temperature on kale growth, and the results are shown in Table 4. Regarding kale physical properties, an overall high growth rate was observed at 20–23 °C, and the lowest growth rate occurred at 14 °C. However, some physical parameters showed greater numbers at 17 and 26 °C. They were plant height (17 °C) and width (26 °C) after two weeks of transplantation, and chlorophyll level (17 °C), leaf length and weight (26 °C) after four weeks of transplantation. The data points of no. of leaves, stem diameter, and leaf parameters (length, width, and weight) were very close to the mean (low standard deviation); however, the data points of other growth variables, especially plant height and width, were spread out over a wide range of values. Standard deviation was greater in samples collected after four weeks of cultivation, compared to the two weeks. According to Duncan's range test results, significant differences were observed for the plant width, weight, and leaf parameters (length, width, and weight) at 2-week sampling time, and the plant height, and leaf parameters at 4-week sampling time, depending on the temperature levels. Contrariwise, the rest of the growth variables (specifically the number of leaves, stem diameter, and chlorophyll level) did not show statistical significance regarding the temperature variations.

Table 3. Correlation matrix of the identified glucosinolate components under the temperature, humidity, and CO_2 treatments.

Variables	T_Pro	T_Sin	T_Glu	T_4-met	T_Neo	H_Pro	H_Sin	H_Glu	H_4-met	H_Neo	C_Pro	C_Sin	C_Glu	C_4-met	C_Neo
T_Pro	1.00														
T_Sin	−0.02	1.00													
T_Glu	−0.12	0.96 ***	1.00												
T_4-met	−0.13	−0.65 ***	−0.59 ***	1.00											
T_Neo	0.51***	0.74 ***	0.76 ***	−0.52 ***	1.00										
H_Pro	−0.52 ***	0.24	0.50 ***	−0.08	0.25	1.00									
H_Sin	−0.13	0.08	−0.06	−0.66 ***	−0.26	−0.35 *	1.00								
H_Glu	−0.01	0.53 ***	0.66 ***	0.14	0.65 ***	0.59 ***	−0.79 ***	1.00							
H_4-met	−0.21	−0.54 ***	−0.36 *	0.88 ***	−0.29	0.37 *	−0.82 ***	0.41 **	1.00						
H_Neo	0.00	0.52 ***	0.35 *	0.11	0.16	−0.38 *	−0.13	0.30 *	−0.17	1.00					
C_Pro	0.70 ***	−0.64 ***	−0.69 ***	0.06	−0.11	−0.50 ***	0.13	−0.54 ***	−0.03	−0.53 ***	1.00				
C_Sin	0.37 *	0.68 ***	0.71 ***	−0.84 ***	0.86 ***	0.31 *	0.17	0.30 *	−0.56 ***	−0.17	−0.01	1.00			
C_Glu	0.70 ***	−0.63**	−0.58 ***	0.27	0.04	−0.22	−0.28	−0.16	0.31 *	−0.57 ***	0.90 ***	0.01	1.00		
C_4-met	0.36 *	−0.91 ***	−0.93 ***	0.36 *	−0.52 ***	−0.45 **	0.11	−0.66 ***	0.22	−0.57 ***	0.90 ***	−0.40 **	0.80 ***	1.00	
C_Neo	0.77 ***	−0.50 ***	−0.56 ***	−0.08	0.03	−0.47 **	0.17	−0.49 **	−0.14	−0.53 ***	0.99 ***	0.16	0.88 ***	0.81 ***	1.00
VIF	1.06	4.72	6.46	2.83	1.57	1.32	1.09	1.14	1.47	1.01	1.56	2.39	1.63	2.92	1.28

Colour scale: −1, −0.6, −0.2, 0, 0.2, 0.6, +1 — Strong negative; Not correlated; Strong positive

*, **, *** indicate the 5%, 1%, and 0.1% significance levels, respectively. T_Pro, T_Sin, T_Glu, T_4-met, T_Neo: progoitrin, sinigrin, glucobrassicin, 4-methoxyglucobrassicin, and neoglucobrassicin observed under experiment 1 (temperature effect); H_Pro, H_Sin, H_Glu, H_4-met, H_Neo: progoitrin, sinigrin, glucobrassicin, 4-methoxyglucobrassicin, and neoglucobrassicin observed under experiment 2 (relative humidity effect); C_Pro, C_Sin, C_Glu, C_4-met, C_Neo: progoitrin, sinigrin, glucobrassicin, 4-methoxyglucobrassicin, and neoglucobrassicin observed under experiment 3 (CO_2 effect), respectively. VIF: variance inflation factor.

Table 4. Effects of different temperature levels on kale growth at different cultivation periods.

Sampling Time	Temp. Level (°C)	Growth Variables								
		P_Height (mm)	P_Width (mm)	P_Weight (g)	No_Leaf	Stem dia. (mm)	Chlor_Level (ppm)	L_Length (mm)	L_Width (mm)	L_Weight (g)
2 weeks	14	95.0 ± 4.5 ^{a}	161.7 ± 8.1 ^{b}	11.9 ± 1.2 ^{b}	7.0 ± 0.0 ^{a}	2.4 ± 0.1 ^{a}	46.9 ± 1.0 ^{a}	8.4 ± 0.5 ^{c}	7.4 ± 0.4 ^{c}	2.9 ± 0.2 ^{b}
	17	99.3 ± 4.7 ^{a}	186.0 ± 24.3 ^{ab}	12.6 ± 0.3 ^{a}	6.7 ± 3.7 ^{a}	2.5 ± 0.2 ^{a}	49.1 ± 0.6 ^{a}	10.6 ± 0.2 ^{b}	9.4 ± 0.4 ^{b}	5.2 ± 0.6 ^{a}
	20	84.7 ± 1.2 ^{a}	167.0 ± 22.7 ^{b}	13.3 ± 0.2 ^{a}	7.0 ± 0.1 ^{a}	2.7 ± 0.1 ^{a}	53.4 ± 5.4 ^{a}	13.0 ± 0.4 ^{a}	1[illegible].3 ± 0.6 ^{a}	6.3 ± 0.4 ^{a}
	23	92.3 ± 11 ^{a}	196.0 ± 14.2 ^{ab}	13.0 ± 0.3 ^{a}	7.0 ± 0.1 ^{a}	2.8 ± 0.9 ^{a}	57.0 ± 2.8 ^{a}	12.8 ± 0.8 ^{a}	11.1 ± 0.6 ^{ab}	6.0 ± 1.1 ^{a}
	26	90.3 ± 8.9 ^{a}	215.0 ± 14.1 ^{a}	13.3 ± 0.3 ^{a}	7.0 ± 0.8 ^{a}	2.8 ± 0.2 ^{a}	52.0 ± 0.3 ^{a}	12.6 ± 0.9 ^{a}	1[illegible].3 ± 0.4 ^{a}	5.8 ± 0.2 ^{a}
4 weeks	14	115.3 ± 8.3 ^{b}	274.3 ± 15.0 ^{a}	21.3 ± 4.1 ^{a}	11.1 ± 0.2 ^{a}	13.0 ± 0.0 ^{a}	55.5 ± 4.3 ^{a}	19.9 ± 0.8 ^{c}	12.6 ± 0.46 ^{b}	9.2 ± 0.9 ^{b}
	17	129.7 ± 5.3 ^{b}	260.3 ± 39.4 ^{a}	23.6 ± 6.2 ^{a}	12.7 ± 0.5 ^{a}	14.3 ± 0.5 ^{a}	61.2 ± 2.4 ^{a}	26.7 ± 0.2 ^{b}	17.1 ± 1.1 ^{ab}	16.0 ± 3.3 ^{ab}
	20	143.3 ± 6.5 ^{ab}	286.3 ± 13.5 ^{a}	28.3 ± 4.3 ^{a}	12.0 ± 0.7 ^{a}	15.7 ± 0.4 ^{a}	48.4 ± 15.1 ^{a}	31.2 ± 2.0 ^{ab}	19.0 ± 1.4 ^{a}	19.1 ± 6.1 ^{a}
	23	137.3 ± 8.6 ^{ab}	278.3 ± 36.5 ^{a}	25.4 ± 3.2 ^{a}	13.5 ± 1.3 ^{a}	17.7 ± 0.4 ^{a}	58.5 ± 1.7 ^{a}	31.8 ± 2.5 ^{ab}	20.2 ± 1.8 ^{a}	20.0 ± 3.8 ^{a}
	26	176.3 ± 27.7 ^{a}	277.0 ± 12.8 ^{a}	22.2 ± 4.8 ^{a}	10.0 ± 0.4 ^{a}	15.6 ± 0.9 ^{a}	53.6 ± 4.3 ^{a}	37.0 ± 2.5 ^{a}	19.8 ± 1.5 ^{a}	20.1 ± 5.7 ^{a}

$^{a, b, c}$ Different letters in the same column indicate a significant difference ($p \leq 0.05$). P_height: plant height, P_width: plant width, P_weight: plant weight, No_leaf: number of leaves, Stem dia.: stem diameter, Chlor_level: chlorophyll level, L_length: leaf length, L_width: leaf width, L_weight: leaf weight.

Figure 5 shows the effects of temperature on the glucosinolate content, based on various components, of harvested kale leaves after two and four weeks of transplantation. Glucobrassicin was found to be a dominant glucosinolate component in both cases. However, an inverse relationship was observed between the contents of all glucosinolate components and increased ambient temperature levels and cultivation period. The total glucosinolate content became lower at each increased temperature level, and the lowest total glucosinolate of kale leaves in each sampling time was observed at 26 °C. A high standard deviation trend was observed for each glucosinolate component due to the low sampling number. Among the five levels of temperature, the total glucosinolate content was higher at 14–17 °C in both cultivation periods. According to Duncan's range test, the concentrations of sinigrin, glucobrassicin, and 4-methoxyglucobrassicin were significantly different for each temperature level in samples collected after two weeks of cultivation; however, no significant differences were observed among the other glucosinolate components (except sinigrin at 4th week) for different temperature levels and cultivation periods.

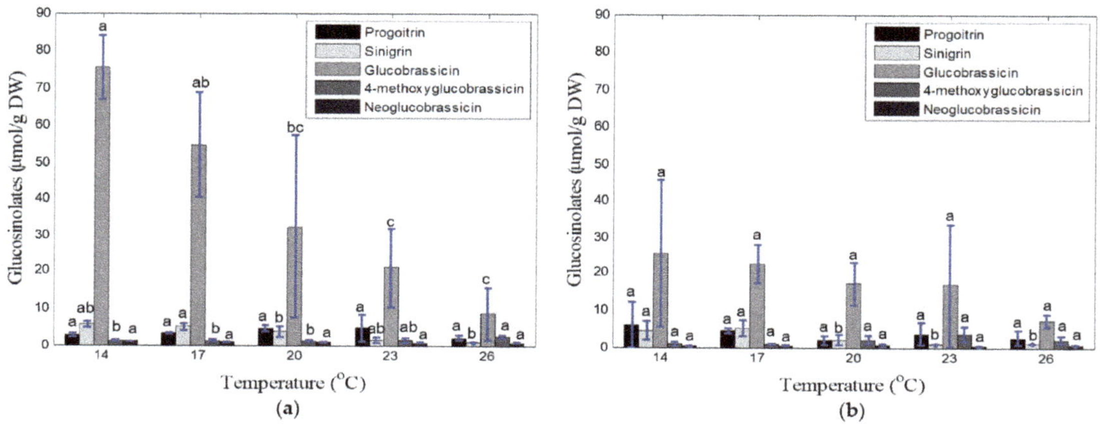

Figure 5. Concentrations of glucosinolate components (μmol/g DW) under different temperature treatments and cultivation periods: two weeks after transplantation (**a**) and four weeks after transplantation (**b**). $^{a, b, c}$ levels of components with the same letters are not significantly different at $p < 0.05$.

The interactions of each growth variable and glucosinolate component under the different temperature treatments after four weeks of transplantation were analyzed using the correlation matrix and the results are summarized in Table 5, where the level of significance and VIF are also mentioned. In many cases, strong positive and negative correlation coefficients were observed. The physical growth variables were found to be strongly correlated with each other (around 0.50–0.98). However, the chlorophyll level showed negative correlations with all the growth variables, and positive correlations were observed with progoitrin, sinigrin, and glucobrassicin. The progoitrin, sinigrin, and glucobrassicin were strongly positively correlated with each other and negatively correlated with 4-methoxyglucobrassicin and neoglucobrassicin. Moreover, the detected glucosinolate components (except 4-methoxyglucobrassicin) were negatively correlated with most of the physical growth variables. According to the VIF analysis, most of the variables were moderately correlated; however, some variables (i.e., leaf length and glucobrassicin) were highly correlated, which might adversely affect other variables.

3.4. Evaluation of Relative Humidity Effects

The effects of relative humidity on kale growth properties are summarized in Table 6. The growth status was evaluated at two different stages (two and four weeks after transplantation). Most of the physical growth variables were prominent at the 85% relative humidity level at both sampling periods, except for the number of leaves, stem diameter, chlorophyll level (55%), and leaf length (75%) in the second week, and the number of leaves, leaf weight (65%), and chlorophyll level (45%) in the fourth week of cultivation. Except for some growth variables, the overall lowest growth performance was observed at the 45% relative humidity level in both sampling times. The data points of some growth variables, such as plant height and width, leaf length and width, were spread out over a wide range compared to other growth variables, and greater standard deviations were observed in samples collected after four weeks of cultivation, compared to the two weeks of cultivation. According to Duncan's range test, all the growth variables (except the chlorophyll level) were significantly different at the 2-week sampling time depending on the relative humidity levels. A similar result was observed (except for the plant weight, chlorophyll level, and leaf width) at the 4-week sampling time.

Table 5. Correlation matrix showing kale growth and glucosinolate variables and their constituents (experiment 1).

Variables	P_Height	P_Width	P_Weight	No_Leaf	Stem dia.	Chlor_Level	L_Length	L_Width	L_Weight	Pro	Sin	Glu	4-Met	Neo
P_height	1													
P_width	0.98 ***	1												
P_weight	0.81 ***	0.73 ***	1											
No_leaf	0.08	−0.08	0.54 ***	1										
Stem dia.	0.76 ***	0.86 ***	0.36 **	−0.57 ***	1									
Chlor_level	−0.36 **	−0.22	−0.63 ***	−0.65 ***	0.12	1								
L_length	0.85 ***	0.85 ***	0.78 ***	−0.04	0.76 ***	−0.50 ***	1							
L_width	0.60 ***	0.60 ***	0.63 ***	−0.1	0.63 ***	−0.51 ***	0.93 ***	1						
L_weight	0.55 ***	0.52 ***	0.66 ***	0.05	0.49 ***	−0.64 ***	0.90 ***	0.98 ***	1					
Pro	−0.60 ***	−0.55 ***	−0.66 ***	−0.13	−0.45 ***	0.78 ***	−0.88 ***	−0.93 ***	−0.96 ***	1				
Sin	−0.68 ***	−0.70 ***	−0.38 **	0.32 *	−0.76 ***	0.43 ***	−0.83 ***	−0.77 ***	−0.72 ***	0.81 ***	1			
Glu	−0.96 ***	−0.96 ***	−0.76 ***	0.06	−0.84 ***	0.41 **	−0.95 ***	−0.77 ***	−0.72 ***	0.75 ***	0.83 ***	1		
4-Met	0.52 ***	0.52 ***	0.38 **	−0.20	0.57 ***	−0.59 ***	0.81 ***	0.86 ***	0.85 ***	−0.92 ***	−0.94 ***	−0.72 ***	1	
Neo	−0.21	−0.33 *	0.33 *	0.68 ***	−0.54 ***	−0.74 ***	0.10	0.31 *	0.47 ***	−0.46 ***	0.10	0.14	0.20	1
VIF	3.47	4.08	3.65	1.06	4.21	2.16	16.67	4.64	2.99	3.09	5.81	11.99	3.46	1.01

Color scale: −1, −0.6, −0.2, 0, 0.2, 0.6, +1 — Strong negative / Not correlated / Strong positive

*, **, *** indicate the 5%, 1%, and 0.1% significance levels, respectively. P_height: plant height, P_width: plant width, P_weight: plant weight, No_leaf: number of leaves, Stem dia.: stem diameter, Chlor_level: chlorophyll level, L_length: leaf length, L_width: leaf width, L_weight: leaf weight, Pro: progoitrin, Sin: sinigrin, Glu: glucobrassicin, 4-Met: 4-methoxyglucobrassicin, and Neo: neoglucobrassicin. VIF: variance inflation factor.

Table 6. Effects of different relative humidity levels on kale growth in different cultivation periods.

Sampling Time	Humi. Level (%)	Growth Variables								
		P_Height (mm)	P_Width (mm)	P_Weight (g)	No_Leaf	Stem Dia. (mm)	Chlor_Level (ppm)	L_Length (mm)	L_Width (mm)	L_Weight (g)
2 weeks	45	68.7 ± 7.6 [b]	171.7 ± 8.1 [b]	5.5 ± 0.1 [c]	8.0 ± 0.0 [ab]	2.9 ± 0.2 [c]	60.5 ± 4.3 [a]	97.7 ± 9.5 [c]	58.0 ± 2.0 [c]	1.1 ± 0.1 [c]
	55	74.3 ± 10.3 [b]	196.0 ± 24.3 [ab]	8.2 ± 1.2 [ab]	8.7 ± 0.6 [a]	3.8 ± 0.3 [a]	66.2 ± 2.4 [a]	111.7 ± 9.5 [bc]	67.7 ± 5.5 [bc]	1.8 ± 0.4 [b]
	65	79.3 ± 5.1 [ab]	177.0 ± 22.6 [b]	6.3 ± 1.4 [bc]	7.7 ± 0.6 [b]	3.3 ± 0.2 [b]	53.4 ± 15.1 [a]	114.0 ± 5.6 [b]	67.3 ± 5.1 [bc]	1.8 ± 0.2 [b]
	75	83.0 ± 1.0 [ab]	206.0 ± 14.2 [ab]	7.7 ± 1.2 [abc]	8.0 ± 0.1 [ab]	3.2 ± 0.2 [bc]	63.5 ± 1.7 [a]	133.3 ± 9.1 [a]	76.7 ± 4.6 [ab]	2.1 ± 0.3 [ab]
	85	93.3 ± 6.8 [a]	225.0 ± 14.1 [a]	8.9 ± 1.5 [a]	8.0 ± 0.0 [ab]	3.5 ± 0.1 [ab]	58.6 ± 4.3 [a]	132.3 ± 8.6 [a]	81.7 ± 8.4 [a]	2.5 ± 0.3 [a]
4 weeks	45	114.3 ± 9.3 [b]	291.7 ± 62.8 [b]	23.4 ± 10.2 [a]	13.0 ± 1.0 [bc]	5.0 ± 0.7 [c]	66.1 ± 1.6 [a]	158.7 ± 25.2 [b]	105.3 ± 21.7 [a]	5.3 ± 1.3 [b]
	55	119.7 ± 15.3 [b]	302.0 ± 18.1 [b]	29.4 ± 3.5 [a]	13.3 ± 0.6 [ab]	5.2 ± 0.3 [bc]	65.3 ± 1.8 [a]	171.0 ± 14.1 [ab]	113.0 ± 7.9 [a]	8.9 ± 1.5 [a]
	65	133.3 ± 16.5 [ab]	314.0 ± 18.0 [b]	33.0 ± 4.9 [a]	14.7 ± 0.6 [a]	5.0 ± 0.2 [c]	47.1 ± 2.8 [a]	172.3 ± 17.1 [ab]	111.3 ± 9.1 [a]	9.1 ± 1.1 [a]
	75	125.3 ± 9.6 [b]	316.3 ± 7.6 [b]	25.7 ± 0.9 [a]	11.7 ± 0.6 [c]	5.7 ± 0.3 [ab]	61.3 ± 3.1 [a]	186.3 ± 6.4 [ab]	114.7 ± 9.7 [a]	8.9 ± 2.2 [a]
	85	166.3 ± 37.9 [a]	383.3 ± 10.1 [a]	34.7 ± 6.9 [a]	13.0 ± 1.0 [bc]	6.1 ± 0.1 [a]	59.2 ± 3.1 [a]	191.0 ± 11.5 [a]	126.0 ± 12.5 [a]	8.7 ± 2.3 [a]

[a, b, c] Different letters in the same column indicate significant differences ($p \leq 0.05$). P_height: plant height, P_width: plant width, P_weight: plant weight, No_leaf: number of leaves, Stem dia.: stem diameter, Chlor_level: chlorophyll level, L_length: leaf length, L_width: leaf width, L_weight: leaf weight.

The results of the glucosinolate analysis for different relative humidity treatments and cultivation periods are shown in Figure 6. The aliphatic glucosinolates (i.e., progoitrin, sinigrin) and indole glucosinolate (i.e., glucobrassicin) were the most prominent components at both of the sampling times. The overall glucosinolate concentrations decreased slightly in the samples collected after the fourth week of cultivation. A high standard deviation was observed, especially for the glucobrassicin, as the sample number was low and sometimes all glucosinolate components were not detected in some samples. According to Duncan's range test, no significant differences were observed among the glucosinolate components (except for the progoitrin, glucobrassicin, and 4-methoxyglucobrassicin at the 2-week sampling time) for different relative humidity treatments and cultivation periods.

Table 7 shows the correlation matrix of physical and glucosinolate properties for different relative humidity treatments after four weeks of cultivation. All physical growth variables, except for the number of leaves and the chlorophyll level, showed strong positive correlations with one another. A fairly good correlation (both positive and negative) was observed between the physical variables and glucosinolate components. However, most of the glucosinolate components were negatively correlated with each other. Moreover, the VIF values were also determined. Except for the leaf length (VIF: 11.93), other predictors were moderately correlated, which resulted in a low influence on other independent variables.

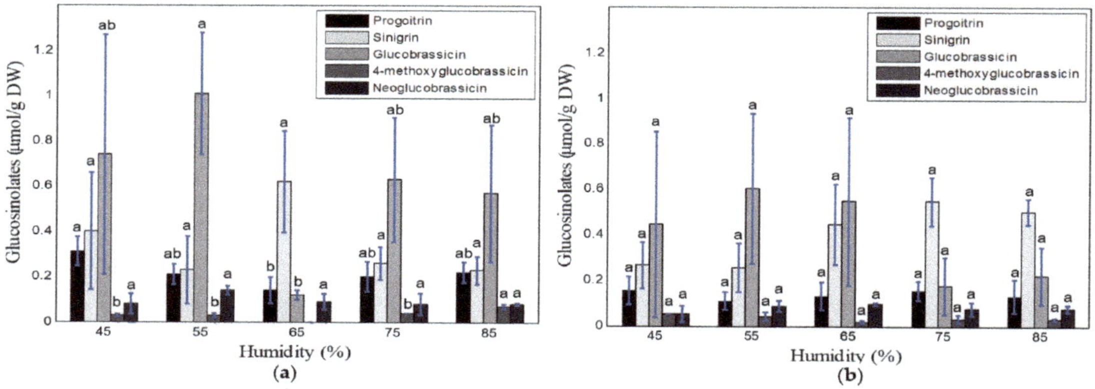

Figure 6. Concentrations of glucosinolate components (μmol/g DW) under different relative humidity treatments and cultivation periods: two weeks after transplantation (**a**) and four weeks after transplantation (**b**). [a, b] levels of components with the same letters are not significantly different at $p < 0.05$.

Table 7. Correlation matrix showing kale growth and glucosinolate variables and their constituents (experiment 2).

Variables	P_Height	P_Width	P_Weight	No_Leaf	Stem Dia.	Chlor_Level	L_Length	L_Width	L_Weight	Pro	Sin	Glu	4-Met	Neo
P_height	1													
P_width	0.99 ***	1												
P_weight	0.81 ***	0.73 ***	1											
No_leaf	0.08	−0.08	0.54 ***	1										
Stem dia.	0.76 ***	0.86 ***	0.36 **	−0.57 ***	1									
Chlor_level	−0.36 **	−0.22	−0.63 ***	−0.65 ***	0.12	1								
L_length	0.76 ***	0.81 ***	0.53 ***	−0.39 **	0.91 ***	−0.21	1							
L_width	0.80 ***	0.85 ***	0.71 ***	−0.14	0.81 ***	−0.02	0.82 ***	1						
L_weight	0.42 ***	0.40 **	0.67 ***	0.12	0.37 **	−0.50 ***	0.69 ***	0.65 ***	1					
Pro	−0.17	−0.13	−0.60 ***	−0.50 ***	0.09	0.09	−0.04	−0.51 ***	−0.54 ***	1				
Sin	0.61 ***	0.62 ***	0.38 **	−0.29 *	0.69 ***	−0.50 ***	0.85 ***	0.42 ***	0.55 ***	0.34 **	1			
Glu	−0.49 ***	−0.59 ***	0.04	0.74 ***	−0.85 ***	−0.13	−0.72 ***	−0.37 **	−0.05	−0.60 ***	−0.76 ***	1		
4-Met	−0.39 **	−0.35 **	−0.47 ***	−0.03	−0.40 **	0.73 ***	−0.68 ***	−0.31 *	−0.79 ***	−0.05	−0.86 ***	0.35 **	1	
Neo	0.28 *	0.20	0.70 ***	0.46 ***	0.03	−0.68 ***	0.41 **	0.42 ***	0.93 ***	−0.64 ***	0.34 **	0.28*	−0.72 ***	1
VIF	3.47	4.07	2.65	1.06	4.21	1.16	11.93	2.56	1.78	1.11	4.97	2.01	1.42	1.22

Color scale: −1, −0.6, −0.2, 0, 0.2, 0.6, +1 — Strong negative, Not correlated, Strong positive.

*, **, *** indicates 5%, 1%, and 0.1% significance levels, respectively. P_height: plant height, P_width: plant width, P_weight: plant weight, No_leaf: number of leaves, Stem dia.: stem diameter, Chlor_level: chlorophyll level, L_length: leaf length, L_width: leaf width, L_weight: leaf weight, Pro: progoitrin, Sin: sinigrin, Glu: glucobrassicin, 4-met: 4-methoxyglucobrassicin, and Neo: neoglucobrassicin. VIF: variance inflation factor.

3.5. Evaluation of CO_2 Effects

A summary of the effects of CO_2 treatments and cultivation periods on kale growth is given in Table 8. The overall growth performance was higher under 700–1000 ppm CO_2. However, a notable growth rate of some parameters (i.e., chlorophyll level and leaf length at the 2-week sampling time, and plant height and chlorophyll level at the 4-week sampling time) was observed under 400 ppm of CO2. Relatively low growth performance was observed under 1300 and 1600 ppm of CO_2 in both sampling periods. Besides this, the spread of standard deviations of the growth variables was almost similar for both of the sampling periods. Comparatively high standard deviations were observed for the plant parameters (height, width, and weight) and chlorophyll level compared to other growth variables. Based on Duncan's range test results, there were no significant differences for the growth parameters (except the plant height at the 4-week sampling time) under different CO_2 concentrations and cultivation periods.

The effects of different CO_2 concentrations on the glucosinolate content are shown in Figure 7. The optimal CO_2 level in relation to the total glucosinolate content was 1300 ppm at both the second and fourth weeks of cultivation. The progoitrin, sinigrin, and neoglucobrassicin contents decreased after the two weeks of cultivation. The low sampling number caused high standard deviations of the detected glucosinolate components. Glucobrassicin was found to be a dominant component in the samples collected after 4-weeks. The results of Duncan's range test showed a significant difference in glucosinolate components (except for the 4-methoxyglucobrassicin and neoglucobrassicin at the 2-week sampling, and sinigrin, glucobrassicin, and neoglucobrassicin at the 4-week sampling time) under different CO_2 concentrations and both cultivation periods.

Table 8. Effects of different carbon dioxide levels on kale growth at different cultivation periods.

Sampling Time	CO_2 Level (ppm)	Growth Variables								
		P_Height (mm)	P_Width (mm)	P_Weight (g)	No_Leaf	Stem Dia. (mm)	Chlor_Level (ppm)	L_Length (mm)	L_Width (mm)	L_Weight (g)
2 weeks	400	85.0 ± 4.5 ^{a}	213.0 ± 9.2 ^{a}	11.0 ± 1.8 ^{a}	4.2 ± 0.3 ^{a}	11.3 ± 0.4 ^{a}	142.0 ± 10.0 ^{a}	53.0 ± 4.2 ^{a}	60.1 ± 1.4 ^{a}	3.5 ± 0.2 ^{a}
	700	89.3 ± 4.7 ^{a}	228.0 ± 11.7 ^{a}	11.3 ± 1.7 ^{a}	4.2 ± 0.2 ^{a}	11.7 ± 0.4 ^{a}	140.7 ± 5.4 ^{a}	47.7 ± 1.2 ^{a}	63.0 ± 4.3 ^{a}	3.6 ± 1.1 ^{a}
	1000	74.7 ± 1.2 ^{a}	196.3 ± 16.8 ^{a}	10.4 ± 1.4 ^{a}	4.3 ± 0.3 ^{a}	11.7 ± 0.4 ^{a}	127.0 ± 6.1 ^{a}	50.0 ± 4.0 ^{a}	64.1 ± 3.2 ^{a}	3.4 ± 0.4 ^{a}
	1300	82.3 ± 11.0 ^{a}	211.3 ± 4.5 ^{a}	8.9 ± 0.4 ^{a}	4.0 ± 0.1 ^{a}	11.0 ± 0.8 ^{a}	125.0 ± 1.4 ^{a}	45.3 ± 4.1 ^{a}	69.8 ± 1.1 ^{a}	3.2 ± 0.6 ^{a}
	1600	80.3 ± 8.9 ^{a}	194.0 ± 17.2 ^{a}	7.5 ± 0.6 ^{a}	4.0 ± 0.3 ^{a}	11.7 ± 0.9 ^{a}	123.0 ± 7.2 ^{a}	52.3 ± 3.2 ^{a}	66.8 ± 1.3 ^{a}	2.9 ± 0.2 ^{a}
4 weeks	400	111.7 ± 6.5 ^{a}	284.3 ± 13.0 ^{a}	22.2 ± 4.8 ^{a}	7.0 ± 0.4 ^{a}	14.7 ± 0.9 ^{a}	186.0 ± 14.3 ^{a}	68.0 ± 2.1 ^{a}	62.7 ± 2.8 ^{a}	5.7 ± 1.7 ^{a}
	700	105.3 ± 4.7 ^{ab}	270.3 ± 49.4 ^{a}	28.4 ± 3.2 ^{a}	7.1 ± 0.3 ^{a}	15.7 ± 0.4 ^{a}	179.0 ± 9.9 ^{a}	67.3 ± 4.7 ^{a}	66.9 ± 1.3 ^{a}	5.3 ± 0.8 ^{a}
	1000	102.7 ± 4.9 ^{ab}	296.3 ± 6.5 ^{a}	25.0 ± 4.3 ^{a}	7.4 ± 0.4 ^{a}	15.7 ± 0.4 ^{a}	175.7 ± 5.2 ^{a}	70.3 ± 2.4 ^{a}	65.6 ± 0.8 ^{a}	4.5 ± 0.1 ^{a}
	1300	99.3 ± 5.5 ^{ab}	288.3 ± 36.5 ^{a}	23.6 ± 6.2 ^{a}	6.7 ± 0.5 ^{a}	14.3 ± 0.5 ^{a}	169.3 ± 12.0 ^{a}	63.0 ± 2.0 ^{a}	67.7 ± 1.2 ^{a}	4.7 ± 1.3 ^{a}
	1600	93.0 ± 5.0 ^{b}	288.0 ± 12.8 ^{a}	25.3 ± 4.1 ^{a}	7.1 ± 0.2 ^{a}	15.0 ± 0.0 ^{a}	179.0 ± 11.3 ^{a}	68.3 ± 4.1 ^{a}	66.4 ± 2.4 ^{a}	4.9 ± 0.9 ^{a}

$^{a, b}$ Different letters in the same column indicate significant differences ($p \leq 0.05$). P_height: plant height, P_width: plant width, P_weight: plant weight, No_leaf: number of leaves, Stem dia.: stem diameter, Chlor_level: chlorophyll level, L_length: leaf length, L_width: leaf width, L_weight: leaf weight.

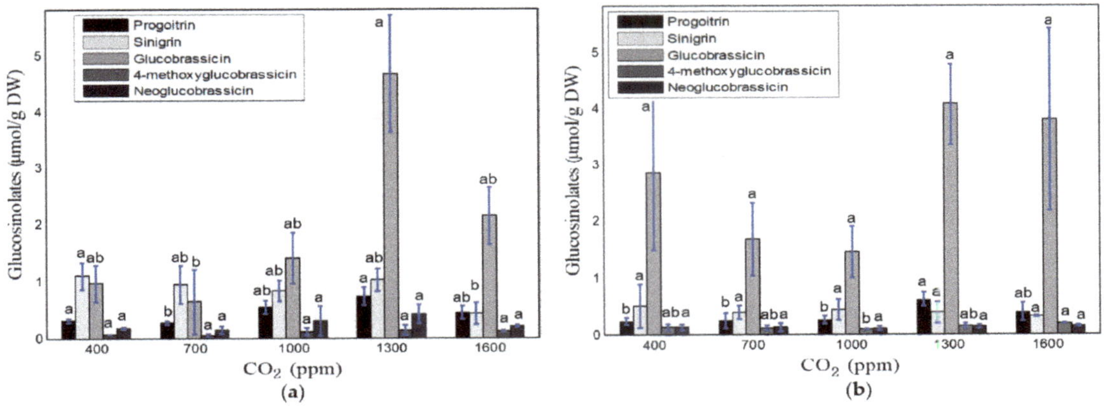

Figure 7. Concentrations of glucosinolate components (μmol/g DW) under different CO_2 treatments and cultivation periods: two weeks after transplantation (**a**) and four weeks after transplantation (**b**). $^{a, b}$ levels of components with the same letters are not significantly different at $p < 0.05$.

Table 9 shows the interactions of each physical and functional parameter with one another, along with the significance levels for different CO_2 treatments after four weeks of transplantation. Weak correlations (both positive and negative) were detected among most of the physical growth variables. However, the glucosinolate components (except sinigrin) were strongly positively correlated with each other and mostly negatively correlated with physical growth variables. According to the VIF analysis, the VIF values of the physical variables varied from 1 to 2, except plant height (4.23), indicating low correlations. Contrariwise, the VIF values of the glucosinolate components (except progoitrin) were comparatively high, which indicated highly correlated relationships and influences on other predictors.

Table 9. Correlation matrix showing kale growth and glucosinolate variables and their constituents (experiment 3).

Variables	P_Height	P_Width	P_Weight	No_Leaf	Stem Dia.	Chlor_Level	L_Length	L_Width	L_Weight	Pro	Sin	Glu	4-Met	Neo
P_height	1													
P_width	−0.32 *	1												
P_weight	−0.16	−0.87 ***	1											
No_leaf	0.06	0.20	−0.37 **	1										
Stem dia.	0.09	−0.19	0.01	0.88 ***	1									
Chlor_level	0.56 ***	−0.31 *	0.08	0.29 *	0.16	1								
L_length	0.15	0.21	−0.38 **	0.94 ***	0.72 ***	0.57 ***	1							
L_width	−0.69 ***	−0.12	0.41 **	−0.22	0.08	−0.84 ***	−0.50 ***	1						
L_weight	−0.16	−0.87 ***	1.00 ***	−0.37 **	0.01	0.08	−0.38 **	0.41 **	1					
Pro	−0.58 ***	0.25	0.10	−0.75 ***	−0.70 ***	−0.78 ***	−0.84 ***	0.63 ***	0.10	1				
Sin	0.92 ***	−0.04	−0.43 ***	0.01	−0.03	0.25	0.03	−0.56 ***	−0.43 ***	−0.36 **	1			
Glu	−0.50 ***	0.19	0.19	−0.80 ***	−0.88 ***	−0.27 *	−0.66 ***	0.2	0.19	0.80 ***	−0.45 ***	1		
4-Met	−0.65 ***	−0.05	0.48 ***	−0.52 ***	−0.55 ***	−0.03	−0.37 **	0.23	0.48 ***	0.53 ***	−0.76 ***	0.85 ***	1	
Neo	−0.54 ***	−0.24	0.64 ***	−0.66 ***	−0.59 ***	−0.05	−0.53 ***	0.29 *	0.64 ***	0.56 ***	−0.68 ***	0.85 ***	0.97 ***	1
VIF	4.23	1.22	1.51	1.75	1.03	1.62	1.79	1.5	1.51	1.82	3.81	6.85	3.15	7.9

Scale: −1 −0.6 −0.2 0 0.2 0.6 +1; Strong negative, Not correlated, Strong positive

*, **, *** indicates the 5%, 1%, and 0.1% significance level, respectively. P_height: plant height, P_width: plant width, P_weight: plant weight, No_leaf: number of leaves, Stem dia.: stem diameter, Chlor_level: chlorophyll level, L_length: leaf length, L_width: leaf width, L_weight: leaf weight, Pro: progoitrin, Sin: sinigrin, Glu: glucobrassicin, 4-Met: 4-methoxyglucobrassicin, and Neo: neoglucobrassicin. VIF: variance inflation factor.

4. Discussion

There is an interaction between plant growth and glucosinolate concentration, which strongly depends on the environmental conditions and water–nutrient consumption rate, along with the plant species, growth method, cultivation period, and cultivation facilities used [12–14,16,49,50]. In this study, the growth rate of kale increased with the cultivation period. The overall maximum growth rate was observed at 20–23 °C, around 85% relative humidity, and 700–1000 ppm CO_2 (Tables 4, 6 and 8). The optimal temperature, relative humidity, and CO_2 range for total glucosinolate content were 14–17 °C, 55–75%, and 1300–1600 ppm. However, the glucosinolate content of kale decreased notably as cultivation period, temperature, and relative humidity level increased (Figures 5 and 6). Contrariwise, it increased with increased CO_2 concentration (Figure 7). All biological processes of plants speed up at higher temperatures [51]. However, the sensitivity of plants to the atmospheric temperature depends on the growth stage. Plants always seek to maintain a balance between the plant-body temperature and air temperature. If the plant is heated up, the transpiration rate increases to cool down plants, which increases water and nutrient uptake, resulting in phonological changes in plants [52,53]. This assimilation process occurs quickly in the early growth stage. We observed a high growth rate at 23–26 °C in the 2nd week, which was reduced along with the temperature range (to 20–23 °C) in the 4th week. However, the rapid transpiration process also ejects many nutrient components, which lowers the concentration of glucosinolate components, as shown in Figure 5 [33]. Conversely, a high level of accumulation of functional components (i.e., glucosinolates) occurs at low temperatures. Steindal et al. [29] explained that low temperatures activate cold acclimatization processes, including many biochemical and physiological changes, to improve the cold tolerance capacity. These procedures reduce the growth and accumulation of osmolytes and the functional component composition. In this study, the lowest rate of physical growth and the highest concentration of glucosinolates were also observed at 14 °C (Table 4, Figure 5). Velasco et al. [49] reported an inverse relationship between low temperatures and the total glucosinolate content. Relative humidity is directly related to CO_2 acclimation through the stomata response, which is connected with plant growth and nutritional levels. Ahmed et al. [54] reviewed several studies and reported that a relative humidity of lower than 40% and higher than 85% causes stomatal malfunctioning, inhibiting the plant growth rate and photosynthesis. They also mentioned that the optimal range of relative humidity for leafy vegetables (i.e., lettuce) is 70–80%. In this study, maximal growth was found at a relative humidity range of 75–85%, and no significant difference was observed at a relative humidity range of 45–85% for the glucosinolate components (Figure 6), which matches the findings of previous studies. In addition to the effects of temperature and relative humidity, a significant impact of the CO_2 concentration was observed on the accumulation of glucosinolate components rather than the growth rate of kale. In open environments, the concentration of CO_2 remains constant (300–400 ppm), but this concentration can be increased in protected cultivation facilities (i.e., greenhouses and plant factories). Usually, the demand for CO_2 increases with the increment of plant growth parameters and biomass [55]. In this study, CO_2 concentrations of 700 to 1000 ppm were associated with better growth performance (Table 8), and higher glucosinolate formation was observed under 1300 to 1600 ppm range of CO_2 (Figure 7). Higher concentrations of CO_2 help to synthesize larger amounts of carbohydrates and other functional components through photosynthesis [13,38,40]. Moreover, lower reduction of the photosynthetic ingredients under elevated CO_2 concentrations improves the glucosinolate content [39,41]. An overaccumulation of glucosinolates was observed under experiment-1 (temperature) compared to experiments-2 and 3 (relative humidity and CO_2). We know glucosinolates are significantly affected by the variety, genetics, plant growth stage, irrigation level, growing media, and environmental variables (i.e., temperature, humidity, CO_2, and light conditions). For example, Chen et al. [56] investigated the variation of glucosinolates in Chinese Brassica campestris vegetables (Chinese cabbage, purple cai-tai, choysum, pakchoi, and turnip) and reported that total glucosinolates varied from 14–130 mg/100 g fresh weight (FW), where He et al. [57] observed the minimum (28.9 µmol/100 g FW) in broccoli and maximum (278 µmol/100 g FW) in Chinese kale. From seedling to early flowering, the total glucosinolate content increased with plant age in *B. oleracea* leaves. After that point, the aliphatic glucosinolate content decreased dramatically over time as the glucosinolates transferred in the flower buds [49]. Qian et al. [58] investigated the effect of light quality on glucosinolate composition and content of Chinese kale sprouts under 23 °C temperature, 80% relative humidity, 16/8 h photoperiod, and red: blue: white light condition, and observed 167.32–288.70 and 72.66–87.48 µmol/g DW of total glucosinolates in shoots and roots, respectively. Similarly, temperature, humidity, and CO_2 have an individual effect on glucosinolate components and accumulation. Rosa and Rodrigues [59] reported that the amount of glucosinolates increases 4–35% in the Brassica species in summer compared to winter seasons. They also observed 386 ± 71 µmol/100 g DW of total glucosinolates in the Chinese cabbage leaves under 20 °C, which increased up to 409 ± 104 µmol/100 g DW under 30 °C. The possible reason behind this increment is the proportional relationship between temperature and the photosynthesis rate. However, glucosinolate components and contents are degraded under both very hot and cold temperatures. Although the light types, intensity, photoperiod, and EC-pH were kept constant in this study (for experiments 1, 2, and 3), the variation of glucosinolate levels was observed due to the individual effect of temperature, humidity, and CO_2. As the experiment 1, 2, and 3 were conducted separately, the overaccumulation of glucosinolates under temperature treatments might have occurred due to the overall growing condition; however, it is very important to maintain consistency between experiments. To minimize the inconsistency between experiments, the following measurements could be considered to handle and minimize the variations. First, similar seedlings could be prepared as much as possible, so that pre-transplanting cannot affect the final harvested product. Moreover, the number of samples could be increased by cultivating kale in bigger and multiple plant beds. Finally, maintenance of the same cultivation condition through more accurate and precise control of the environmental variables is necessary.

In the correlation matrixes, strong positive correlations were observed among all the physical growth variables, except for the chlorophyll level. Negative correlations with other growth variables were shown. The most likely reason for this phenomenon is that chlorophyll is an indicator of the health of the photosynthetic apparatus, and the concentration (amount per mass) is a function of the leaf area. As the midrib and petiole of kale (depending on the cultivar) are large, the midrib might become enlarged, diluting the concentration of chlorophyll in the lamina during the growth period, which results in negative correlations with other growth variables [60]. Moreover, the efficiency of chlorophyll varies over time due to the engagement–disengagement of assorted photoprotective mechanisms under fluctuating light conditions. This results in energy loss (absorbed by chlorophyll as heat) and affects carbohydrate (glucose) accumulation. This might be another reason for the negative correlation with kale growth [61,62]. Besides this, glucosinolate components were strongly positively correlated with each other under elevated CO_2 concentrations, because glucosinolate synthesis is proportionally related to photosynthesis [38]. The mechanisms of biochemical reactions are, in fact, very complex, and in many processes, the biochemical pathways are only hypothetical or assumed, and the intermediate reactions and products are not fully known. At any stage in the biochemical chain, double bonds, which are very reactive, may be affected by temperature, relative humidity, and CO_2 as well as by free radicals in the environment. Particularly, glucosinolate synthesis can be illustrated according to the following steps: (1) radical substitution and the addition of water occurs at elevated temperatures and/or in the presence of radicals; (2) addition reactions to carbon–nitrogen double bonds, resulting in carbonic acid esters; (3) electrophilic addition of water to double bonds that creates two new sigma bonds, resulting in the formation of alcohol; and (4) the occurrence of rearrangement, transposition, and isomerization involving double bonds, allyl radicals, and the glucose cycle. The increased reactivity of double bonds makes them very susceptible to environmental factors, specifically temperature, relative humidity, and CO_2.

Cartea and Velasco [6] reported that the concentrations of glucosinolate components vary depending on genetics and environmental factors, along with the crop cultivation methods, harvest, storage, and even the processes of meal preparation. Velasco et al. [49] specifically showed that the concentrations of aliphatic glucosinolate components gradually increase in vegetative tissues (i.e., leaves) and are transferred to the reproductive tissues (i.e., flowers and seeds) during the flowering period. In addition, the indole glucosinolate components of leaves and flower buds gradually decrease after a certain period of cultivation. However, the concentrations of aromatic glucosinolates do not vary significantly with the cultivation period. In this study, kale was cultivated in the plant factory using an aeroponic method (one type of hydroponics). The fast growth rate due to proper ambient environment and nutrient management might be a reason for high glucosinolate accumulation in the early stage (two weeks after transplantation), and it gradually declined with the cultivation period (four weeks after transplantation). Determination of the proper harvesting time of brassicaceous plants has been investigated in several studies [63,64]. Based on the environmental factors and cultivation methods used in this study, early harvesting (2–3 weeks after transplantation) is suggested as a possible strategy to achieve glucosinolate-rich kale.

5. Conclusions

This study was conducted to investigate the effects of temperature, relative humidity, and CO_2 on the growth and glucosinolate content of kale plants hydroponically grown in a plant factory, where five different treatments of each environmental variable were applied separately, and samples were collected after two different periods of cultivation. According to the results, the optimal temperature, relative humidity, and CO_2 range for growth and total glucosinolate content were 20–23 °C, 85%, and 700–1000 ppm, and 14–17 °C, 55–75%, and 1300–1600 ppm, respectively. The glucosinolate content of kale was high in the early growth stage, with low temperature and humidity levels, and elevated CO_2 concentrations. Strong positive correlations were observed among the physical growth variables, and weak correlations were found between the growth and glucosinolate parameters, which indicated that high physical growth might not ensure the high concentration of glucosinolates. According to the findings of this study, early harvesting (i.e., after 2 weeks of transplantation) could be preferred. As the optimum level of temperature, humidity, and CO_2 was different in two- and four-week sampling times, dynamic ambient environment management might be adopted. Farmers could maintain the optimum range of each environmental variable separately based on their target (growth or glucosinolate level), or preferred combined management of the temperature, relative humidity, and CO_2 during kale cultivation within protected cultivation facilities.

Author Contributions: Conceptualization, S.-O.C. and M.C.; methodology, S.-O.C. and M.C.; software, M.C.; validation, M.C.; formal analysis, M.C., S.K., M.N.I., M.A., M.N.R., W.-H.L. and S.-O.C.; investigation, S.-O.C. and W.-H.L.; resources, S.-O.C.; data curation, M.C., S.K., M.N.I., M.N.R. and

M.A.; writing-original draft preparation, M.C.; writing-review and editing, S.-O.C. and W.-H.L.; visualization, M.C., S.K., M.N.I., M.A. and M.N.R.; supervision, S.-O.C.; project administration, S.-O.C.; funding acquisition, S.-O.C. All authors have read and agreed to the published version of the manuscript.

Funding: This work was supported by the Korea Institute of Planning and Evaluation for Technology in Food, Agriculture and Forestry (IPET) through the Agriculture, Food and Rural Affairs Convergence Technologies Program for Educating Creative Global Leaders, funded by Ministry of Agriculture, Food and Rural Affairs (MAFRA) (project No. 320001-4), Republic of Korea.

Institutional Review Board Statement: Not applicable.

Informed Consent Statement: Not applicable.

Data Availability Statement: All the data reported here are available from the authors upon request.

Conflicts of Interest: The authors declare no conflict of interest.

References

1. Chang, J.; Wang, M.; Jian, Y.; Zhang, F.; Zhu, J.; Wang, Q.; Sun, B. Health-promoting phytochemicals and antioxidant capacity in different organs from six varieties of Chinese kale. *Sci. Rep.* **2019**, *9*, 20344. [CrossRef]
2. Abellán, Á.; Domínguez-Perles, R.; Moreno, D.A.; García-Viguera, C. Sorting out the value of cruciferous sprouts as sources of bioactive compounds for nutrition and health. *Nutrients* **2019**, *11*, 429. [CrossRef]
3. Jeon, J.; Kim, J.K.; Kim, H.; Kim, Y.J.; Park, Y.J.; Kim, S.J.; Kim, C.; Park, S.U. Transcriptome analysis and metabolic profiling of green and red kale (*Brassica oleracea* var. *Acephala*) seedlings. *Food Chem.* **2018**, *241*, 7–13. [CrossRef]
4. Giorgetti, L.; Giorgi, G.; Cherubini, E.; Gervasi, P.G.; Croce, C.M.D.; Longo, V.; Bellani, L. Screening and identification of major phytochemical compounds in seeds, sprouts and leaves of tuscan black kale *Brassica oleracea* (L.) ssp Acephala (DC) Var. Sabellica L. *Nat. Prod. Res.* **2018**, *32*, 1617–1626. [CrossRef]
5. Akram, W.; Saeed, T.; Ahmad, A.; Yasin, N.A.; Akbar, M.; Khan, W.U.; Ahmed, S.; Guo, J.; Luo, W.; Wu, T.; et al. Liquiritin elicitation can increase the content of medicinally important glucosinolates and phenolic compounds in chinese kale plants. *J. Sci. Food Agric.* **2020**, *100*, 1616–1624. [CrossRef] [PubMed]
6. Cartea, M.E.; Velasco, P. Glucosinolates in Brassica Foods: Bioavailability in Food and Significance for Human Health. *Phytochem. Rev.* **2008**, *7*, 213–229. [CrossRef]
7. Agerbirk, N.; Olsen, C.E. Glucosinolate Structures in Evolution. *Phytochemistry* **2012**, *77*, 16–45. [CrossRef]
8. Holst, B.; Fenwick, G.R. Glucosinolates. In *Encyclopedia of Food Sciences and Nutrition*, 2nd ed.; Caballero, B., Ed.; Academic Press: Oxford, UK, 2003; pp. 2922–2930. ISBN 978-0-12-227055-0.
9. Mithen, R.F.; Dekker, M.; Verkerk, R.; Rabot, S.; Johnson, I.T. The nutritional significance, biosynthesis and bioavailability of glucosinolates in human foods. *J. Sci. Food Agric.* **2000**, *80*, 967–984. [CrossRef]
10. Johnson, I.T. Glucosinolates: Bioavailability and importance to health. *Int. J. Vitam. Nutr. Res.* **2002**, *72*, 26–31. [CrossRef] [PubMed]
11. Piotrowski, M.; Schemenewitz, A.; Lopukhina, A.; Müller, A.; Janowitz, T.; Weiler, E.W.; Oecking, C. Desulfoglucosinolate sulfotransferases from arabidopsis thaliana catalyze the final step in the biosynthesis of the glucosinolate core structure. *J. Biol. Chem.* **2004**, *279*, 50717–50725. [CrossRef]
12. Neugart, S.; Baldermann, S.; Hanschen, F.S.; Klopsch, R.; Wiesner-Reinhold, M.; Schreiner, M. The intrinsic quality of brassicaceous vegetables: How secondary plant metabolites are affected by genetic, environmental, and agronomic factors. *Sci. Hortic.* **2018**, *233*, 460–478. [CrossRef]
13. Singh, S.K.; Reddy, V.R.; Sharma, M.P.; Agnihotri, R. Dynamics of plant nutrients, utilization and uptake, and soil microbial community in crops under ambient and elevated carbon dioxide. In *Nutrient Use Efficiency: From Basics to Advances*; Rakshit, A., Singh, H.B., Sen, A., Eds.; Springer India: New Delhi, India, 2015; pp. 381–399. ISBN 978-81-322-2169-2.
14. Lee, G.J.; Heo, J.W.; Jung, C.R.; Kim, H.H.; Jo, J.S.; Lee, J.G.; Lee, G.J.; Nam, S.Y.; Hong, E.Y. Effects of artificial light sources on growth and glucosinolate contents of hydroponically grown kale in plant factory. *Prot. Hort. Plant Fac.* **2016**, *25*, 77–82. [CrossRef]
15. Lee, H.H.; Yang, S.C.; Lee, M.K.; Ryu, D.K.; Park, S.; Chung, S.O.; Park, S.U.; Lim, Y.P.; Kim, S.J. Effect of developmental stages on glucosinolate contents in kale (*Brassica oleracea* var. acephala). *Hortic. Sci. Technol.* **2015**, *33*, 177–185. [CrossRef]
16. Kim, K.H.; Chung, S.O. Comparison of plant growth and Glucosinolates of Chinese cabbage and Kale crops under three cultivation conditions. *J. Biosyst. Eng.* **2018**, *43*, 30–36. [CrossRef]
17. Kozai, T. *Smart Plant Factory: The Next Generation Indoor Vertical Farms*; Springer: Singapore, 2018; ISBN 9789811310652.
18. Chowdhury, M.; Kabir, M.S.N.; Kim, H.T.; Chung, S.O. Method of pump, pipe, and tank selection for aeroponic nutrient management systems based on crop requirements. *J. Agric. Eng.* **2020**, *51*, 119–128. [CrossRef]
19. Kozai, T.; Niu, G.; Takagaki, M. *Plant Factory: An Indoor Vertical Farming System for Efficient Quality Food Production*; Academic Press: London, UK, 2019; ISBN 978-0-12-816692-5.

20. Chowdhury, M.; Jang, B.E.; Kabir, M.S.N.; Lee, D.H.; Kim, H.T.; Park, T.S.; Chung, S.O. Performance Evaluation of Commercial Ion-Selective Electrodes for Hydroponic Cultivation System. *Acta Hortic.* **2020**, 831–838. [CrossRef]
21. Jung, D.H.; Kim, H.J.; Choi, G.L.; Ahn, T.I.; Son, J.E.; Sudduth, K.A. Automated lettuce nutrient solution management using an array of ion-selective electrodes. *Trans. ASABE* **2015**, *58*, 1309–1319.
22. Chowdhury, M.; Jang, B.E.; Kabir, M.S.N.; Kim, Y.J.; Na, K.D.; Park, S.B.; Chung, S.O. Factors affecting the accuracy and precision of ion-selective electrodes for hydroponic nutrient supply systems. *Acta Hortic.* **2020**, 997–1004. [CrossRef]
23. Cho, W.J.; Kim, H.J.; Jung, D.H.; Kim, D.W.; Ahn, T.I.; Son, J.E. On-site ion monitoring system for precision hydroponic nutrient management. *Comput. Electron. Agric.* **2018**, *146*, 51–58. [CrossRef]
24. Benton Jones, J. *Hydroponics: A Practical Guide for the Soilless Grower*; CRC Press: Boca Raton, FL, USA, 2016; ISBN 978-1-4200-3770-8.
25. Li, Q.; Li, X.; Tang, B.; Gu, M. Growth responses and root characteristics of lettuce grown in aeroponics, hydroponics, and substrate culture. *Horticulturae* **2018**, *4*, 35. [CrossRef]
26. Wu, X.; Huang, H.; Childs, H.; Wu, Y.; Yu, L.; Pehrsson, P.R. Glucosinolates in Brassica vegetables: Characterization and factors that influence distribution, content, and intake. *Annu. Rev. Food Sci. Technol.* **2021**, *12*, 43–73. [CrossRef]
27. Alegre, S.; Pascual, J.; Trotta, A.; Gollan, P.J.; Yang, W.; Yang, B.; Aro, E.-M.; Burow, M.; Kangasjärvi, S. Growth under high light and elevated temperature affects metabolic responses and accumulation of health-promoting metabolites in kale varieties. *bioRxiv* **2019**, 816405. [CrossRef]
28. McClung, C.R.; Lou, P.; Hermand, V.; Kim, J.A. The Importance of ambient temperature to growth and the induction of flowering. *Front. Plant Sci.* **2016**, *7*. [CrossRef]
29. Steindal, A.L.H.; Rødven, R.; Hansen, E.; Mølmann, J. Effects of photoperiod, growth temperature and cold acclimatisation on glucosinolates, sugars and fatty acids in Kale. *Food Chem.* **2015**, *174*, 44–51. [CrossRef]
30. Lee, J.H.; Oh, M.M. Short-term low temperature increases phenolic antioxidant levels in kale. *Hortic. Environ. Biotechnol.* **2015**, *56*, 588–596. [CrossRef]
31. Maibam, A.; Nisar, S.; Zargar, S.M.; Mahajan, R. High-temperature response and tolerance in agronomic crops. In *Agronomic Crops: Volume 3: Stress Responses and Tolerance*; Hasanuzzaman, M., Ed.; Springer: Singapore, 2020; pp. 173–190. ISBN 9789811500251.
32. Soengas, P.; Rodríguez, V.M.; Velasco, P.; Cartea, M.E. Effect of temperature stress on antioxidant defenses in *Brassica oleracea*. *Acs Omega* **2018**, *3*, 5237–5243. [CrossRef]
33. Anjum, S.A.; Xie, X.; Wang, L.; Saleem, M.F.; Man, C.; Lei, W. Morphological, physiological and biochemical responses of plants to drought stress. *Afr. J. Agric. Res.* **2011**, *6*, 2026–2032. [CrossRef]
34. Han, W.; Yang, Z.; Huang, L.; Sun, C.; Yu, X.; Zhao, M. Fuzzy comprehensive evaluation of the effects of relative air humidity on the morpho-physiological traits of Pakchoi (*Brassica Chinensis* L.) under high temperature. *Sci. Hortic.* **2019**, *246*, 971–978. [CrossRef]
35. Avotins, A.; Gruduls, J.; Apse-Apsitis, P.; Bicāns, J. Development and testing results of iot based air temperature and humidity measurement system for industrial greenhouse. *Agron. Res.* **2018**, *16*, 943–951.
36. Amani, M.; Foroushani, S.; Sultan, M.; Bahrami, M. Comprehensive review on dehumidification strategies for agricultural greenhouse applications. *Appl. Therm. Eng.* **2020**, *181*, 115979. [CrossRef]
37. Islam, M.N.; Iqbal, M.Z.; Ali, M.; Jang, B.E.; Chowdhury, M.; Kabir, M.S.N.; Jang, S.H.; Chung, S.O. Performance evaluation of a suspension-type dehumidifier with a heating module for smart greenhouses. *J. Biosyst. Eng.* **2020**, *45*, 155–166. [CrossRef]
38. La, G.; Fang, P.; Teng, Y.; Li, Y.; Lin, X. Effect of CO_2 enrichment on the Glucosinolate contents under different nitrogen levels in bolting stem of Chinese Kale (*Brassica Alboglabra* L.). *J. Zhejiang Univ. Sci. B* **2009**, *10*, 454–464. [CrossRef] [PubMed]
39. Ainsworth, E.A.; Long, S.P. What have we learned from 15 years of free-air CO_2 enrichment (FACE)? A Meta-Analytic Review of the Responses of Photosynthesis, Canopy Properties and Plant Production to Rising CO_2. *New Phytol.* **2005**, *165*, 351–372. [CrossRef] [PubMed]
40. Madhu, M.; Hatfield, J.L. Dynamics of plant root growth under increased atmospheric Carbon Dioxide. *Agron. J.* **2013**, *105*, 657–669. [CrossRef]
41. Kimball, B.A. Crop responses to elevated CO_2 and interactions with H_2O, N, and temperature. *Curr. Opin. Plant Biol.* **2016**, *31*, 36–43. [CrossRef]
42. Chung, S.O.; Kang, S.W.; Bae, K.S.; Ryu, M.J.; Kim, Y.J. The potential of remote monitoring and control of protected crop production environment using mobile phone under 3G and Wi-Fi communication conditions. *Eng. Agric. Environ. Food* **2015**, *8*, 251–256. [CrossRef]
43. Zhang, Y.; Ji, J.; Song, S.; Su, W.; Liu, H. Growth, nutritional quality and health-promoting compounds in Chinese Kale grown under different ratios of red:blue LED lights. *Agronomy* **2020**, *10*, 1248. [CrossRef]
44. Lefsrud, M.G.; Kopsell, D.A.; Sams, C.E. Irradiance from distinct wavelength light-emitting diodes affect secondary metabolites in Kale. *HortScience* **2008**, *43*, 2243–2244. [CrossRef]
45. Naznin, M.T.; Lefsrud, M.; Gravel, V.; Azad, M.O.K. Blue light added with red leds enhance growth characteristics, pigments content, and antioxidant capacity in lettuce, spinach, kale, basil, and sweet pepper in a controlled environment. *Plants* **2019**, *8*, 93. [CrossRef]
46. Doheny-Adams, T.; Redeker, K.; Kittipol, V.; Bancroft, I.; Hartley, S.E. Development of an efficient glucosinolate extraction method. *Plant Methods* **2017**, *13*, 17. [CrossRef]

47. ISO 9167:2019 Rapeseed and Rapeseed Meals—Determination of Glucosinolates Content—Method Using High-Performance Liquid Chromatography. Available online: https://www.iso.org/standard/72207.html (accessed on 15 June 2021).
48. Hair, J.F., Jr.; Black, W.C.; Babin, B.J.; Anderson, R.E. *Multivariate Data Analysis: A Global Perspective*, 7th ed., Prentice Hall: Upper Saddle River, NJ, USA, 2009.
49. Velasco, P.; Cartea, M.E.; González, C.; Vilar, M.; Ordás, A. Factors affecting the glucosinolate content of kale (*Brassica oleracea Acephala* Group). *J. Agric. Food Chem.* **2007**, *55*, 955–962. [CrossRef]
50. Bohinc, T.; Trdan, S. Environmental factors affecting the glucosinolate content in brassicaceae. *J. Food Agric. Environ.* **2012**, *10*, 357–360.
51. Hatfield, J.L.; Prueger, J.H. Temperature extremes: Effect on plant growth and development. *Weather Clim. Extrem.* **2015**, *10*, 4–10. [CrossRef]
52. Keenan, T.F.; Richardson, A.D.; Hufkens, K. On Quantifying the apparent temperature sensitivity of plant phenology. *New Phytol.* **2020**, *225*, 1033–1040. [CrossRef] [PubMed]
53. Lapenis, A.; Henry, H.; Vuille, M.; Mower, J. Climatic factors controlling plant sensitivity to warming. *Clim. Chang.* **2014**, *122*, 723–734. [CrossRef]
54. Ahmed, H.A.; Yu-Xin, T.; Qi-Chang, Y. Optimal Control of Environmental Conditions Affecting Lettuce Plant Growth in a Controlled Environment with Artificial Lighting: A Review. *S. Afr. J. Bot.* **2020**, *130*, 75–89. [CrossRef]
55. Kaiser, E.; Morales, A.; Harbinson, J.; Kromdijk, J.; Heuvelink, E.; Marcelis, L.F.M. Dynamic photosynthesis in different environmental conditions. *J. Exp. Bot.* **2015**, *66*, 2415–2426. [CrossRef] [PubMed]
56. Chen, X.; Zhu, Z.; Gerendás, J.; Zimmermann, N. Glucosinolates in Chinese Brassica Campestris Vegetables: Chinese Cabbage, Purple Cai-Tai, Choysum, Pakchoi, and Turnip. *HortScience* **2008**, *43*, 571–574. [CrossRef]
57. He, H.; Liu, L.; Song, S.; Tang, X.; Wang, Y. Evaluation of glucosinolate composition and contents in chinese brassica vegetables. *Acta Hortic.* **2003**, 85–92. [CrossRef]
58. Qian, H.; Liu, T.; Deng, M.; Miao, H.; Cai, C.; Shen, W.; Wang, Q. Effects of Light Quality on Main Health-Promoting Compounds and Antioxidant Capacity of Chinese Kale Sprouts. *Food Chem.* **2016**, *196*, 1232–1238. [CrossRef]
59. Rosa, E.A.S.; Rodrigues, P.M.F. The Effect of Light and Temperature on Glucosinolate Concentration in the Leaves and Roots of Cabbage Seedlings. *J. Sci. Food Agric.* **1998**, *78*, 208–212. [CrossRef]
60. Lin, K.H.; Shih, F.C.; Huang, M.Y.; Weng, J.H. Physiological characteristics of photosynthesis in yellow-green, green and dark-green Chinese kale (*Brassica oleracea* L. var. *Alboglabra Musil.*) under varying light intensities. *Plants* **2020**, *9*, 960. [CrossRef]
61. Blankenship, R.E. *Molecular Mechanisms of Photosynthesis*; John Wiley & Sons: St. Louis, MO, USA, 2014; ISBN 978-1-4051-8976-7.
62. Eberhard, S.; Finazzi, G.; Wollman, F.A. The dynamics of photosynthesis. *Annu. Rev. Genet.* **2008**, *42*, 463–515. [CrossRef]
63. Hagen, S.F.; Borge, G.I.A.; Solhaug, K.A.; Bengtsson, G.B. Effect of cold storage and harvest date on bioactive compounds in curly kale (*Brassica oleracea L.* var. *Acephala*). *Postharvest Biol. Technol.* **2009**, *51*, 36–42. [CrossRef]
64. Yoon, H.I.; Kim, J.S.; Kim, D.; Kim, C.Y.; Son, J.E. Harvest strategies to maximize the annual production of bioactive compounds, glucosinolates, and total antioxidant activities of kale in plant factories. *Hortic. Environ. Biotechnol.* **2019**, *60*, 883–894. [CrossRef]

foods

MDPI

Article

Seasonal Variation of Glucosinolate Hydrolysis Products in Commercial White and Red Cabbages (*Brassica oleracea* var. *capitata*)

Nicole S. Wermter [1,2], Sascha Rohn [2,3] and Franziska S. Hanschen [1,*]

1 Plant Quality and Food Security, Leibniz-Institute of Vegetable and Ornamental Crops (IGZ), Theodor-Echtermeyer-Weg 1, 14979 Grossbeeren, Germany; nicolewermter@yahoo.de

2 Hamburg School of Food Science, Institute of Food Chemistry, University of Hamburg, Grindelallee 117, 20146 Hamburg, Germany; rohn@chemie.uni-hamburg.de

3 Department of Food Chemistry and Analysis, Institute of Food Technology and Food Chemistry, Technische Universität Berlin, TIB 4/3-1, Gustav-Meyer-Allee 25, 13355 Berlin, Germany

* Correspondence: hanschen@igzev.de

Received: 26 October 2020; Accepted: 12 November 2020; Published: 17 November 2020

Abstract: *Brassica* vegetables contain glucosinolates, which are well-known for their potential to form health-promoting isothiocyanates. Among those crucifers, white and red cabbage are commonly consumed vegetables, exhibiting different glucosinolate and hydrolysis profiles thereof. Regarding the health beneficial effects from these vegetables, more information, especially concerning the seasonal variation of glucosinolate profiles and the formation of their bioactive hydrolysis products in commercial cabbages, is needed. In this study, glucosinolates and glucosinolate hydrolysis product profiles in red and white cabbages from three different food retailers were monitored over six different sampling dates across the selling season in autumn. For the first time, it was shown that, while glucosinolate profiles were similar in each cabbage variety, glucosinolate hydrolysis product profiles and hydrolysis behavior varied considerably over the season. The highest total isothiocyanate concentrations were observed in conventional red (1.66 µmol/g FW) and organic white (0.93 µmol/g FW) cabbages purchased at the first sampling date in September. Here, red cabbage was with up to 1.06 µmol/g FW of 4-(methylsulfinyl)butyl isothiocyanate (sulforaphane), an excellent source for this health-promoting isothiocyanate. Cabbages purchased 11 weeks later in autumn released lower levels of isothiocyanates, but mainly nitriles and epithionitriles. The results indicate that commercial cabbages purchased in early autumn could be healthier options than those purchased later in the year.

Keywords: glucosinolate; cabbage; isothiocyanate; epithionitrile; nitrile; *Brassica*; seasonal variation; food retailer

1. Introduction

With a consumption of 5.2 kg/person and year in 2017/2018, red cabbage (*Brassica oleracea* var. *capitata* f. *rubra*) and white cabbage (*Brassica oleracea* var. *capitata* f. *alba*) are the most consumed Brassicaceae vegetables in Germany, contributing to 5% of total vegetable intake [1]. These vegetables are rich in glucosinolates (GLSs), secondary, sulfurous plant constituents, which are particularly present in vacuoles of plant cells of the Brassicaceae family. The chemical GLS structure is determined by the nature of the side chain, depending on the amino acid inserted during biosynthesis [2,3] and GLSs typically contain a glucose unit, bound with a central carbon atom with nitrogen grouping via a thioether bridge. The carbon atom, in turn, is linked to a sulfate group and an organic aglycone residue, possessing an alkyl, alkenyl, aryl, or indole group [2]. This organic residue is inherent to the GLS, its chemical properties and its flavor, respectively [4,5]. Upon attack by herbivores or due to cutting or chopping of vegetables rich in GLSs, GLSs, which were previously present in spatially separated cell vacuoles, are hydrolyzed by myrosinase to various herbivore-toxic degradation products [6,7]. Hydrolysis by myrosinase occurs due to enzymatic cleavage of the thioglycoside bond, first resulting in an unstable aglucone (thiohydroximate-*O*-sulfate). The aglucone can then undergo a LOSSEN-like rearrangement to form isothiocyanates (ITCs) or decompose into nitriles and molecular sulfur. Moreover, in the presence of certain proteins such as the epithiospecifier protein (ESP), an aglycone with a terminal double bond favors epithionitrile (ETN) formation [7].

Consumption of ITCs can positively affect human health as they have antimicrobial, antidiabetogenic, chemopreventive, and anticarcinogenic properties [8,9]. Previous studies have shown a positive correlation between ITC uptake and cancer prevention [8,10,11], and especially 4-(methylsulfinyl)butyl ITC (sulforaphane; 4MSOB-ITC) is valued for its anticarcinogenic potential [8]. Simple nitriles and ETNs on the other hand, seem to have less health beneficial effects [12,13]. Studies have elucidated that *Brassica* vegetables, not only ITCs, but also nitriles and ETNs can be the most predominant degradation products [14,15]. Consequently, in order to estimate the health beneficial potential of *Brassica* vegetables, it is of great importance to not only analyze intact GLSs, but also their behavior during hydrolysis.

The natural GLS content in vegetables varies in accordance with genotype, plant developmental step, soil and cultivation conditions, and other ecophysiological influences, but it is also affected by storage [14,16,17]. Moreover, GLS levels vary over the growth season and several studies reported higher GLS levels in *B. oleracea* plants grown in spring compared to plants grown in autumn [17], while in broccoli (*B. oleracea* var. *italica*), GLS levels were higher when grown in the summer season compared to the spring season [18]. Similarly, Nuñez-Gómez et al. (2020) recently reported higher GLS levels in broccoli grown in autumn compared to broccoli grown in the spring season. Moreover, when comparing two spring seasons, the one with less rainy days and higher temperature resulted in higher GLS levels [19]. Experiments performed under controlled conditions indicate that temperature, as well as day length affect GLS biosynthesis in *B. oleracea* in a structure-dependent way [20–22].

In order to predict GLS-based health beneficial properties, more information especially concerning GLS profiles and especially on their hydrolysis behavior in *Brassica* foods available for the consumer, is needed. Possessing a long harvesting period—typically ranging from June to November—red and white cabbages are facing seasonal changes and different ecophysiological influences, which can have a great impact on the GLS profile [16,20] and might also affect the potential to form health-preventive ITC. To date, little is known about how GLS hydrolysis products are affected by different cultivation conditions. No revealing insight has been given on how GLS, and even more importantly, the formation behavior of their hydrolysis products varies in commercial cabbages across the whole season. Therefore, the objective of this study was to monitor the variation of GLS levels and the formation of their bioactive hydrolysis products in commercial red and white cabbage heads obtained from three local retailers in Germany and to link the data with the cultivation practices and post-harvest storage conditions.

2. Materials and Methods

2.1. Chemicals

Allyl GLS (Allyl; ≥99%, reference compound), 4-hydroxybenzyl GLS (4OHbenzyl; purity ≥99%, internal standard), and methylene chloride (GC Ultra Grade, solvent) were purchased from Carl Roth GmbH and Co. KG (Karlsruhe, Germany). Allyl ITC (Allyl-ITC; ≥99%, reference compound), benzonitrile (≥99.9%, internal standard), DEAE-Sephadex A-25 (anion exchanger), and the reference compounds 3-butenenitrile (Allyl-CN; ≥98%), 4-pentenenitrile (3But-CN; ≥97%), 3-phenylpropanenitrile (≥99%), were obtained from Sigma-Aldrich Chemie GmbH (Steinheim, Germany). The reference compounds 3-butenyl ITC (3But-ITC; ≥95%) and 4-pentenyl ITC (≥95%) were purchased from TCI Deutschland GmbH (Eschborn, Germany). 3-(Methylsulfinyl)propyl ITC (3MSOP-ITC) and 4-(methylsulfanyl)butyl ITC (4MTB-ITC; ≥98%)

were purchased from Santa Cruz Biotechnology (Heidelberg, Germany). 4-(Methylsulfinyl)butyl ITC (4MSOB-ITC) was purchased from Enzo Life Sciences GmbH (Lörrach, Germany). The ETN 1-cyano-2,3-epithiopropane (CETP; ≥95%) was synthesized by Taros Chemicals GmbH and Co. KG (Dortmund, Germany) and 1-cyano-3,4-epithiobutane (CETB; ≥95%) was synthetized by ASCA GmbH Angewandte Synthesechemie Adlershof (Berlin, Germany). 5-(Methylsulfanyl)pentanenitrile (4MTB-CN) and 5-(methylsulfinyl)pentanenitrile (4MSOB-CN) were purchased from Enamine (SIA Enamine, Latvia, Riga). 3-Butenyl GLS (3But; ≥95%), 2(*R*)-hydroxy-3-butenyl GLS (2OH3But; ≥98%), 4-(methylsulfanyl)butyl GLS (4MTB; ≥98%), 4-(methylsulfinyl)butyl GLS (4MSOB; ≥98%), 3-(methylsulfinyl)propyl GLS (3MSOP; ≥98%), and 2-phenylethyl GLS (2PE; ≥98%) were acquired from Phytolab GmbH and Co. KG, Vestenbergsgreuth, Germany. The solvents methanol (≥99.95%), acetonitrile (LC-MS grade), and arylsulfatase (enzyme) were purchased from Th. Geyer GmbH and Co. KG (Renningen, Germany).

2.2. Plant Material

Three red cabbages (*Brassica oleracea* convar. *capitata* var. *rubra* L.) and three white cabbages (*Brassica oleracea* convar. *capitata* var. *alba)* were purchased at regular intervals from the same two local conventional supermarkets (CON_1, CON_2) and from the same organic supermarket (ORG_1) in Brandenburg, Germany over a period of 3 months (September–November 2019). The supermarkets were selected with regard to different German food trading companies. The two conventional supermarkets selected belong to the two biggest food trading companies in Germany and the organic supermarket also belongs to a big organic food supermarket chain. The exact sampling dates (S1–S6) are given in Table 1. Additionally, red and white cabbage heads, grown in the field at the Leibniz-Institute of Vegetable and Ornamental Crops (IGZ) in Grossbeeren, Germany, were harvested freshly in order to compare GLSs and their hydrolysis products with commercial cabbages. Therefore, red (cultivar 'Redma RZ F1') and white (cultivar 'Dottenfelder Dauer') cabbage seeds were sown (13.06.2019 and 20.06.2019) on loamy soil (pH 7.3) and then grown for 3 months at the IGZ (52°20′59.0″ N 13°18′57.5″ E). The red cabbage was cultivated with 100% of the required nitrogen level and fertilized using calcium ammonium nitrate and Patentkali® (419 kg N/ha). Before cultivation, the field was fertilized once with calcium ammonium nitrate (CAN) and Patentkali® and later fertilized a second time, with CAN only during the cultivation period. In total, 377.78 kg CAN/ha and 40.74 kg Patentkali® were applied for fertilization.

For white cabbage, Aminofert® Vinasse fertilizer (BayWa AG, Munich, Germany) was applied (60 kg N/ha) with 30% of the required nitrogen level. Here, fertilization occurred before the sowing of the seeds on 06 June 2019 and a second time on 25 June 2019 for head formation. The origin and cultivation background of commercial cabbages were investigated by interrogating service staff at supermarkets and by contacting growers. Cabbages sold at CON_1 and ORG_1 were procured from different farming areas in northern Germany (CON_1: Neuenkirchen and Helse (Schleswig-Holstein), Germany; ORG_1: Blankensee (Mecklenburg-Western Pomerania), Hedwigenkoog (Schleswig-Holstein), Vierlinden and Seeblick (Brandenburg, Germany), whereas cabbages purchased from CON_2 could continuously be procured from the same farming area, but on different fields within a 30 km perimeter in Neuenkirchen, Schleswig-Holstein, Germany) (Table 2). With regard to the cultivars, mainly white cabbage varieties such as "Storema", "Lennox", "Marcello", and "Impala" and red cabbage varieties, especially "Futurima", "Rodima", "Bandolero", and "Klimaro" were cultivated in the Dithmarschen region (Helse, Neuenkirchen) for CON_1 and CON_2. Cultivars "Rodynda"(red cabbage) and "Dowinda" (white cabbage) were mainly cultivated in Hohennauen, Germany for ORG_1. Early cabbage cultivars ("Marcello", "Bandolero") were likely purchased between 04 and 05 September 2019, whereas later cabbage cultivars ("Storema", "Lennox", "Impala", "Futurima", "Klimaro", "Rodynda", "Dowinda") could be purchased between 09 September 2019 and 06 November 2019.

Table 1. Overview of the dates of the purchased and harvested red and white cabbage heads.

Supplier	Abbreviation	Date of Purchase	
		White Cabbage	Red Cabbage
CON_1	S1	04.09.2019	04.09.2019
	S2	16.09.2019	16.09.2019
	S3	30.09.2019	30.09.2019
	S4	21.10.2019	21.10.2019
	S5	04.11.2019	04.11.2019
	S6	18.11.2019	18.11.2019
CON_2	S1	05.09.2019	05.09.2019
	S2	16.09.2019	16.09.2019
	S3	30.09.2019	30.09.2019/01.10.2019
	S4	21.10.2019	21.10.2019/23.10.2019
	S5	04.11.2019	4.11.2019/06.11.2019
	S6	18.11.2019	18.11.2019
ORG_1	S1	09.09.2019	09.09.2019
	S2	19.09.2019	19.09.2019
	S3	01.10.2019	01.10.2019
	S4	23.10.2019	23.10.2019
	S5	06.11.2019	06.11.2019
	S6	20.11.2019	20.11.2019
Fresh harvest from field (IGZ)		**Date of harvest**	
		White Cabbage	**Red Cabbage**
		29.10.2019	10.10.2019

Table 2. Total overview of the German origin, cultivation, and storage conditions of commercial red and white cabbages and red and white cabbages from the field experiment. Purchase dates of S1–S6 can be found in Table 1.

Supplier	Origin of Cultivation	Samples/Date of Harvest (IGZ)	Harvest Season	Cabbage Type	Cabbage Genotype	Soil Type and Field	Fertilizer	Certifica-Tion Mark	Storage Conditions *1	Storage Conditions *2
CON_1	25792, Neuenkirchen Schleswig-Holstein	S1–S3	Summer, autumn	(1), (2)	White cabbage hybrids	Sea marsh	Urea, CAN, phosphate and potash	QS	(i)	20 °C day/night
		S1–S4			Red cabbage hybrids					
CON_1	25709, Helse Schleswig-Holstein	S4–S6	Autumn, winter	(2), (3)	White cabbage hybrids	Sea marsh, pH-value 7.0–7.4	PKS fertilizer (blends), ammonium nitrate and urea, calcium cyanamide	QS	(ii)	20 °C day/night
		S5, S6			Red cabbage hybrids					
CON_2	25792, Neuenkirchen Schleswig-Holstein	S1–S6	Summer, autumn, winter	(1), (2), (3)	Red and white cabbage hybrids	Sea marsh	Urea, CAN, phosphate and potash	QS	(iii)	7–10 °C day/night in cooling counter, max. 1 week
ORG_1	17237, Blankensee Mecklenburg-Western Pomerania	S1, S4, S5	Autumn	(2)	White cabbage hybrids	Sandy loam	Hair-meal pellets	Bioland	(iv)	10 °C day/night in cooling counter, max. 1 week
		S1			Red cabbage hybrids					
ORG_1	25761, Hedwigenkoog Schleswig-Holstein	S2, S3	Autumn	(2)	White cabbage, non-hybrid	Sea Marsh	Compost	Demeter	(iv)	Day/night in cooling counter, max. 1 week
		S2, S3, S4, S5			Red cabbage, non-hybrid					
ORG_1	15306, Vierlinden Brandenburg	S6	Autumn	(2)	White cabbage hybrids	Sandy loam	Hair-meal pellets	Bioland	(v)	Day/night in cooling counter, max. 1 week
ORG_1	14715, Seeblick Brandenburg	S6	Autumn	(2)	Red cabbage, non-hybrids	Sandy loam	Compost	Demeter	(iv)	Day/night in cooling counter, max. 1 week
IGZ	14979, Großbeeren Brandenburg	10 October 2019	Autumn	(2)	Red cabbage (Redma RZ F1), hybrid	Silty loam	CAN, patentkali	-	No storage	No storage
IGZ	14979, Großbeeren Brandenburg	29 October 2019	Autumn	(2)	White cabbage (Dottenfelder Dauer), non-hybrid	Silty loam	Vinasse	-	No storage	No storage

*1 Before selling, *2 during selling; CAN-calcium ammonium nitrate. (i) 6–8 °C for 1–2 days after harvest, storage for 1–2 days (intermediate trade), then storage at 6–8 °C (cold storage warehouse). (1) Early cabbage (fresh harvest). (ii) 6–7 °C for 1–2 days in central warehouse. (2) Late cabbage (fresh harvest). (iii) 6–8 °C for 1–2 days after harvest, then storage at 6–8 °C in cold storage warehouse. (3) Long-term stored (stored). (iv) 2–6 °C for 1–2 days after harvest, then storage at 4–7 °C for 1–3 days in cold storage warehouse (max 48 h). (v) 0.1–0.3 °C cold storage warehouse.

Long-term stored white cabbages ('Storema", "Lennox", "Impala", "Dowinda") were purchased as of 18 November 2019 and stored at 0.1–0.3 °C in warehouses at the wholesaler. Cabbages from CON_1 and CON_2 were fertilized using a combination of urea, calcium ammonium nitrate, phosphate and potassium, whilst organically cultivated cabbage ORG_1 was fertilized using *hair-meal pellets* (200 kg N/ha) or compost (Table 2). According to growers in Blankensee, Neuenkirchen, Vierlinden, and Helse, only hybrid cultivars such as "Storema", "Impala", "Lennox", and "Bandolero" were grown, harvested, and later sold as ripe cabbages to CON_1 and CON_2, whereas non-hybrid cultivars such as "Rodynda" and "Dowinda" (grown for ORG_1) were generally grown in Hedwigenkoog and Hohennauen. According to growers in Helse, seeds were sown from the 16th to the 20th calendar week of 2019, and cabbages were harvested from the 23rd to the 46th calendar week on heavy, sea marsh soil. Similar sowing and harvesting dates also applied for cabbages cultivated in Neuenkirchen, which were harvested between weeks 23 to 45 and also grown on heavy, sea marsh soil. According to growers in Helse, which supplied CON_1 with cabbage, cabbage heads grown for the conventional market generally grew slower in the Dithmarschen region and were grown for 150 d in Helse on sea marsh soil (late cabbage). The growth of early cultivars (S1 sampling of CON_1 and CON_2) was accelerated with non-woven fibre barriers. The alleged storage conditions, according to all growers before the selling period and conditions in the supermarket during selling time, according to salespersons in CON_1, CON_2, and ORG_1 are listed in Table 2, while in Supplemental Table S1 all information collected during this study on the analyzed cabbages are given for each cabbage separately and in more detail.

In order to give a better understanding of how GLSs and their hydrolysis products can differ in commercial cabbages and how they might change within the season, GLSs and their hydrolysis products were additionally monitored by comparing cabbage heads from an organic and two conventional supermarkets (ORG_1, CON_1, CON_2) with cabbage heads grown at the Leibniz-Institute of Vegetable and Ornamental Crops (IGZ), Grossbeeren, Germany. A further aim of this work was to link the results with the common cultivation practices and the alleged storage conditions (Table 2).

2.3. Sample Preparation

Fresh cabbage heads were chopped in half. Of the two obtained halves, one of the halves was halved again, and two quarters were obtained. Afterwards, one of the obtained quarters was divided into 2–3 strips (weight: 70–180 g, width: 1–1.5 cm) along the middle, and the strips were frozen at −20 °C overnight before lyophilization (11 d) and were later ground. The remaining plant material of the same quarter, from which the strips were obtained, was then cut into small pieces of 1 cm width. The chopped plant material was thoroughly mixed by hand, and 15–20 g fresh cabbage was then given into a round bottom glass vessel for homogenization. Afterwards, 15–20 mL of distilled water was added, in order to obtain a 1:1 ratio of plant material and water. Then, samples were homogenized for 1 min at a rate of 20,000 rpm using a mixer (H04, Edmund Bühler GmbH, Tübingen, Germany) and incubated for 1 h at room temperature (22 °C).

2.4. Analysis of Glucosinolates

To determine the profiles and concentrations of GLS in red and white cabbages, 10 mg of lyophilized powder was extracted and GLS was analyzed as their desulfo-form [23]. Briefly, 10 mg of dry plant powder was extracted thrice using 70% of hot methanol in the presence of 0.025 µmol 4-hydroxybenzyl GLS as an internal standard. The extracts were combined and desulfated on a DEAE-Sephadex A-25 ion-exchanger column using aryl sulfatase. Afterwards, desulfo-GLSs were eluted with 1 mL of water and analyzed using an Agilent UHPLC-DAD-ToF-MS system equipped with a Poroshell 120 EC-C18 column (100 × 2.1 mm, 2.7 µm; Agilent Technologies), a gradient of water, and 40% acetonitrile, as described previously [23]. Desulfo-GLSs were quantified at 229 nm via the internal standard and the calibration factor reported in the DIN EN ISO 9167-1 and calculated on this basis for 4-hydroxybenzyl GLS.

2.5. Determination of Glucosinolate Breakdown Products

For the analysis of GLS hydrolysis products released from red and white cabbage tissue, the protocol described by Hanschen and Schreiner (2017) was followed with small modifications [14]: Briefly, 500 mg of the homogenized fresh samples (containing 50% of water) were weighed into solvent resistant centrifuge tubes. During the first two samplings, 1 g of sample homogenate was used, which might have led to reduced recoveries for nitriles and epithionitriles, due to higher water to solvent ratio. Then, the internal standard benzonitrile (0.2 μmol) was added and GLS hydrolysis products were extracted 3 times using methylene chloride: 2 mL during the first extraction and 1.5 mL of methylene chloride in the second and third extraction. Then, samples were analyzed as described previously [14], except that in the present study an Agilent J&W VF-5ms GC-MS column (30 m × 0.25 mm × 0.25 μm) coupled to a 10 m EZ-Guard P/N:CP9013 column was used for analyte separation.

2.6. Statistical Analysis

To investigate differences between different sampling dates (S1–S6), means were compared using ANOVA and Tukey's HSD test and STATISTICA version 13.5.0.17 software (TIBCO Software Inc., Palo Alto, CA, USA) with a significance level of $p \leq 0.05$. All analyses were carried out in triplicate by analyzing three biological replicates.

3. Results

Samples (three cabbage heads) from each of the three retailers (conventional supermarkets CON_1, CON_2, and organic supermarket ORG_1) were collected between September and November 2019 every 2 to 3 weeks, summing up to a total of six sampling dates (S1–S6). GLS-profiles and the corresponding GLS hydrolysis products were monitored over the six sampling periods and additionally compared to fresh samples (four red and white cabbage heads) harvested from a field at the IGZ in Grossbeeren, Germany. Most analyzed cabbage heads differed in their regional origin and harvest time (Table 2). The dates of the purchased or harvested cabbages over the six sampling periods (S1–S6) are listed in Table 1.

3.1. Glucosinolates in White and Red Cabbage from Local Food Retailers

The GLS profile of the most abundant GLS of white cabbage purchased from the different food retailers over the 3-month period is given in Figure 1A–C, whilst the GLS profile for red cabbage is displayed in Figure 1D–F. Table 3 shows the chemical structures of the most abundant cabbage GLS, as well as their GLS hydrolysis product names including the abbreviations. In the heads of the analyzed white and red cabbage cultivars, a total of 12 chemically different GLSs were detected (Table S2). The main GLSs were allyl GLS (Allyl), 3-butenyl GLS (3But), 2-hydroxy-3-butenyl GLS (2OH3But), 3-(methylsulfinyl)propyl GLS (3MSOP), 4-(methylsulfinyl)butyl GLS (4MSOB), and indol-3-ylmethyl GLS (I3M) (Figure 1). The GLS profile in white and red cabbage cultivars between different supermarkets was often found to be similar within the same cabbage variety. In that way, Allyl, 3MSOP, and I3M were found to be most dominant in white cabbage, with maximum Allyl concentrations reaching 0.71 ± 0.02 μmol/g FW in S5 (Figure 1A), up to 0.64 ± 0.17 μmol/g FW 3MSOP in S1 of ORG_1, and up to 0.49 ± 0.12 μmol/g FW I3M in S4 of ORG_1 (Figure 1C). Red cabbage was often the richest in 2OH3But and 4MSOB (Figure 1D–F), with up to 0.99 ± 0.19 μmol/g FW 2OH3But in S1 of CON_1 (Figure 1D) and 0.93 ± 0.05 μmol/g FW 4MSOB in S1 of CON_1 (Figure 1D), respectively. However, Allyl, 3But, 3MSOP, and I3M were also formed in considerable amounts in red cabbage (Figure 1D–F).

Table 3. Structures of main cabbage glucosinolates, their abbreviations, and their hydrolysis products analyzed in the present study. N.d.: Not detected.

Glucosinolates (GLSs)			Corresponding Breakdown Products					
			Isothiocyanate (ITC)		Nitrile		Epithionitrile (ETN)	
Structure	Abbreviation	Name (trivial name)	Abbreviation	Name	Abbreviation	Name	Abbreviation	Name
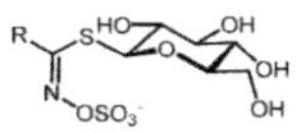								
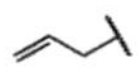	Allyl	allyl GLS (sinigrin)	Allyl-ITC	2-propenyl ITC	Allyl-CN	3-butenenitrile	CETP	1-cyano-2,3-epithiopropane
	3But	3-butenyl GLS (gluconapin)	3But-ITC	3-butenyl ITC	3But-CN	4-pentenenitrile	CETB	1-cyano-3,4-epithiobutane
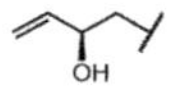	2OH3But	2-(*R*)-2-hydroxy-3- butenyl GLS (progoitrin)	OZT	5-vinyl-1,3-oxazolidine-2-thione		3-hydroxy-pentenenitrile	CHETB A CHETB B	1-cyano-2-hydroxy-3,4-epithiobutane
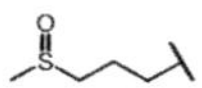	3MSOP	3-(methylsulfinyl)propyl GLS (glucoiberin)	3MSOP-ITC	3-(methylsulfinyl)- propyl ITC	3MSOP-CN	4-(methylsulfinyl)-butanenitrile	-	-
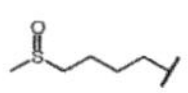	4MSOB	4-(methylsulfinyl)butyl GLS (glucoraphanin)	4MSOB-ITC	4-(methylsulfinyl)butyl ITC	4MSOB-CN	5-(methylsulfinyl)-pentanenitrile	-	-
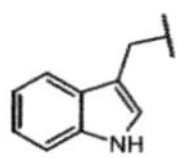	I3M	indol-3-ylmethyl GLS (glucobrassicin)		n.d.		indole-3-acetonitrile	-	-

In general, red cabbage heads produced higher levels of GLSs (Figure 1D–F) compared to white cabbage heads (Figure 1A–C). The highest total GLS levels in red cabbage heads were detected in S1 from CON_1 (3.12 ± 0.27 μmol/g FW) (Figure 1D), whereas the highest total GLS content for white cabbage was observed in S5 from CON_1 (1.61 ± 0.11 μmol/g FW) (Figure 1A).

Whilst general increases in total GLS concentration were found from sampling S1 to S6 (especially due to Allyl) in white cabbages from CON_2 (Figure 1B), a general decreasing trend in total GLS concentration was detected in white cabbages procured from ORG_1 (Figure 1C). In red cabbage, total GLSs varied for the individual samples from the same food retailers in all purchased red cabbage heads (Figure 1D–F), and no specific trend was noted between purchase dates from the different retailers. With regard to individual GLSs, many GLSs in cabbages from the three supermarkets did not significantly change over the different sampling dates, such as GLSs in red cabbage from CON_2 (Figure 1E), while others were affected (Figure 1).

The two major GLSs detected in white cabbage heads were 3MSOP and Allyl. In white cabbages procured from CON_1 (S1-S6, Figure 1A), Allyl increased by 2.1-fold from S1 to S5 and then decreased in S6 to levels similar to S1. In white cabbages, which were purchased from CON_2, a general increase in Allyl content was observed from S1 to S6 (0.17 ± 0.03 to 0.61 ± 0.07 μmol/g FW) (Figure 1B), while in white cabbages purchased from ORG_1 (Figure 1C), Allyl did not significantly change over time. In red cabbage, no significant changes were observable for Allyl (Figure 1D–F). 3MSOP as the other main GLS of white cabbage stayed the same over the sampling period in cabbages from CON_2 and ORG_1, but displayed increased levels in S4 and S5 of cabbages from CON_1 compared to the S1-S3 samples (Figure 1A). In red cabbage, 3MSOP did not change across the consecutive sampling dates (Figure 1D–F). 4MSOB as a major GLS in red cabbage displayed reduced content in S2-S4 compared to S1 in CON_1-cabbage (Figure 1D), but did not change in cabbages from CON_2 and ORG_1. Likewise, in white cabbage of ORG_1, 4MSOB was not affected. In white cabbage from CON_1, 4MSOB slightly decreased from S1 to S5 but was highest in S6 (Figure 1A), and in CON_2-cabbages, this GLS generally displayed similar levels over the sampling dates (Figure 1B). Similarly, 2OH3But varied significantly only in white and red cabbage procured from CON_1 (Figure 1A,D), but not in cabbages from the other supermarkets. In CON_1-cabbages, 2OH3But was highest in S1 (red cabbage) or S2 (white cabbage) samples, then decreased until S4 (red cabbage) or S5 (white cabbage) and then again increased in the last samples (by tendency in red cabbage; significantly in white cabbage). The indole GLS I3M generally had similar levels over the different sampling dates in white and red cabbages, as well (Figure 1).

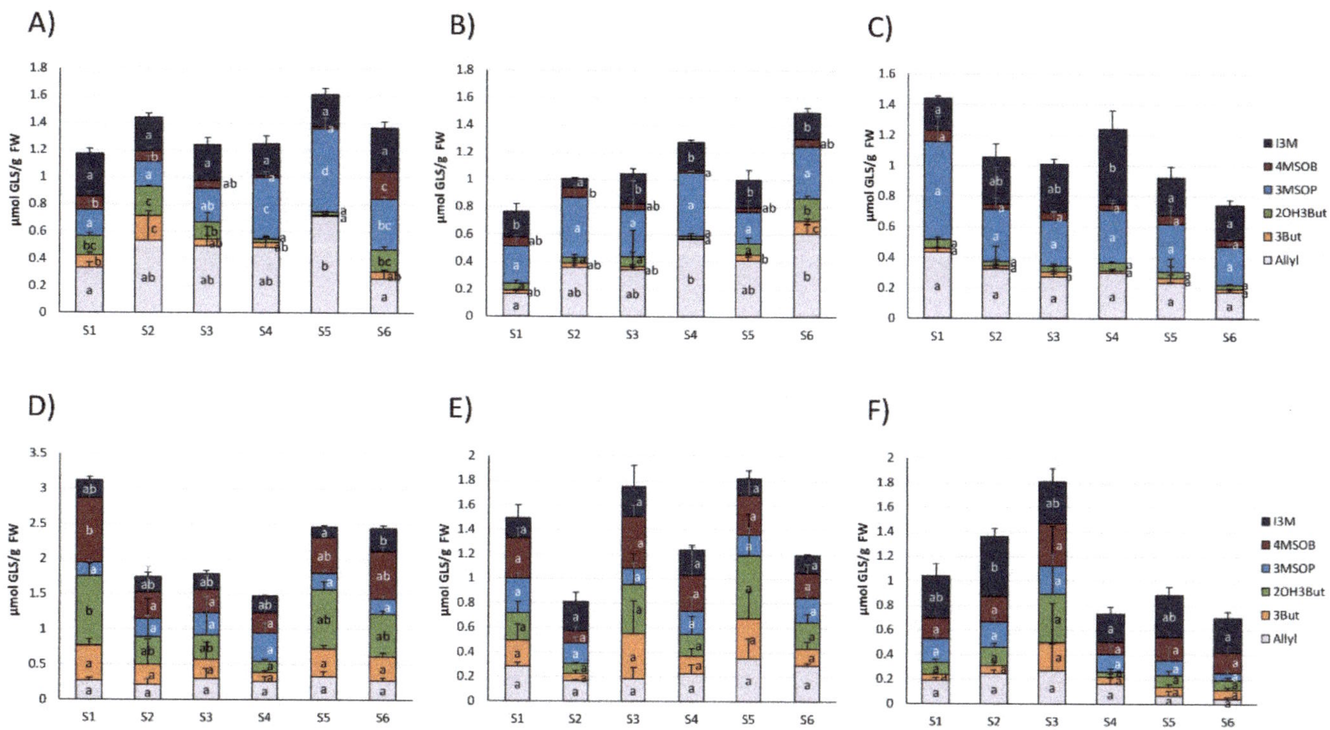

Figure 1. Glucosinolate (GLS) profile in white (**A–C**) and red cabbages (**D–F**) from conventional (**A**,**B**,**D**,**E**) and organic supermarkets (**C**,**F**) on different sampling dates (S1–S6). Exact sampling dates can be found in Table 1. A) and D) Represent conventional supermarket 1 (CON_1), B) and E) stand for conventional supermarket 2 (CON_2), and C) and F) show results from organic supermarket 1 (ORG_1). Each color in the bar of the given bar chart represents the mean plus standard deviation (SD) of the GLSs from three cabbage heads from the same supermarket ($n = 3$). Lower case letters indicate significant differences in means between the levels of a GLS on different sampling dates, as tested by ANOVA and Tukey HSD test at the $p \leq 0.05$ level. Abbreviations: FW: Fresh weight; abbreviations of compounds as listed in Table 3.

3.2. Glucosinolate Hydrolysis Product Formation in White and Red Cabbages from Local Food Retailers

Resulting from the homogenization of the fresh cabbage material, GLSs in cabbages from the different food retailers were degraded. The most pronounced GLS hydrolysis products released from white and red cabbages included the ITCs (or follow-up products from ITC) 3-(methylsulfinyl)propyl ITC (3MSOP-ITC), 4-(methylsulfinyl)butyl ITC (4MSOB-ITC), 3-butenyl ITC (3But-ITC), and 5-vinyloxazolidine-2-thione (OZT), the nitriles 5-(methylsulfinyl)pentanenitrile (4MSOB-CN) and 4-(methylsulfinyl)butanenitrile (3MSOP-CN) and the ETNs 1-cyano-2,3-epthiopropane (CETP), 1-cyano-3,4-epithiobutane (CETB), and isomers of 1-cyano-2-hydroxy-3,4-epithiobutane (CHETB A, CHETB B) (Table S3). Overall, the main GLS hydrolysis products in white cabbages were 3-MSOP-CN and 3MSOP-ITC, which were formed from 3MSOP and CETP, originating from the GLS Allyl (Figure 2A–C).

Red cabbages released mainly 4MSOB-CN and 4MSOB-ITC, originating from 4MSOB and CETP, but also 3MSOP-products and CETB and CHETB were often released in higher amounts (Figure 2D–F). Usually, homogenized red cabbages released more GLS-hydrolysis products compared to white cabbages (Figures 2 and 3). The formation of the cancer-preventive ITC 4MSOB-ITC was highest in red cabbages procured from CON_1 in the S1 sample (1.06 ± 0.25 µmol/g FW), where it was also the main GLS-hydrolysis product. Although still being the main GLS-hydrolysis product in some samples, less 4MSOB-ITC was released in CON_2- and ORG_1- cabbages (up to 0.33 ± 0.18 µmol/g FW in S3 from CON_2 and up to 0.44 ± 0.16 µmol/g FW in S2 from ORG_1) (Figure 2D–F). The formation of GLS hydrolysis products in cabbages purchased from the three different supermarkets generally varied over the course of the sampling period (Figures 2 and 3). Overall, the S5 sample of red cabbages with 2.68 ± 0.57 µmol/g FW displayed the highest total level of released GLS-hydrolysis products.

Regarding white cabbages, total ITC concentrations were highest with up to 0.50 ± 0.19 µmol/g FW in the later samples S4 and S5 from CON_1, where they were also the main GLS hydrolysis product type. In addition, total nitrile levels were higher in these later samples. Total ETN levels were also highest in S5 (Figure 3A). In white cabbages from CON_2, total ITC levels did not change during the sampling season, while ETN levels increased in later samples (S3–S6) and nitriles only displayed increased levels in S4 (Figure 3B). In organic white cabbages from ORG_1, cabbages from the first sampling date (S1) released with up to 0.96 ± 0.14 µmol/g FW mainly ITCs, which was generally the highest observed ITC level in white cabbages. On the other hand, samples S3, S5, and S6 only showed low levels of ITC formation (0.02–0.06 µmol/g FW) with nitriles and ETNs as the most dominant GLS-hydrolysis products (Figure 3C).

Red cabbages from the first samples (S1–S3) usually mainly released ITCs, while later samples (S3–S6) mainly released nitriles or ETNs (Figure 3D–F). More specifically, the total ITC formation was highest in the first samples and peaked in cabbages from S1 (CON_1 and CON_2) or S2 (ORG_1), while later samples released much lower total ITCs (Figure 3D–E). In red cabbages from CON_1, ETNs and nitriles displayed increased levels in S5 and S6 cabbages compared to S1–S4 (Figure 3D). Comparably, in CON_2, the red cabbages total ETN formation was also higher in S5, but nitriles were not affected (Figure 3E). In organic red cabbages from ORG_1, total nitriles were also highest in S5, while ETN-formation fluctuated and displayed increased levels in S2 and S5 compared to S3 and S6 (Figure 3F).

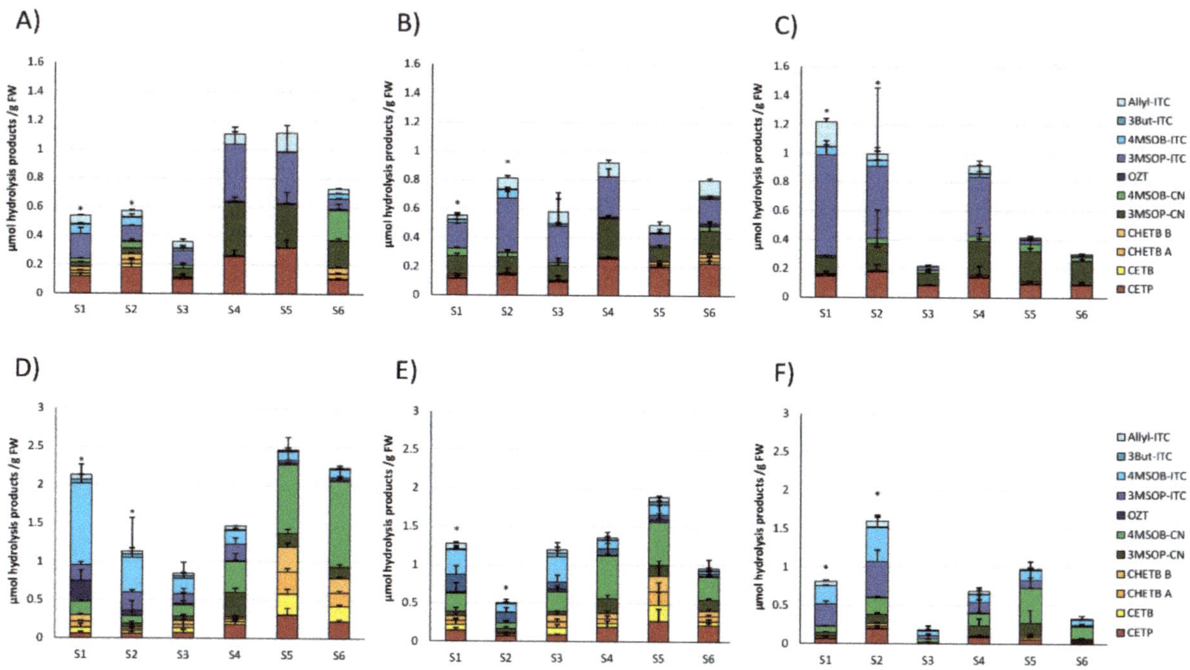

Figure 2. Main glucosinolate (GLS) hydrolysis products released from white (**A–C**) and red cabbages (**D–F**) from conventional (**A,B,D,E**) and organic supermarkets (**C,F**) on different sampling dates (S1–S6). Exact sampling dates can be found in Table 1. (**A,D**) Represent conventional supermarket 1 (CON_1), (**B,E**) stand for conventional supermarket 2 (CON_2), and (**C,F**) show results from organic supermarket 1 (ORG_1). Each color in the bar of the given bar chart represents the mean plus standard deviation (SD) of the GLS hydrolysis products from three cabbage heads from the same supermarket (n = 3). Abbreviations: FW: Fresh weight; abbreviations of compounds are listed in Table 3. * Samples were analyzed from 1 g of sample homogenate (instead of 0.5 g).

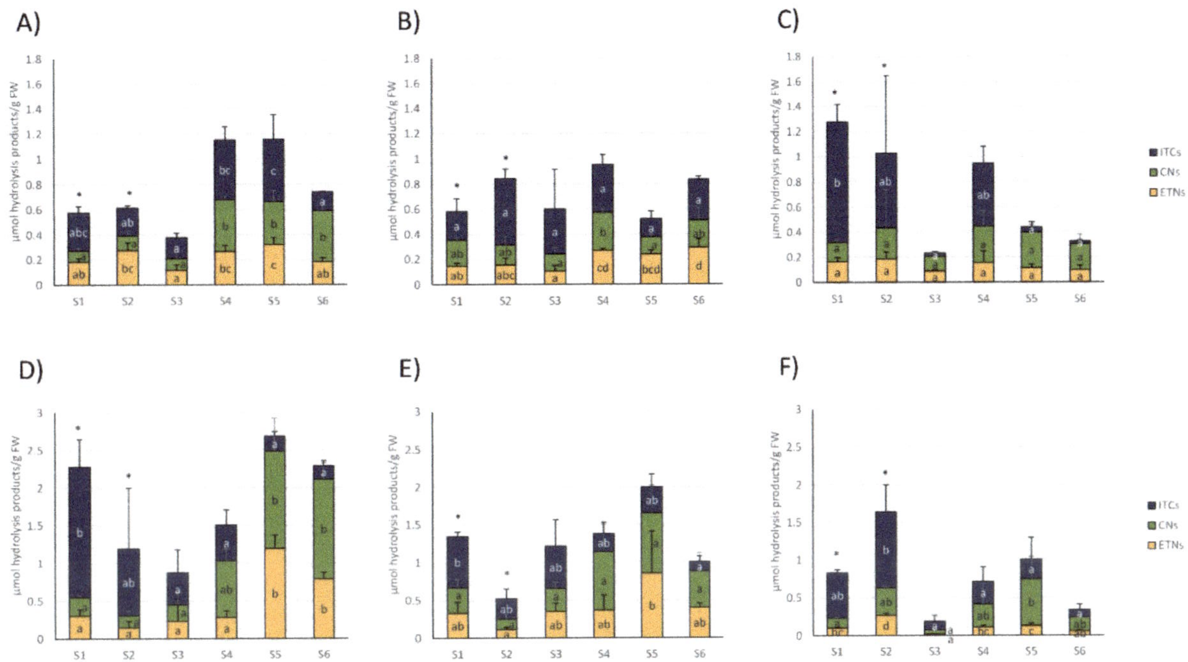

Figure 3. Formation of total isothiocyanates (ITCs), nitriles (CNs), and epithionitriles (ETNs) from white (**A**–**C**) and red cabbages (**D**–**F**) from conventional (**A**,**B**,**D**,**E**) and organic supermarkets (**C**,**F**) on different sampling dates (S1–S6). Exact sampling dates can be found in Table 1. (**A**,**D**) Represent conventional supermarket 1 (CON_1), (**B**,**E**) stand for conventional supermarket 2 (CON_2), and (**C**,**F**) show results from organic supermarket 1 (ORG_1). Each color in the bar of the given bar chart represents the mean minus standard deviation (SD) of the total ITC, CN, or ETN levels from three cabbage heads from the same supermarket (n = 3). Different lower case letters indicate significant differences in means between total ITCs, CNs, and ETNs levels in different samples, as tested by ANOVA and Tukey's HSD test at the $p \leq 0.05$ level. FW: Fresh weight. * Samples were analyzed from 1 g of sample homogenate (instead of 0.5 g).

Regarding the ratios (%) of total EPTs, nitriles, and ITCs relative to the total amount of formed GLS hydrolysis products, usually the relative ITC formation was higher in the first samples of cabbages, where they were often the main GLS hydrolysis products, but ITC formation decreased towards later sampling dates. Relative nitrile formation often behaved the other way around and was higher in later samples compared to early samples (Figure 4). More specifically, in white cabbages from CON_1, the % of ITCs more than halved, while relative nitrile levels more than tripled from S1 to S6 and relative ETN formation was slightly increased in S2 compared to the other samples (Figure 4A). In white cabbages from CON_2, the relative ITC formation was highest in S2 with 63 ± 11% and then decreased to S5 by 56% to 28 ± 9% of ITC formation. While relative nitrile formation was hardly affected, relative ETN formation increased from S1 to S5 to 46 ± 10% of ETN formation in S5 (Figure 4B). In organic white cabbages from ORG_1, GLSs also mainly released ITCs with 75 ± 3% in S1, while cabbages from later samples (S3, S5, and S6) mainly released nitriles with up to 65 ± 7% (in S5) (Figure 4C). Red cabbages showed a very similar GLS hydrolysis behavior: With up to 75 ± 7% of ITC formation (S1 from CON_1) the first samples (S1-S3) released mainly ITCs and the formation decreased towards later sampling dates, while nitrile formation increased in reverse with later samples to up to 64 ± 8% (S5 from ORG_1).

The relative ETN release was not affected in red cabbages from CON_2 and ORG_1, but increased with later samples in CON_1 up to 45 ± 9% of all GLS products (in S5) (Figure 4D–F). As an indicator of ESP-activity, the relative formation of CETP, Allyl-ITC, and Allyl-CN were monitored, as well. In red cabbages from CON_1 and ORG_1, the relative release of CEPT increased from first to last samples, while in white cabbages, an increase from S1 to S2–S5 was found (Supplemental Figure S1). The relative CETP formation was unaffected in red cabbages from CON_2 and conventional white cabbages.

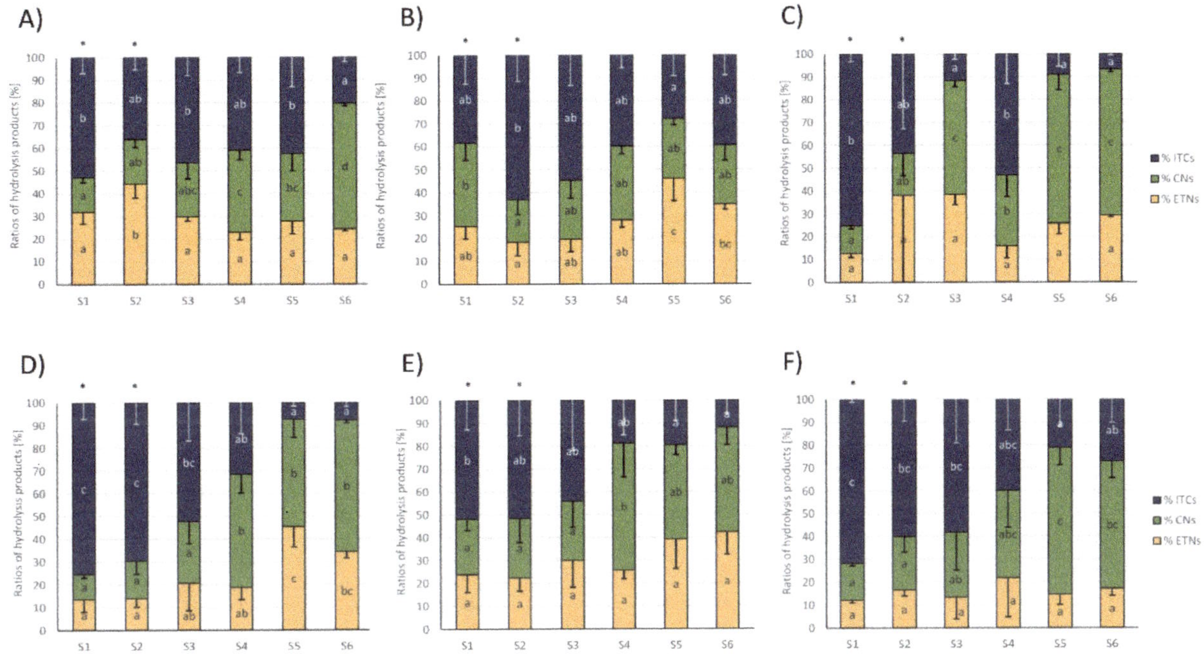

Figure 4. Total glucosinolate (GLS) hydrolysis product ratios (percentage of total isothiocyanates (ITCs), nitriles (CNs), and epithionitriles (ETNs)) from white (**A–C**) and red cabbages (**D–F**) from conventional (**A,B,D,E**) and organic supermarkets (**C,F**) on different sampling dates (S1–S6). Exact sampling dates can be found in Table 1. (**A,D**) Represent conventional supermarket 1 (CON_1), (**B,E**) stand for conventional supermarket 2 (CON_2), and (**C,F**) show results from organic supermarket 1 (ORG_1). Each color in the bar of the given bar chart represents the mean minus standard deviation (SD) of the GLS hydrolysis products from three cabbage heads from the same supermarket ($n = 3$). Different lower case letters indicate significant differences in means between the ratios of total ITCs, CNs, and ETNs in different samples, as tested by ANOVA and Tukey's HSD test at the $p \leq 0.05$ level. FW: Fresh weight. * Samples were analyzed from 1 g of sample homogenate (instead of 0.5 g).

3.3. Glucosinolates and Glucosinolate Hydrolysis Products Formation in Freshly Harvested White and Red Cabbages

The GLS profile of freshly harvested white and red cabbages was similar to the commercial ones, and 3MSOP and Allyl were most dominant in white cabbage (Figure 5A), whilst 2OH3But and 4MSOB contributed the most towards the total GLS content of freshly harvested red cabbage (Figure 5D). The total GLS and GLS-hydrolysis product level of red cabbages was higher compared to the freshly harvested white cabbages. With regard to individual GLS hydrolysis product formation, the main GLS hydrolysis products released from homogenized freshly harvested white cabbages were CETP and 3MSOP-CN (Figure 5B) and from freshly harvested red cabbages 4MSOB-CN and 4MSOB-ITC (Figure 5E). Of the detected GLS hydrolysis products from freshly harvested white cabbages, 36 ± 13% were nitriles (mainly 3MSOP-CN), 34 ± 7% were ETNs, and 30 ± 9% were ITCs (Figure 5C), while in red cabbages GLSs were degraded to 52 ± 3% nitriles (mainly 4MSOB-CN), 28 ± 2% ETNs, and 20 ± 4% ITCs (Figure 5F).

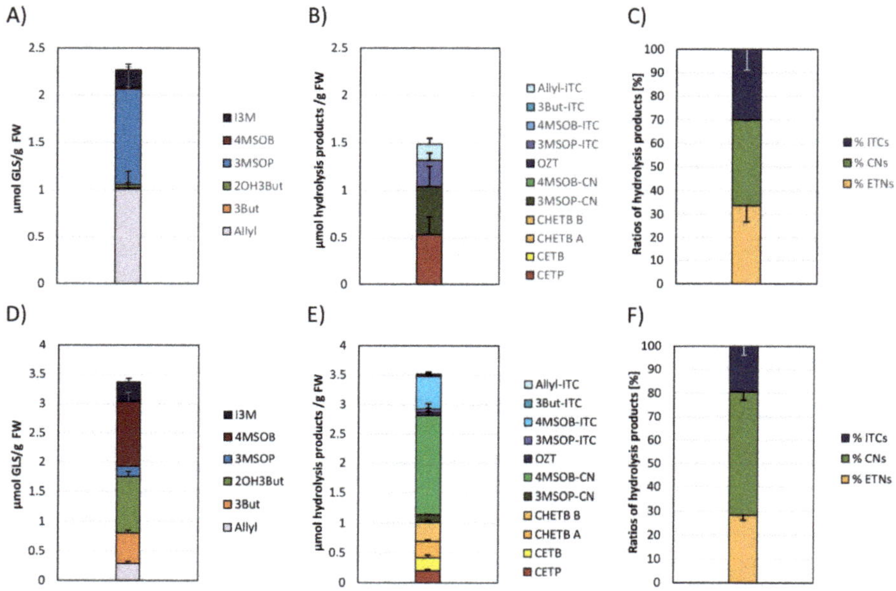

Figure 5. Glucosinolate (GLS) (**A**,**D**) and their absolute (**B**,**E**) and relative (**C**,**F**) hydrolysis product formation in white (**A–C**) and red cabbages (**D–F**) harvested freshly from the field. Each color in the bar of the given bar chart represents the mean plus standard deviation (SD) of the GLS (**A**,**D**), their respective hydrolysis products (**B**,**E**), or the ratio of relative isothiocyanates (ITCs), nitriles (CNs), and epithionitriles (ETNs) (**C**,**F**) from three cabbage heads freshly harvested from the field (n = 3). Abbreviations: FW: Fresh weight; abbreviations of compounds are listed in Table 3.

4. Discussion

In this study, the GLS content and formation of GLS hydrolysis products was evaluated in commercial white and red cabbages purchased from two conventional and one organic supermarket in Germany over a period of 3 months and compared to freshly harvested cabbages. In general, the composition of individual GLSs in red and white cabbages among different food retailers displayed only slight fluctuations and the GLS profile and levels were similar over the six sampling periods (Figure 1). With Allyl and 3MSOP being the main GLS in commercial and freshly harvested white cabbages and red cabbages being rich in 4MSOB and 2OH3But, (but also of Allyl, 3But, 3MSOP, and I3M), the GLS profiles and levels were in accordance to previous reports [14,24]. The small variability in GLS levels is an unexpected observation, as the analyzed white and red cabbages differed in genotype, came from different regions, were cultivated on different soil types using different fertilizers and storage practices, and were also purchased from different food retailers that belonged to different food trading companies (Table 2, Supplemental Table S1). Previous studies showed that GLS levels in *Brassica oleracea* vegetables are affected by cultivar (genotype) [14,25], nutrient supply [26,27], climatic conditions [17,20,28], as well as storage conditions [29,30]. As the variability of the GLSs was relatively low, it is suspected that genotypes were similar in their initial GLS concentrations and that also cultivation practices and storage conditions had no major effect on the GLS content of the cabbages, when they were finally sold in the supermarket. On the other hand, long-term storage (2 °C, 95% of relative humidity, up to 100 days) was shown to decrease the GLS content in Chinese cabbage (*Brassica rapa* L. spp. *pekinensis*), with GLSs being more stable in cabbages stored under a controlled atmosphere (CA) (2% O_2 and 2% CO_2) [31]. Accordingly, Osher et al. (2018) even reported an increase for aliphatic ITC-formation in cabbage (*B. oleracea*) stored at 1 °C under CA for 60 days (CA: 2% O_2, 5% CO_2), while ITC formation declined when stored under normal atmosphere (up to 45 and 72% decline in Allyl-ITC after 60 and 90 days, respectively) [32]. In the present study, I3M levels of organic white and red cabbages from ORG_1 were often higher, compared to the cabbages from the conventional food retailers CON_1 and CON_2 (Figure 1C). Likewise, using NMR spectroscopy, Lucarini et al. (2020) recently also found nearly 3 times as much I3M in organic broccoli compared to conventionally grown ones [33]. In that study, the main difference in both farming practices was the fertilization practice, as no pesticides were used. While the same amount of nitrogen was supplied to the soil, for conventional broccoli with 0.2 t/ha urea and 15 t/ha bovine manure were applied, while organic grown broccoli was fertilized with 28 t/ha [33]. Nevertheless, increased I3M biosynthesis could be also explained due to the absence of chemical pesticides in organic cultivation practices, resulting in the stimulation of indole GLS biosynthesis upon herbivory damage via the methyl jasmonate signalling pathway [34].

Upon homogenization, GLSs were hydrolyzed in cabbages, yielding nitriles, ITC (or breakdown product thereof), and ETNs. Usually, the recovery of aliphatic GLS hydrolysis products was good with recoveries in a range of 60–130%. However, in some samples, low recoveries of aliphatic GLS hydrolysis samples were also observed (for example, S3 and S6 white cabbage samples and the S3 red cabbage sample from ORG_1) (Figures 1 and 2). Regarding this observation, myrosinase activity in these samples was probably low, therefore, resulting in a low recovery of hydrolysis products. In pre-experiments performed for the current study, the recovery of GLS hydrolysis products did not benefit from longer incubation times. Probably due to chemical instabilities of the products [35,36], it is likely that the myrosinase activity decreases with incubation time and that the initial myrosinase activity might be a major factor for the recovery of products. This hypothesis is further supported by the observation that sulfate, which is released during GLS hydrolysis, is a competitive inhibitor of myrosinase activity [37].

With regard to the differences in GLS hydrolysis products in the different samples, in contrast to GLS, the formation of GLS hydrolysis products varied strongly between the purchase dates (S1–S6; Figure 2). Especially during the first samples (S1–S3) mainly ITCs were released from red cabbages, while in later samples nitriles were preferentially formed (S5, S6) (Figure 3). Likewise, the relative ITC formation in white and red cabbages generally decreased until the last sample, while nitriles were usually the major hydrolysis products (Figure 4). These results show that in contrast to some previous reports [14,38], ITCs can be the main hydrolysis products in cabbages, as this was the case for cabbages purchased in early autumn. Here, red cabbage could be an excellent source for cancer-preventive 4MSOB-ITC (sulforaphane) (up to 1.06 µmol/g FW in red cabbage from CON_1 at S1), releasing levels, which were 6-times higher compared to mature broccoli and comparable to the 4MSOB-ITC release from broccoli sprouts [14].

With regard to the high nitrile and ETN formation at later purchase dates, the ESP protein activity is made responsible for ETN-release from alkenyl GLS and for increased formation in simple nitriles from non-alkenyl GLS [39,40]. Therefore, it was suspected that the ESP-activity increased towards later samples, while it remained low in the first samples. When regarding the relative formation of Allyl hydrolysis products, as an indicator of ESP activity, it can be further supported that the ESP activity increased with later purchase dates in red cabbage from CON_1 and organic cabbage, while it was not considerably affected in conventional white cabbage (Figure S1). As the relative nitrile formation significantly increased in white cabbage from CON_1 in later samples (Figure 4A) (but relative CETP release not, Figure S1), it is suspected, that next to the ESP activity, also other factors influence GLS hydrolysis, which could have resulted in changes in hydrolysis product behavior from S1 to S6 due to their variation. As especially nitriles increased (Figure 4A), it is suspected that white cabbage contains nitrile specifier proteins, which are involved in nitrile formation in *Arabidopsis thaliana* [41]. This suspicion is strengthened by the observation that *Brassica oleracea* contains a gene with a homology of 80% compared to the nitrile specifier protein 1 of *A. thaliana* [42]. Further, this hypothesis is supported by a recent study, which could neither prove the nitrile specifier protein activity for three *B. oleracea* ESP isoforms in vitro nor in vivo, but nitrile formation from alkyl GLS was observed [43].

To date, there is little data how pre- and postharvest factors affect glucosinolate hydrolysis. Freshly harvested white and red cabbages showed a similar GLS hydrolysis behavior compared to the supermarket cabbages purchased during similar dates (S5 white cabbage, S4 red cabbage) (Figure 5; Figures 2–4). Due to this finding and due to the different storage conditions of growers and retailers, in the present study storage does not seem to be the factor that caused a reduced ITC release in cabbages purchased in later autumn. With regard to preharvest factors, nitrogen and sulfur supply affected the release of GLS hydrolysis products in the ETN-producer Chinese cabbage (*Brassica rapa* L. ssp. *pekinensis*) and ITCs were reduced in response to the increasing N and decreasing S supply [44]. In pak choi (*B. rapa* subsp. *chinensis* (L.) Hanelt), which also mainly released ETN, the ITC/CN and ITC/ETN ratio increased with the increasing sulfur supply [45]. With regard to the present study, it is unlikely that differences in fertilization were responsible for the observed changes in the GLS hydrolysis behavior as also the red cabbage from CON_2 which originated from the same grower (Table 2, Table S1) showed reduced % ITC release at later purchase dates (Figure 4E). Moreover, herbivore feeding can affect the GLS hydrolysis behavior, and simple nitrile formation was shown to increase in response to the specialist insect feeding (*Pieris rapae*) in *Arabidopsis thaliana* Col-0 [46], while ITC-emitting plants appear to be better defended against generalist herbivores [47]. However, as organic and conventional cabbages displayed a similar hydrolysis behavior (Figure 4D–F), it is suspected that climatic conditions such as reduced radiation or decreasing temperatures across the autumn season might be responsible for the observed shifts. Moreover, all of the conventional cabbages originated from the Dithmarschen region (Schleswig-Holstein, Germany; Table S1), which is characterized by a coastal climate and marsh soil (being ideal conditions for growing cabbages). It is the largest coherent cabbage-growing area in Europe. In Table S4, the climatic data of the presumable growing season in 2019 is presented. Consequently, it is likely that temperature and radiation interact with regard to the GLS hydrolysis behavior. Recently, Jasper et al. (2020) showed that at higher temperatures during growth, more GLS hydrolysis products were formed from rocket (*Eruca sativa*), while GLS levels were less affected [48]. Likewise, Ku et al. (2013) reported different ITC conversion rates in broccoli grown in 2 different years and linked this to different climatic conditions. Unfortunately, in that study, nitriles were not analyzed and therefore, conversion rates could have been also affected by changes of the myrosinase activity [49]. As organic cabbages which originated from different regions in Germany (Brandenburg, Mecklenburg-Western Pomerania, Schleswig-Holstein; Table 2 and Table S1) also showed similar changes in the GLS hydrolysis behavior compared to the conventional cabbages originating from the Dithmarschen region, it can be assumed that the results obtained in this study might be also valid for other countries and regions with similar climatic conditions. The specific role of climatic growth conditions on GLS hydrolysis in *B. oleracea* vegetables will need to be evaluated in future studies.

5. Conclusions

Current findings in this study have highlighted the great diversity, particularly of the GLS hydrolysis behavior in white and red cabbages between the different supermarkets over the six sampling periods: Whilst the GLS composition and content remained similar between the different food retailers, the composition and content of the individual hydrolysis products formed varied across the season and high ITC levels were generally noted in early sampling periods (early September) and decreased, particularly in red cabbages over time. Here, the increased specifier protein activity is made responsible for the reduced ITC-release.

In conclusion, with regard to their potential to release more ITCs, consumption of commercial cabbages purchased in early autumn could be healthier options than those purchased in later autumn months. The fact that ITCs can be preferentially formed in earlier autumn months, but hardly towards the end of autumn, underlines the need to unravel the factors that affect the GLS-hydrolysis outcome. The results of this study might also help growers and food companies produce cabbages and products with more pungency due to a higher ITC formation. Due to the potential of red cabbage to form high rates of health-promoting 4MSOB-ITC in cabbages purchased in early autumn, red cabbage consumption could be an alternative for people who dislike broccoli.

Supplementary Materials: The following are available online at http://www.mdpi.com/2304-8158/9/11/1682/s1, Figure S1: Allyl glucosinolate hydrolysis product ratios [percentage of allyl isothiocyanate (Allyl-ITC), 3-butenenitrile (Allyl-CN) and 1-cyano-2,3-epithiopropane (CETP)] from white (A–C) and red cabbages (D–F) from conventional (A, B, D, E) and organic supermarkets (C, F) on different sampling dates (S1–S6). Table S1: Overview of white and red cabbages purchased from three different food retailers or harvested freshly from the field (Großbeeren, D). Table S2: Glucosinolates (GLSs) content in commercial and cabbages harvested freshly from the field in µmol/g fresh weight. Table S3: Glucosinolate (GLS) hydrolysis product formation in commercial and cabbages harvested freshly from the field in µmol/g fresh weight. Table S4: Climatic data of the Dithmarschen region in northern Germany in 2019.

Author Contributions: Conceptualization, N.S.W., F.S.H., and S.R.; methodology, N.S.W. and F.S.H.; validation, F.S.H.; investigation, N.S.W. and F.S.H.; resources, F.S.H.; writing—original draft preparation, N.S.W. and F.S.H.; writing—review and editing, F.S.H. and S.R.; visualization, N.S.W. and F.S.H.; supervision, F.S.H. and S.R.; project administration, F.S.H.; funding acquisition, F.S.H. All authors have read and agreed to the published version of the manuscript.

Funding: Franziska Hanschen and this research study are part of the Leibniz-Junior Research Group OPTIGLUP, funded by the Leibniz-Association (J16/2017).

Acknowledgments: The excellent technical assistance of Andrea Jankowsky, Andrea Maikath, and Jessica Eichhorn is gratefully acknowledged.

Conflicts of Interest: The authors declare no conflict of interest.

References

1. BMEL-Bundesministerium für Ernährung und Landwirtschaft. Available online: https://www.bmel-statistik.de/fileadmin/daten/GBT-0070004-2018.pdf (accessed on 15 July 2020).
2. Blažević, I.; Montaut, S.; Burčul, F.; Olsen, C.E.; Burow, M.; Rollin, P.; Agerbirk, N. Glucosinolate structural diversity, identification, chemical synthesis and metabolism in plants. *Phytochemistry* **2020**, *169*, 112100. [CrossRef] [PubMed]
3. Katsarou, D.; Omirou, M.; Liadaki, K.; Tsikou, D.; Delis, C.; Garagounis, C.; Krokida, A.; Zambounis, A.; Papadopoulou, K.K. Glucosinolate biosynthesis in Eruca sativa. *Plant Physiol. Biochem.* **2016**, *109*, 452–466. [CrossRef] [PubMed]
4. Hanschen, F.S.; Lamy, E.; Schreiner, M.; Rohn, S. Reactivity and stability of glucosinolates and their breakdown products in foods. *Angew. Chem. Int. Ed.* **2014**, *53*, 11430–11450. [CrossRef]
5. Bell, L.; Oloyede, O.O.; Lignou, S.; Wagstaff, C.; Methven, L. Taste and flavor perceptions of glucosinolates, isothiocyanates, and related compounds. *Mol. Nutr. Food Res.* **2018**, *62*, e1700990. [CrossRef] [PubMed]
6. Koroleva, O.A.; Cramer, R. Single-cell proteomic analysis of glucosinolate-rich S-cells in Arabidopsis thaliana. *Methods* **2011**, *54*, 413–423. [CrossRef]

7. Wittstock, U.; Burow, M. Glucosinolate breakdown in Arabidopsis: Mechanism, regulation and biological significance. *Arab. Book* **2010**, *8*, e0134. [CrossRef]
8. Palliyaguru, D.L.; Yuan, J.-M.; Kensler, T.W.; Fahey, J.W. Isothiocyanates: Translating the power of plants to people. *Mol. Nutr. Food Res.* **2018**, *62*, e1700965. [CrossRef]
9. Romeo, L.; Iori, R.; Rollin, P.; Bramanti, P.; Mazzon, E. Isothiocyanates: An overview of their antimicrobial activity against human infections. *Molecules* **2018**, *23*, 624. [CrossRef]
10. Chou, Y.-C.; Chang, M.-Y.; Lee, H.-T.; Shen, C.-C.; Harnod, T.; Liang, Y.-J.; Wu, R.S.-C.; Lai, K.; Hsu, F.; Chung, J.-G. Phenethyl isothiocyanate inhibits in vivo growth of xenograft tumors of human glioblastoma cells. *Molecules* **2018**, *23*, 2305. [CrossRef]
11. Veeranki, O.L.; Bhattacharya, A.; Tang, L.; Marshall, J.R.; Zhang, Y. Cruciferous vegetables, isothiocyanates, and prevention of bladder cancer. *Curr. Pharmacol. Rep.* **2015**, *1*, 272–282. [CrossRef]
12. Hanschen, F.S.; Herz, C.; Schlotz, N.; Kupke, F.; Rodríguez, M.M.B.; Schreiner, M.; Rohn, S.; Lamy, E. The Brassica epithionitrile 1-cyano-2,3-epithiopropane triggers cell death in human liver cancer cells in vitro. *Mol. Nutr. Food Res.* **2015**, *59*, 2178–2189. [CrossRef] [PubMed]
13. Matusheski, N.V.; Jeffery, E. Comparison of the bioactivity of two glucoraphanin hydrolysis products found in broccoli, sulforaphane and sulforaphane nitrile. *J. Agric. Food Chem.* **2001**, *49*, 5743–5749. [CrossRef] [PubMed]
14. Hanschen, F.S.; Schreiner, M. Isothiocyanates, nitriles, and epithionitriles from glucosinolates are affected by genotype and developmental stage in Brassica oleracea varieties. *Front. Plant Sci.* **2017**, *8*, 1095. [CrossRef] [PubMed]
15. Klopsch, R.; Witzel, K.; Börner, A.; Schreiner, M.; Hanschen, F.S. Metabolic profiling of glucosinolates and their hydrolysis products in a germplasm collection of Brassica rapa turnips. *Food Res. Int.* **2017**, *100*, 392–403. [CrossRef]
16. Verkerk, R.; Schreiner, M.; Krumbein, A.; Ciska, E.; Holst, B.; Rowland, I.; De Schrijver, R.; Hansen, M.; Gerhäuser, C.; Mithen, R.; et al. Glucosinolates in Brassica vegetables: The influence of the food supply chain on intake, bioavailability and human health. *Mol. Nutr. Food Res.* **2008**, *53*, S219. [CrossRef]
17. Cartea, M.E.; Velasco, P.; Obregón, S.; Padilla, G.; De Haro, A. Seasonal variation in glucosinolate content in Brassica oleracea crops grown in northwestern Spain. *Phytochemistry* **2008**, *69*, 403–410. [CrossRef]
18. Rosa, E.A.; Rodrigues, A.S. Total and individual glucosinolate content in 11 broccoli cultivars grown in early and late seasons. *HortScience* **2001**, *36*, 56–59. [CrossRef]
19. Nuñez-Gómez, V.; Baenas, N.; Navarro-González, I.; García-Alonso, F.J.; Moreno, D.A.; González-Barrio, R.; Periago, M.J. Seasonal variation of health-promoting bioactives in broccoli and methyl-jasmonate pre-harvest treatments to enhance their contents. *Foods* **2020**, *9*, 1371. [CrossRef] [PubMed]
20. Charron, C.S.; Sams, C.E. Glucosinolate content and myrosinase activity in rapid-cycling Brassica oleracea grown in a controlled environment. *J. Am. Soc. Hortic. Sci.* **2004**, *129*, 321–330. [CrossRef]
21. Mølmann, J.A.; Steindal, A.L.; Bengtsson, G.B.; Seljåsen, R.; Lea, P.; Skaret, J.; Johansen, T.J. Effects of temperature and photoperiod on sensory quality and contents of glucosinolates, flavonols and vitamin C in broccoli florets. *Food Chem.* **2015**, *172*, 47–55. [CrossRef]
22. Pereira, F.M.V.; Rosa, E.; Fahey, J.W.; Stephenson, K.K.; Carvalho, R.; Aires, A. Influence of temperature and ontogeny on the levels of glucosinolates in nroccoli (Brassica oleracea var. italica) sprouts and their effect on the induction of mammalian phase 2 enzymes. *J. Agric. Food Chem.* **2002**, *50*, 6239–6244. [CrossRef] [PubMed]
23. Hanschen, F.S. Domestic boiling and salad preparation habits affect glucosinolate degradation in red cabbage (Brassica oleracea var. capitata f. rubra). *Food Chem.* **2020**, *321*, 126694. [CrossRef] [PubMed]
24. Ciska, E.; Martyniak-Przybyszewska, B.; Kozlowska, H. Content of glucosinolates in cruciferous vegetables grown at the same site for two years under different climatic conditions. *J. Agric. Food Chem.* **2000**, *48*, 2862–2867. [CrossRef]
25. Brown, A.F.; Yousef, G.G.; Jeffery, E.H.; Klein, B.P.; Wallig, M.A.; Kushad, M.M.; Juvik, J.A. Glucosinolate profiles in broccoli: Variation in levels and implications in breeding for cancer chemoprotection. *J. Am. Soc. Hortic. Sci.* **2002**, *127*, 807–813. [CrossRef]
26. Schonhof, I.; Blankenburg, D.; Müller, S.; Krumbein, A. Sulfur and nitrogen supply influence growth, product appearance, and glucosinolate concentration of broccoli. *J. Plant Nutr. Soil Sci.* **2007**, *170*, 65–72. [CrossRef]

27. Omirou, M.D.; Papadopoulou, K.K.; Papastylianou, I.; Constantinou, M.; Karpouzas, D.G.; Asimakopoulos, I.; Ehaliotis, C. Impact of nitrogen and sulfur fertilization on the composition of glucosinolates in relation to sulfur assimilation in different plant organs of broccoli. *J. Agric. Food Chem.* **2009**, *57*, 9408–9417. [CrossRef]
28. Charron, C.S.; Saxton, A.M.; Sams, C.E. Relationship of climate and genotype to seasonal variation in the glucosinolate-myrosinase system. I. Glucosinolate content in ten cultivars of Brassica oleracea grown in fall and spring seasons. *J. Sci. Food Agric.* **2005**, *85*, 671–681. [CrossRef]
29. Vallejo, F.; Tomás-Barberán, F.; García-Viguera, C. Health-promoting compounds in broccoli as influenced by refrigerated transport and retail sale period. *J. Agric. Food Chem.* **2003**, *51*, 3029–3034. [CrossRef]
30. Rodrigues, A.S.; Rosa, E.A.S. Effect of post-harvest treatments on the level of glucosinolates in broccoli. *J. Sci. Food Agric.* **1999**, *79*, 1028–1032. [CrossRef]
31. Kang, J.-H.; Woo, H.-J.; Park, J.-B.; Chun, H.H.; Park, C.W.; Bin Song, K. Effect of storage in pallet-unit controlled atmosphere on the quality of Chinese cabbage (Brassica rapa L. spp. pekinensis) used in kimchi manufacturing. *LWT* **2019**, *111*, 436–442. [CrossRef]
32. Osher, Y.; Chalupowicz, D.; Maurer, D.; Ovadia-Sadeh, A.; Lurie, S.; Fallik, E.; Kenigsbuch, D. Summer storage of cabbage. *Postharvest Biol. Technol.* **2018**, *145*, 144–150. [CrossRef]
33. Lucarini, M.; Di Cocco, M.E.; Raguso, V.; Milanetti, F.; Durazzo, A.; Lombardi-Boccia, G.; Santini, A.; Delfini, M.; Sciubba, F. NMR-based metabolomic comparison of Brassica oleracea (var. italica): Organic and conventional farming. *Foods* **2020**, *9*, 945. [CrossRef] [PubMed]
34. Textor, S.; Gershenzon, J. Herbivore induction of the glucosinolate–myrosinase defense system: Major trends, biochemical bases and ecological significance. *Phytochem. Rev.* **2008**, *8*, 149–170. [CrossRef]
35. Hanschen, F.S.; Kaufmann, M.; Kupke, F.; Hackl, T.; Kroh, L.W.; Rohn, S.; Schreiner, M. Brassica vegetables as sources of epithionitriles: Novel secondary products formed during cooking. *Food Chem.* **2018**, *245*, 564–569. [CrossRef] [PubMed]
36. Fechner, J.; Kaufmann, M.; Herz, C.; Eisenschmidt, D.; Lamy, E.; Kroh, L.W.; Hanschen, F.S. The major glucosinolate hydrolysis product in rocket (Eruca sativa L.), sativin, is 1,3-thiazepane-2-thione: Elucidation of structure, bioactivity, and stability compared to other rocket isothiocyanates. *Food Chem.* **2018**, *261*, 57–65. [CrossRef] [PubMed]
37. Shikita, M.; Fahey, J.W.; Golden, T.R.; Holtzclaw, W.D.; Talalay, P. An unusual case of 'uncompetitive activation' by ascorbic acid: Purification and kinetic properties of a myrosinase from Raphanus sativus seedlings. *Biochem. J.* **1999**, *341*, 725–732. [CrossRef] [PubMed]
38. Kyung, K.; Fleming, H.; Young, C.; Haney, C. 1-Cyano-2,3-epithiopropane as the primary sinigrin hydrolysis product of fresh cabbage. *J. Food Sci.* **1995**, *60*, 157–159. [CrossRef]
39. Matusheski, N.V.; Swarup, R.; Juvik, J.A.; Mithen, R.; Bennett, M.; Jeffery, E. Epithiospecifier protein from broccoli (Brassica oleracea L. ssp. italica) inhibits formation of the anticancer agent sulforaphane. *J. Agric. Food Chem.* **2006**, *54*, 2069–2076. [CrossRef]
40. Burow, M.; Markert, J.; Gershenzon, J.; Wittstock, U. Comparative biochemical characterization of nitrile-forming proteins from plants and insects that alter myrosinase-catalysed hydrolysis of glucosinolates. *FEBS J.* **2006**, *273*, 2432–2446. [CrossRef]
41. Wittstock, U.; Meier, K.; Dörr, F.; Ravindran, B.M. NSP-dependent simple nitrile formation dominates upon breakdown of major aliphatic glucosinolates in roots, seeds, and seedlings of Arabidopsis thaliana Columbia-0. *Front. Plant Sci.* **2016**, *7*, 1821. [CrossRef]
42. Román, J.; González, D.; Inostroza-Ponta, M.; Mahn, A. Molecular modeling of epithiospecifier and nitrile-specifier proteins of broccoli and their interaction with aglycones. *Molecules* **2020**, *25*, 772. [CrossRef] [PubMed]
43. Witzel, K.; Abu Risha, M.; Albers, P.; Börnke, F.; Hanschen, F.S. Identification and characterization of three epithiospecifier protein isoforms in Brassica oleracea. *Front. Plant Sci.* **2019**, *10*, 1552. [CrossRef] [PubMed]
44. Geilfus, C.-M.; Hasler, K.; Witzel, K.; Gerendás, J.; Mühling, K.H. Interactive effects of genotype and N/S-supply on glucosinolates and glucosinolate breakdown products in Chinese cabbage (Brassica rapa L. ssp. pekinensis). *J. Appl. Bot. Food Qual.* **2016**, *89*, 279–286.
45. Meschede, C.A.C.; Abdalla, M.A.; Mühling, K.H. Sulfur but not nitrogen supply increases the ITC/nitrile ratio in pak choi (Brassica rapa subsp. chinensis (L.) Hanelt). *J. Appl. Bot. Food Qual.* **2020**, *93*, 95–104.

46. Burow, M.; Losansky, A.; Müller, R.; Plock, A.; Kliebenstein, D.J.; Wittstock, U. The genetic basis of constitutive and herbivore-induced ESP-independent nitrile formation in Arabidopsis. *Plant Physiol.* **2008**, *149*, 561–574. [CrossRef]
47. Mumm, R.; Burow, M.; Bukovinszkine'Kiss, G.; Kazantzidou, E.; Wittstock, U.; Dicke, M.; Gershenzon, J. Formation of simple nitriles upon glucosinolate hydrolysis affects direct and indirect defense against the specialist herbivore, Pieris rapae. *J. Chem. Ecol.* **2008**, *34*, 1311–1321. [CrossRef]
48. Jasper, J.; Wagstaff, C.; Bell, L. Growth temperature influences postharvest glucosinolate concentrations and hydrolysis product formation in first and second cuts of rocket salad. *Postharvest Biol. Technol.* **2020**, *163*, 111157. [CrossRef]
49. Ku, K.-M.; Jeffery, E.; Juvik, J.A. Influence of seasonal variation and methyl jasmonate mediated induction of glucosinolate biosynthesis on quinone reductase activity in broccoli florets. *J. Agric. Food Chem.* **2013**, *61*, 9623–9631. [CrossRef]

foods

MDPI

Article

Influence of Cabbage (*Brassica oleracea*) Accession and Growing Conditions on Myrosinase Activity, Glucosinolates and Their Hydrolysis Products

Omobolanle O. Oloyede *, Carol Wagstaff and Lisa Methven

Department of Food and Nutritional Sciences, Harry Nursten Building, University of Reading, Whiteknights, Reading RG6 6DZ, UK; c.wagstaff@reading.ac.uk (C.W.); l.methven@reading.ac.uk (L.M.)

* Correspondence: bola.oloyede@reading.ac.uk; Tel.: +44-(0)118-378-3606

Abstract: Glucosinolates are secondary plant metabolites present in *Brassica* vegetables. The endogenous enzyme myrosinase is responsible for the hydrolysis of glucosinolates, yielding a variety of compounds, including health-promoting isothiocyanates. The influence of cabbage accession and growing conditions on myrosinase activity, glucosinolates (GSL) and their hydrolysis products (GHPs) of 18 gene-bank cabbage accessions was studied. Growing conditions, cabbage morphotype and accession all significantly affected myrosinase activity and concentration of glucosinolates and their hydrolysis products. In general, cabbages grown in the field with lower growth temperatures had significantly higher myrosinase activity than glasshouse samples. Profile and concentration of glucosinolates and their hydrolysis products differed across the accessions studied. Aliphatic glucosinolates accounted for more than 60 % of total glucosinolates in most of the samples assessed. Nitriles and epithionitriles were the most abundant hydrolysis products formed. The results obtained showed that consumption of raw cabbages might reduce the amount of beneficial hydrolysis products available to the consumer, as more nitriles were produced from hydrolysis compared to beneficial isothiocyanates. However, red and white cabbages contained high concentrations of glucoraphanin and its isothiocyanate, sulforaphane. This implies that careful selection of accessions with ample concentrations of certain glucosinolates can improve the health benefits derived from raw cabbage consumption.

Keywords: *Brassica oleracea*; cabbage; growing condition; myrosinase activity; glucosinolates; glucosinolate hydrolysis products; isothiocyanates; nitriles; epithionitriles

Citation: Oloyede, O.O.; Wagstaff, C.; Methven, L. Influence of Cabbage (*Brassica oleracea*) Accession and Growing Conditions on Myrosinase Activity, Glucosinolates and Their Hydrolysis Products. *Foods* **2021**, *10*, 2903. https://doi.org/10.3390/foods10122903

Academic Editors: Franziska S. Hanschen and Sascha Rohn

Received: 11 October 2021
Accepted: 18 November 2021
Published: 23 November 2021

1. Introduction

Cabbage (*Brassica oleracea*) belongs to the *Brassicaceae* family and comprises eight distinct cultivar groups, all descended from wild cabbage (*B. oleracea* var. *oleracea*) [1]. Epidemiological studies have shown that the consumption of *Brassica* vegetables reduces the risks of cardiovascular diseases and cancer [2] and is reported to have a cytoprotective effect against tissue damage associated with oxidative stress as well as antimicrobial activity against bacterial and fungal pathogens [3,4].

Brassica vegetables are unique in comparison to other vegetables because they contain the enzyme myrosinase and a group of thioglucosides called glucosinolates (GSLs). GSLs are sulphur and nitrogen containing biologically active secondary metabolites found in plants of the order Capparales, which includes the *Brassicaceae* family and other economically important agricultural crops [5–7]. In plants, GSLs act as plant defense mechanisms against stress, insect, and pest attack [8]. GSLs have been grouped into three main classes based on the structure of their different amino acid precursors; these groups are aliphatic, aromatic and indole GSLs. Aliphatic GSLs are derived from alanine, leucine, isoleucine, methionine or valine; aromatic GSLs are from phenylalanine or tyrosine, while tryptophan-derived GSLs are called indole GSLs [9,10]. A recent review by Blažević et al. [11] stated that between 88–137 glucosinolates (GSLs) have been characterised in plants to date.

GSLs and myrosinase enzymes coexist in separate compartments in the plants; while glucosinolates exists in the vacuoles of various cells [6], myrosinase enzymes are localised inside the myrosin cells. When plant tissue is disrupted, GSLs are hydrolysed by plant myrosinase enzymes, resulting in the formation of various hydrolysis products such as isothiocyanates (ITCs), thiocyanates, nitriles and epithionitriles [5]. The extent of glucosinolate hydrolysis and the type of hydrolysis compound produced is dependent on a number of factors, which include coexisting myrosinase enzyme, presence of epithiospecifier protein (ESP), ascorbic acid, Fe^{2+} and $MgCl_2$, structure of the glucosinolate side chain, the plant species, as well as reaction conditions such as pH and temperature [9,12,13].

ITCs, the primary products of GSL hydrolysis from myrosinase, are responsible for the well-documented health-promoting properties of *Brassica* vegetables, such as reduced risk of cardiovascular diseases (CVD) and cancer [2,5]. For example, sulforaphane (SFP), the hydrolysis product of glucoraphanin present in high concentrations in broccoli and cabbage, has been reported to possess chemoprotective, antimicrobial, anti-inflammatory, and antithrombotic properties [14,15]. Allyl isothiocyanate (AITC), another common ITC present in cabbages and produced upon myrosinase hydrolyses of the glucosinolate sinigrin (SIN), was reported to be potent against human breast cancer cells [16], human erythroleukemic K562 cells [17], and more potent on human A549 and H1299 non-small cell lung cancer cells in vitro than 2-phenylethyl-ITC (PEITC; ITC from gluconasturtin) [18]. However, in the presence of epithiospecifier proteins (ESPs), nitriles and epithionitriles (EPTs), which have not been shown to proffer any beneficial characteristics for health, are formed [19]. GSLs and ITCs are also partly responsible for the bitter taste and pungent aromas of *Brassica* vegetables, which limits consumer acceptance and liking of *Brassica* vegetables [20–23].

There are several factors that affect the GSL-myrosinase system in *Brassicas;* these factors include climatic factors, location, and growing conditions [24–27], morphotype and the variety of plant [28,29]; with the impact of these factors varying between studies. For example, while some authors have suggested that the effect of plant genotype on GSL concentrations is greater than that of environmental factors [30,31], others have reported higher variations in GSL concentrations as a result of environmental conditions than genetic factors [32,33].

To date, most studies on myrosinase activity have focused on single cultivars of B. *oleracea* species [28,34–37], with studies on the myrosinase activity of different varieties within a species limited [29,38]. Variations in myrosinase activities were reported in different varieties of Brussels sprouts, broccoli, cauliflower, Chinese cabbage, and white cabbage [38]. The authors found a two-fold difference in the myrosinase activities of five broccoli varieties as well as two cauliflower and Chinese cabbage varieties. Low temperature conditions are reported to increase the myrosinase activity of *B. oleracea* species (Brussels sprout, broccoli, cauliflower, cabbage, and kale) grown in the autumn season [25].

Several studies have been undertaken on the formation of GSLs in cabbage varieties, some of which have investigated GSL concentrations in cabbages grown under different conditions; with , most focused on GSL concentrations of cabbages grown in different locations or different seasons [29,32,39–41]. However, none of the studies analysed the glucosinolate hydrolysis products (GHPs) of cabbages under different plant growth conditions and instead made suggestions on potential GHP concentrations of the samples based on the concentrations of GSLs observed. These suggestions may be problematic, as studies have shown that GSL concentration is not necessarily correlated with the abundance of GHPs formed [42,43].

Little is known of the GHPs in cabbages, as most studies have focused on a specific cabbage variety [44] or ITCs in other *B. oleracea* such as broccoli [45]. A recent study analysed the GSLs and GHPs of cabbages with a focus on red, white and savoy cabbages, but the samples were grown under the same conditions [43]. To fully understand the health benefits that can be derived from cabbage consumption, however, there is a need to characterize the GHPs produced from GSL hydrolysis and understand the factors affecting the type and concentrations of GHPs formed. With growing health campaigns promoting the consumption of more fruits and vegetables, and consumers wanting to include more fresh vegetables like cabbage in their diet, many people now grow their own cabbages at home in pots, either in green/glasshouses or in the garden [46,47]. It is therefore important to investigate the effect of these plant growth conditions on the GSL-myrosinase system to ensure that the health benefits desired from their consumption are not lost.

In light of this gap in present knowledge, the purpose of this study was to investigate the influence of growing conditions and accession identity on myrosinase activity as well as the GSL and GHP content of cabbage. A total of 18 cabbage accessions across six different cabbage morphotypes were selected from a genetic resources unit and grown under two different conditions. In addition to wild cabbage, this study used red, white, and green (savoy) cabbage (*B. oleracea* var. *capitata*), kale (*B. oleracea* var. *acephala*) and sea kale (*B. oleracea* var. *tronchuda*). The primary hypothesis of the study was that cabbage growth conditions will affect myrosinase activity as well as the GSL and GHP contents of cabbage. The secondary hypothesis was that while cabbage morphotype and accession would affect myrosinase activity, cabbage morphotype rather than accession will affect the profile and concentrations of the GSLs and GHPs formed. The results of myrosinase activity and variations in the amount and profile of GSLs and GHPs in cabbage accessions across both plant conditions studied are presented.

2. Materials and Methods

2.1. Plant Material

Cabbage accessions were selected from the University of Warwick Crop Centre Genetic Resources Unit (Wellesbourne, UK). Eighteen cabbage accessions comprising six cabbage morphotypes (wild (*B. oleracea* var. *oleracea*), black kale (*B. oleracea* var. *acephala*), tronchuda (*B. oleracea* var. *tronchuda*), savoy, red and white (*B. oleracea* var. *capitata*)) were used for the experiment. Cabbages were selected based on their geographical origin, whether or not they were of hybrid descent, and morphology of head formation (closed heart or open leaf), as shown in Table 1 and Supplementary Figure S1. Seeds of one white cabbage accession (WC-DLI) did not germinate when sown and thus will not be discussed further. Out of the remaining 17 accessions planted, RC-RM (red cabbage) and SC-SDG (savoy cabbage) did not survive in the glasshouse.

A total of 15 biological replicates of each accession were germinated in seedling trays using potting compost under controlled environmental conditions (Saxcil cabinets). A 16 h photo period was used (16 h light, 8 h dark); relative humidity was set to 60%, with day and night temperatures of 22 °C and 16 °C, respectively. Seedlings were allowed to grow in seedling trays until the appearance of 3–4 true leaves, before being transplanted to individual 2.5 L pots containing loam-based compost (7–8 May 2014) and left to grow in the glasshouse (minimum night temperature 13 °C). After 50 days (26–27 June 2014), five replicates of each accession were transplanted to larger pots (10 L) containing loam-based compost and allowed to grow until commercial maturity in the glasshouse, while seven replicates of each accession were transplanted to the field and allowed to grow to commercial maturity. On the field, each accession was planted on 7 metre beds with 0.6 metres between plants and rows. Both glasshouse and field cabbages were fertilized twice weekly with nitrogen phosphate potassium (NPK) (100 kg/ha N, 100 kg/ha P and 200 kg/ha K) fertilizer. Standard agricultural practices were employed in the cultivation of the cabbages, including a programme of pest management using insecticides and fungicides. Cabbages were grown between 7 March–25 November 2014 in the plant growth facilities, Whiteknights campus of the University of Reading, UK (Supplementary Figure S2).

Table 1. Origin and botanical and common names of cabbage accessions planted between May and November 2015.

Genus/Morphotype [a]	Accession Name	Accession Code	Common Name	Origin	Head Formation
Black kale					
Brassica oleracea var. *acephala*	Cavolo nero di toscana o senza palla	BK-CNDTP (BK1)	Fodder black kale	Italy	Open leaf
Brassica oleracea var. *acephala*	Cavolo palmizio	BK-CPNT (BK2)	Black kale	Italy	Open leaf
Brassica oleracea var. *acephala*	Cavolo nero di toscana o senza testa	BK-CNDTT (BK3)	Fodder black kale	Italy	Open leaf
Wild					
Brassica oleracea var. *oleracea*	Wild cabbage	WD-8707 (WD1)	Wild cabbage	Great Britain	Open leaf
Brassica oleracea var. *oleracea*	Wild cabbage	WD-GRU (WD2)	Wild cabbage	New Zealand	Open leaf
Brassica oleracea var. *oleracea*	Wild cabbage	WD-8714 (WD3)	Wild cabbage	Great Britain	Open leaf
Tronchuda					
Brassica oleracea var. *tronchuda*	Penca mistura	TC-PCM (TC1)	Tronchuda cabbage	Portugal	Open leaf
Brassica oleracea var. *tronchuda*	Penca povoa	TC-CPDP (TC2)	Tronchuda cabbage	Portugal	Open leaf
Brassica oleracea var. *tronchuda*	Tronchuda	TC-T (TC3)	Tronchuda cabbage	Portugal	Open leaf
Savoy					
Brassica oleracea var. *capitata*	Hybrid savoy wirosa cabbage	SC-HSC (SC1)	Hybrid savoy cabbage	Great Britain	Closed heart
Brassica oleracea var. *capitata*	Pointed winter	SC-PW (SC2)	Savoy cabbage	Great Britain	Closed heart
Brassica oleracea var. *capitata*	Dark green	SC-SDG (SC3)	Savoy cabbage	Italy	Closed heart
Red					
Brassica oleracea var. *capitata*	Red langendijker	RC-RL (RC1)	Red cabbage	Great Britain	Closed heart
Brassica oleracea var. *capitata*	Rocco marner (Hybrid)	RC-RM (RC2)	Hybrid red cabbage	Great Britain	Closed heart
Brassica oleracea var. *capitata*	Red Danish	RC-RD (RC3)	Red cabbage	Netherlands	Closed heart
White					
Brassica oleracea var. *capitata*	Early market	WC-FEM (WC1)	White spring cabbage	Great Britain	Closed heart
Brassica oleracea var. *capitata*	Couve repolho	WC-CRB (WC2)	White cabbage	Portugal	Closed heart
Brassica oleracea var. *capitata*	De louviers	WC-DLI (WC3)	Hybrid white cabbage	Great Britain	Closed heart

[a] Names in bold refer to cabbage morphotype

Cabbages were harvested over a period of two days upon reaching commercial maturity based on visual inspection. Though some accessions attained commercial maturity earlier than others, they had sufficiently good field holding capacity to be left until all accessions were mature before harvesting, so that all plants experienced equivalent environmental conditions. Harvested plants were placed on ice in freezer bags and immediately stored in a cold room at 4 °C for 24 h before processing. The average weight of each field cabbage head per plant was 700 g (closed heart) and 300 g (open leaf), while the glasshouse cabbages were smaller (400 g for closed heart and 250 g for open leaf cabbages) (Supplementary Figure S1). Climatic data for both growing conditions are presented in Supplementary Table S1.

2.2. Reagents and Chemicals

Sinigrin standard was purchased from Santa Cruz Biotechnology (Heidelberg, Germany) and D-glucose determination kit was from R-Biopharm Rhone (Heidelberg, Germany). All other chemicals used were purchased from Sigma–Aldrich (Dorset, UK).

2.3. Sample Preparation

The outer leaves and central core of 4–5 cabbage heads (biological replicates) were removed and discarded in order to remove senescent leaves and achieve a representative sample spanning similar leaf ages for each morphotype. Cabbages were chopped into pieces of approximately 1 cm in width using a kitchen knife (representing how cabbages would normally be sliced by consumers), mixed together, and washed under running tap water; excess water was drained using a salad spinner (OXO Good Grips Clear Manual Salad Spinner, Chambersburg, PA, USA). A total of 120 g of cabbage samples was put into sterile sterilin tubes, immediately placed on ice, and transferred to a −80 °C freezer. Frozen samples were freeze-dried (Stokes freeze drier, Philadelphia, PA, USA), ground using a tissue grinder (Thomas Wiley® Mini-Mill, Thomas Scientific, Swedesboro, NJ, USA) and stored at −20 °C until further analysis.

2.4. Myrosinase Enzyme Extraction and Assay

Myrosinase enzyme was extracted using the method described by Ghawi et al. [48]. A sample of 0.1 g was suspended in 0.15 g polyvinylpolypyrrolidone (PVPP) and 10 mL of Tris-HCL buffer (200 mM, pH 7.5) containing 0.5 mM ethylenediaminetetracetic acid (EDTA) and 1.5 mM dithiothreitol (DTT). The mixture was stirred for 15 min at 5 °C and centrifuged (11,738× *g*) for 15 min at 5 °C. The final volume of supernatant was made up to 10 mL using the Tris-HCL buffer. Then, 6.2 g ammonium sulphate was added to the supernatant to achieve 90% saturation and stirred at 5 °C for 30 min. The samples were then centrifuged (13,694× *g*) for 15 min at 5 °C. The resulting pellet was suspended in 2 mL Tris-HCl buffer (10 mM, pH 7.5) and assayed for myrosinase activity.

Myrosinase activity was measured using the coupled enzyme method described by Gatfield and Sand [49] and Wilkinson et al. [50], with slight modifications. The procedure depends on the glucose released from the reaction between myrosinase enzymes and the substrate (sinigrin). The mixture for the reaction consisted of 0.9 mL of 5 mM ascorbic acid, 0.5 mL ATP/NADP+ solution, 10 µL hexokinase/glucose-6-phosphate dehydrogenase and 50 µL crude enzyme extract. The mixture was homogenized and allowed to stand for 3 min, and then 50 µL sinigrin substrate (0.6 M) added. The change in absorbance due to NADPH formation was read on a spectrophotometer at 340 nm. Myrosinase enzyme activity was determined by taking the slope of the linear part of the curve of absorbance versus the time of reaction. One unit of myrosinase activity is defined as the amount of enzyme that produces 1 µmol of glucose from sinigrin substrate per minute at pH 7.5.

2.5. Protein Assay

Protein content was measured using the Bradford method [51]. The procedure is based on formation of a complex between dye (brilliant Blue G, Sigma-Aldrich, Dorset, UK) and the protein present in the sample, and absorbance is read at 595 nm using a spectrophotometer. 50 μL filtered crude enzyme extract was added to 1.5 mL of concentrated dye reagent, vortexed and allowed to stand for 20 min before taking the absorbance reading. Bovine serum albumin (BSA) (Sigma-Aldrich, Dorset, UK) was used to construct a standard curve, and the protein concentration of sample was calculated from the standard curve obtained. Protein content was used to calculate the specific activity of myrosinase enzymes (U/mg protein).

2.6. Glucosinolate Extraction and LC-MS^2 Analysis

The method used for GSL extraction is as described by Bell et al. [52], with modifications. Briefly, 40 mg ground cabbage powder was heated in a heat block at 75 °C for two minutes. Then, 1 mL 70% (v/v) methanol preheated to 70 °C was added to each sample, vortexed and placed in a preheated (70 °C) water bath for 20 min. Samples were centrifuged at full speed for five minutes (18 °C), and supernatant was collected in fresh Eppendorf tubes. The volume was adjusted to 1 mL with 70% (v/v) methanol and frozen at −80 °C until further analysis.

Samples were filtered using 0.22 μm Millex syringe filters with a low protein binding Durapore polyvinylidene fluoride (PVDF) membrane (Fisher scientific, Loughborough, UK) and diluted with 9 mL HPLC-grade water. LC-MS analysis of GSL extracts was performed in negative ion mode on an Agilent 1200 Series LC system (Agilent, Stockport, UK) equipped with a variable wavelength detector and coupled to a Bruker HCT ion trap (Bruker, Coventry, UK). Sample separation was achieved on a Gemini 3 μm C_{18} 110 Å (150 × 4.6 mm) column (with Security Guard column, C_{18}; 4 mm × 3 mm; Phenomenex, Macclesfield, UK). GSLs were separated during a 40 min chromatographic run, with a 5 min post-run sequence. Mobile phases consisted of 95% of 0.1% ammonium formate solution and 5% acetonitrile. The flow rate was optimised for the system at 0.4 mL/min, with a column temperature of 30 °C and with 5 μl of sample injected into the system. GSLs were quantified at a wavelength of 229 nm.

MS analysis settings were as follows: electrospray ionization (ESI) was carried out at atmospheric pressure in negative ion mode (scan range m/z 100–1500 Da). Nebulizer pressure was set at 50 psi, gas-drying temperature at 350 °C, and capillary voltage at 2000 V. GSLs were quantified using sinigrin hydrate standard. Five concentrations of sinigrin hydrate (14–438 μg/mL) were prepared with 70% methanol and used to prepare an external calibration curve (r^2 = 0.996). Compounds were identified using their parent mass ion and characteristic ion fragments as well as comparing with literature ion data (Table 2). Compounds were quantified using Bruker Daltonics HyStar software (Bruker, Coventry, UK). Relative response factors (RRFs) were used in the calculation of GSL concentrations where available [53]. Where such data could not be found for intact GSLs, RRFs were assumed to be 1.0.

Table 2. Intact glucosinolates identified in cabbage accessions analysed by LC-MS.

Common Name	Chemical Name	Abbreviation	Mass Parent Ion	MS^2 Spectrum Ion (Base Ion in Bold) [a]	Reference
sinigrin	2-propenyl(allyl) GSL	SIN	358	278, 275, **259**, 227, 195, 180, 162	[54,55]
gluconapin	3-butenyl GSL	GPN	372	292, 275, **259**, 195, 194, 176	[54,56]
epi/progoitrin	(R, S)-2-hydroxy-3-butenyl GSL	PROG	388	332, 308, 301, 275, **259**, 210, 195, 146, 136	[54–56]
glucoiberverin	3-(methylthio)propyl GSL	GIBVN	406	326, 275, **259**, 288, 228,195	[52,54,55]
glucoerucin	4-(methylthio)butyl GSL	GER	420	340, 291, 275, **259**, 227, 195, 178, 163	[52,54,55]
glucoiberin	3-(methylsulfinyl) propyl GSL	GIBN	422	407, **358**, 259	[54–56]
glucoraphanin	4-(methylsulfinyl) butyl GSL	GRPN	436	422, **372**, 291, 259, 194	[52,54,55]
glucobrassicin	3-indolylmethyl GSL	GBSN	447	275, **259**, 251, 205	[54–56]
4-hydroxyglucobrassicin	4-hydroxy-3-indolylmethyl GSL	4-HOH	463	383, **285**, 267, 259, 240, 195	[54–56]

Key: GSL = glucosinolate; [a] Base ion highlighted in bold

2.7. Extraction of Glucosinolate Hydrolysis Products

GHPs were extracted and analysed following the method described by Bell et al. [57]. A total of 0.5 g of lyophilized cabbage was mixed with 10 mL deionized water, vortexed and allowed to incubate for three hours at 30 °C. The mixture was then centrifuged at 5000× *g* (18 °C) for ten minutes, and the supernatant collected. The pellet was extracted two more times with 10 mL deionized water, and the supernatants were combined and filtered (0.45 µm syringe filters, Epsom, UK) into glass centrifuge tubes. GHPs were extracted by adding an equal volume of dichloromethane (DCM) to the supernatant, vortexed for one minute and centrifuged at 3000× *g* for ten minutes. After centrifugation, the organic phase was collected, and the extraction step repeated twice. The organic phase collected was combined, 2 g sodium sulphate salt was added to remove any excess liquid present, and the mixture was filtered into a round-bottom flask. The filtrate was dried using a rotatory evaporator (37 °C), re-dissolved in 1 mL DCM, and filtered (0.22 µm filter; Fisher scientific, Loughborough, UK) in GC-MS glass vials (VWR, Lutterworth, UK) for GC-MC analysis.

2.8. GC-MS Analysis

GC–MS analysis was performed on an Agilent 7693/5975 GC–MS autosampler system (Agilent, Manchester, UK). The sample was injected onto a HP-5MS 15 m non-polar column DB-5MS (J and W scientific, Santa Clara, CA, USA) (0.25-µm film thickness, 0.25 mm I.D.). The injection temperature was 250 °C in split mode (1:20). The oven temperature was programmed from 40 to 320 °C at a rate of 5 °C/min until 250 °C. The carrier gas was helium, with flow rate of 1.1 mL/min and pressure of 7.1 psi. Mass spectra were obtained by electron ionization at 70 eV, and mass scan from 35 to 500 amu. A total of 1 µL of the sample was injected, and compounds were separated during a 42 min run. Compounds were identified using the National Institute of Standards and Technology (NIST) library and literature ion data (Table 3; see Figure S3 for GC-MS chromatograms) and quantified based on an external standard calibration curve. Five concentrations (0.15–0.5 mg/mL) of sulforaphane standard (Sigma Aldrich, Dorset, UK) were prepared in DCM (r^2 = 0.99). Data analysis was performed using ChemStation for GC-MS (Agilent, Manchester, UK).

Table 3. Glucosinolate hydrolysis products identified in cabbage accessions analysed by GC-MS.

Precursor Glucosinolate	Glucosinolate Hydrolysis Product		Abbreviation	LRI [a]	ID [b]	MS^2 Spectrum Ion (Base Ion in Bold) [c]	Reference
	Common name	**Chemical Name**					
	allyl thiocyanate	2-propenyl thiocyanate	ATC	871	B	99, 72, 45, 44, **41**, 39	[58]
sinigrin	allyl-ITC	2-propenyl isothiocyanate	AITC	884	B	**99**, 72, 71, 45, 41, 39	[58,59]
	1-cyano-2,3-epithiopropane	3,4-epithiobutane nitrile	CETP	1004	B	**99**, 72, 66, 59, 45, 41, 39	[58]
gluconapin	3-butenyl-ITC	1-butene, 4-isothiocyanate	3BITC	983	B	113, 85, **72**, 64, 55, 46, 45, 41	[58–60]
	4,5-epithiovaleronitrile	1-cyano-3,4-epithiobutane	EVN	1121	B	**113**, 86, 80, 73, 60, 45	[60]
progoitrin	goitrin	5-vinyloxazolidin-2-thione	GN	1545	B	**129**, 86, 85, 68, 57, 45, 43, 41, 39	
	1-cyano-2-hydroxy-3,4-epithiobutane isomer 1	2-hydroxy-3,4-epithiobutylcyanide diastereomer-1	CHETB-1	1225	B	129, 111, 89, 84, 68, **61**, 58, 55, 45	[61]
	1-cyano-2-hydroxy-3,4-epithiobutane isomer 2	2-hydroxy-3,4-epithiobutylcyanide diastereomer-2	CHETB-2	1245	B	129, 111, 89, 84, 68, **61**, 58, 55, 45	
glucoiberverin	4-methylthiobutyl nitrile	4-methylthio butanenitrile	4MBN	1085	B	115, 74, 68, **61**, 54, 47, 41	[58]
glucoerucin	erucin	4-(methylthio)-butyl-ITC	ER	1427	B	161, 146, **115**, 85, 72, 61, 55	[58,59]
	erucin nitrile	1-cyano-4-(methylthio) butane	ERN	1200	B	129, 87, 82, **61**, 55, 48, 41, 47	[58,59]
glucoiberin	iberin	3-methylsulfinylpropyl-ITC	IB	1617	B	163, 130, 116, 102, 100, 86, **72**, 63, 61,41	[58]
	iberin nitrile	4-methylsulfinylbutanenitrile	IBN	1384	B	**131**, 78, 64, 47, 41	[58]
gluconasturtin	2-phenylethyl-ITC	2-isothiocyanatoethyl benzene	PEITC	1458	B	163, 105, **91**, 65, 51, 40	[58]
	benzenepropanenitrile	2-phenylethyl cyanide	BPN	1238	B	131, **91**, 85, 65, 63, 57, 44, 51	[60]
glucoraphanin	sulforaphane	4-methylsulfinylbutyl-ITC	SFP	1757	A	160, 114, 85, **72**, 64, 63, 61, 55. 41, 39	[57,59]
	sulforaphane nitrile	5-(methylsulfinyl) pentanenitrile	SFN	1526	B	145, 128, 82, 64, **55**, 41	[57,59]
glucobrassiccin	indole-3-carbinol	1H-indole-3-methanol	I3C	1801	B	**144**, 145, 116, 108, 89	[61]
	indoleacetonitrile	1H-indole-3-acetonitrile	1IAN	1796	B	**155**, 145, 144, 130, 116, 89, 101, 63	[62]
pentyl glucosinolate	pentyl-ITC	1-isothiocyanato-pentane	PITC	1165	B	129, 114, 101, 96, 72, 55, **43**, 41, 39	[63]
glucotropaeolin	benzeneacetonitrile	2-phenylacetonitrile	BAN	1137	A	**117**, 90, 89, 77, 63, 51	[64]

Key: ITC—isothiocyanate. [a] Linear retention index on a HP-5MS non-polar column. [b] A, mass spectrum and LRI agree with those of authentic compound; B, mass spectrum agrees with reference spectrum in the NIST/EPA/NIH mass spectra database and that in the literature. [c] Base ion highlighted in bold.

2.9. Statistical Analysis

The results are the average of three biological replicates (each replicate consists of leaves from 4–5 cabbage heads) and two technical replicates ($n = 6$). Data obtained were analysed using 2-way ANOVA, with both cabbage accession (or morphotype) and growing condition (glasshouse and field) fitted as treatment effects, and Tukey's HSD multiple pairwise comparison test used to determine significant differences ($p < 0.05$) between samples. Multifactor analysis (MFA) was used to visualise the GSL and GHP data in a minimum number of dimensions (two or three). All statistical analyses were performed in XLSTAT (version 2019.4.2, Addinsoft, Paris, France).

3. Results and Discussion

3.1. Effect of Growing Conditions, Cabbage Morphotype and Accession on Myrosinase Activity

The myrosinase activity of cabbages grown on the field and in the glasshouse is shown in Figure 1. Myrosinase activity ranged from 12.2 U/g DW (BK-CPNT) to 127.4 U/g DW (SC-PW) in glasshouse samples and from 31.5 U/g DW (BK-CPNT and RC-RL) to 154.8 U/g DW (SC-PW) in field samples. Growing condition (glasshouse versus field), cabbage morphotype, cabbage accession and the interactions between these parameters significantly ($p < 0.0001$) affected myrosinase activity. The myrosinase activity of cabbage accessions within a cabbage morphotype differed significantly for all cabbage morphotypes studied. This agrees with previous reports that myrosinase activity varies within varieties and plant species [65]. Singh et al. [38] and Penas et al. [29] also reported variations in the myrosinase activity of different cabbage varieties within and between cabbage morphotypes. There were significant differences in the myrosinase activity of field and glasshouse grown cabbages across most of the accessions studied. Field grown cabbages had significantly higher myrosinase activity than glasshouse cabbages, except for WC-FEM, where the myrosinase activity of the glasshouse sample was significantly ($p < 0.003$) higher than that of the field grown counterpart.

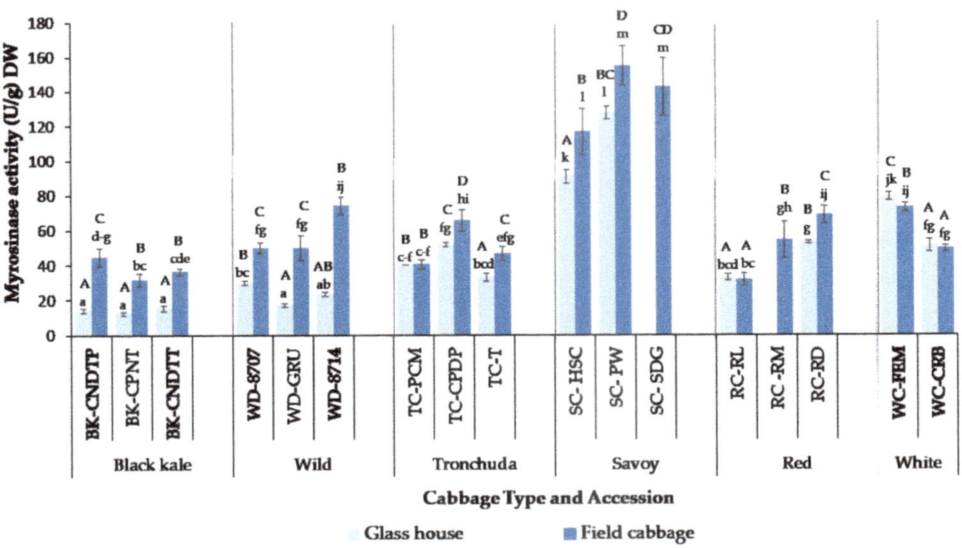

Figure 1. Myrosinase activity of field and glasshouse grown cabbages. Values are means of three biological replicates (each replicate comprising 4–5 cabbage heads) and two separately extracted technical replicates ($n = 6$). Error bars represent standard deviation from mean values. Missing data points implies cabbage accession did not survive under glasshouse growing conditions. Letters "A-D": bars not sharing a common uppercase letter indicates significant differences ($p < 0.0001$) between accessions and growing conditions within a cabbage morphotype. Letters "a-k": bars not sharing a common lowercase letter indicates significant differences ($p < 0.0001$) between accessions and growing conditions between cabbage morphotypes. See Table 1 for full names of cabbage accessions.

The myrosinase activity of TC-PCM, RC-RL and WC-CRB accessions did not differ significantly between field and glasshouse grown cabbages. Authors have previously reported that growing/environmental conditions affect myrosinase activity in *B. oleracea* species [24–26,29], and the results obtained from this study agree with their reports. The lower myrosinase activity of glasshouse cabbages might have been due to higher growth temperatures than those grown in the field. Minimum and maximum glasshouse temperatures were 14 and 43 °C, respectively, while minimum and maximum field temperatures were 6 and 24 °C, respectively (Supplementary Table S1). There are several possible reasons for the differences observed. One hypothesis could be that high temperatures reduced myrosinase enzyme synthesis or led to its more rapid denaturation. Another possible reason may have been that the process of synthesis and degradation of the enzyme (turn-over rate) was occurring faster at the higher growth temperatures, meaning that the plant did not accumulate a pool of enzymes at any one time. However, given that we can only see a snapshot in time when plants are sampled for enzyme assays, and each accession was harvested just once at a consistent time of day, it is not possible to infer the kinetics of these reactions occurring within the plant from the data in the present study. The kinetics of myrosinase synthesis and degradation within the plant is an area that warrants further study. Penas et al. [29], in their study of cabbages grown in different parts of Spain, reported that myrosinase activity was lower in cabbages grown in eastern Spain that were exposed to a higher growing temperature when compared to those grown in northern Spain with lower growing temperatures. It is, however impossible to say unequivocally that the lower myrosinase activity observed in the glasshouse samples is as a result of higher growth temperatures and not due to other stress factors, as we were unable to grow the plants in the glasshouse under lower temperatures similar to those observed in the field due to unavailability of cooling facilities within the glasshouse used in the study.

Another possible reason for the significantly lower enzyme activity in glasshouse cabbages could be due to stress factors during growth. Glasshouse cabbages were grown in pots, which may have led to stress from restricted root volume and reduced the amounts of nutrients (sulphur and nitrogen) available, potentially resulting in fewer enzymes and substrates being synthesized. Cabbage grown in the glasshouse achieved a lower above ground biomass than the field grown ones, indicating some form of stress. This was also evident in the differences in size of the closed heart cabbage heads, with the glasshouse plants having smaller heads than the field plants, as reported in Section 2.1. Their leaves appeared to be thinner and less robust than the field cabbages, as is often found in plants grown in protected environments that are not exposed to stimuli, such as wind, which for decades has been known to encourage the formation of thicker cell walls and smaller cells [66]. Hirai et al. [67] found that under nitrogen and/or sulfur limiting growth conditions, genes encoding myrosinase enzyme synthesis were down-regulated in Arabidopsis in order to facilitate storage of these elements in the form of glucosinolates in the leaf tissue. Yuan et al. [68] and Rodríguez-Hernández et al. [69] showed that salt stress reduced myrosinase activity in radish sprouts and broccoli, respectively. Pests and insect attack in field cabbages may have also led to higher myrosinase synthesis and/or accumulation in the cabbages. Accessions that did not show significantly different myrosinase activities between the two growing environments, or in the case of WC-FEM, higher myrosinase activity in glasshouse samples, might have been able to tolerate the glasshouse conditions and may have found it conducive for growth, while accessions that did not survive in the glasshouse may have found the conditions too harsh. Increased myrosinase activity as a result of abiotic stress, such as salt, temperature and drought, has been reported in various *Brassicaceae* species [70–72]. Increased myrosinase activity would result in enhanced glucosinolate hydrolysis to beneficial isothiocyanates, which would not only be beneficial to consumers but would also serve as defence compounds for the plants, thereby protecting them against insect and pest attacks.

3.2. Protein Content and Specific Myrosinase Activity of Glasshouse and Field Grown Cabbages

The protein content and specific activity of myrosinase for all accessions and growing conditions studied are presented in Table 4. The protein content and specific activity of samples studied were significantly ($p < 0.05$) affected by growing conditions and cabbage accession. Protein content did not correlate with myrosinase activity.

Table 4. Protein content ((mg/g ± SD) DW) and specific activity ((U/mg soluble protein ± SD) DW) of cabbage accessions grown in the glasshouse and on the field.

Cabbage Morpho-type/Accession	Protein Content (mg/g ± SD) DW Glasshouse	Protein Content (mg/g ± SD) DW Field	Specific activity (U/mg Soluble Protein ± SD) DW Glasshouse	Specific activity (U/mg Soluble Protein ± SD) DW Field
Black Kale				
BK-CNDTP	$29.1 \pm 0.4^{gh, B}$	$33.7 \pm 0.6^{l, C}$	$0.5 \pm 0.0^{a, A}$	$1.3 \pm 0.2^{d\text{-}h, C}$
BK-CPNT	$24.5 \pm 0.1^{e, A}$	$35.4 \pm 1.0^{m, D}$	$0.5 \pm 0.1^{a, A}$	$0.9 \pm 0.1^{a\text{-}d, B}$
BK-CNDTT	$25.4 \pm 3.9^{e, A}$	$36.7 \pm 0.7^{m, E}$	$0.6 \pm 0.1^{ab, A}$	$1.0 \pm 0.0^{b\text{-}e, B}$
Wild				
WD-8707	$27.4 \pm 0.7^{f, C}$	$31.4 \pm 0.1.2^{jk, E}$	$1.1 \pm 0.1^{c\text{-}f, B}$	$1.6 \pm 0.1^{ghi, C}$
WD-GRU	$25.3 \pm 0.1^{e, B}$	$29.9 \pm 0.6^{hi, D}$	$0.7 \pm 0.1^{abc, A}$	$1.7 \pm 0.2^{hij, C}$
WD-8714	$18.4 \pm 0.1^{a, A}$	$30.6 \pm 0.8^{ij, DE}$	$1.3 \pm 0.1^{d\text{-}h, B}$	$2.4 \pm 0.2^{l, D}$
Tronchuda				
TC-PCM	$32.8 \pm 0.1^{kl, D}$	$33.6 \pm 0.2^{l, E}$	$1.2 \pm 0.0^{d\text{-}h, AB}$	$1.2 \pm 0.1^{d\text{-}g, AB}$
TC-CPDP	$21.2 \pm 0.2^{b, A}$	$27.8 \pm 0.6^{fg, B}$	$2.4 \pm 0.1^{l, C}$	$2.4 \pm 0.3^{l, C}$
TC-T	$30.5 \pm 0.2^{hij, C}$	$33.1 \pm 0.8^{l, DE}$	$1.1 \pm 0.1^{cde, A}$	$1.4 \pm 0.1^{e\text{-}h, B}$
Savoy				
SC-HSC	$24.5 \pm 1.0^{e, A}$	$24.6 \pm 1.43^{e, A}$	$3.7 \pm 0.1^{m, A}$	$4.7 \pm 0.3^{n, B}$
SC-PW	$24.1 \pm 0.1^{cde, A}$	$24.3 \pm 0.3^{de, A}$	$5.3 \pm 0.1^{o, BC}$	$6.4 \pm 0.5^{q, D}$
SC-SDG	dng	$24.4 \pm 0.5^{de, A}$	dng	$5.8 \pm 0.7^{p, CD}$
Red				
RC-RL	$21.0 \pm 0.5^{b, A}$	$33.6 \pm 0.6^{l, C}$	$1.6 \pm 0.1^{ghi, B}$	$0.9 \pm 0.1^{a\text{-}d, A}$
RC-RM	dng	$35.4 \pm 1.0^{m, D}$	dng	$1.5 \pm 0.3^{f\text{-}i, B}$
RC-RD	$25.3 \pm 0.1^{e, B}$	$36.7 \pm 0.7^{m, E}$	$2.1 \pm 0.0^{jkl, C}$	$1.9 \pm 0.1^{ijk, C}$
White				
WC-FEM	$21.2 \pm 0.9^{b, A}$	$21.3 \pm 0.4^{b, A}$	$3.8 \pm 0.2^{m, C}$	$3.4 \pm 0.2^{m, B}$
WC-CRB	$22.8 \pm 0.6^{c, B}$	$23.0 \pm 1.2^{cd, B}$	$2.2 \pm 0.2^{kl, A}$	$2.1 \pm 0.1^{kl, A}$

Values are means of three processing replicates and two technical replicates ($n = 6 \pm$ SD). SD: standard deviation from mean; dng: did not grow. Letters "A-E": mean values not sharing a common uppercase letter differ significantly ($p < 0.05$) between accessions and growing condition within a cabbage type for each parameter (i.e., protein content and specific activity). Letters "a-q": mean values not sharing a common lowercase letter differ significantly ($p < 0.05$) between cabbage types, accessions, and growing condition for each parameter (i.e., protein content and specific activity). See Table 1 for full names of cabbage accessions.

Savoy and white cabbage accessions, which had the highest myrosinase activity, had the lowest protein contents. Just like myrosinase activity, the protein content of glasshouse samples was significantly lower than the field samples. This might be as a result of plant stress during growth, which prevents the plant from producing more nutrients than required or using up its stored nutrients in order to survive, as previously discussed in Section 3.1. Plant proteins have been reported to react negatively to environmental stress [26]. The results obtained are in agreement with Rosa and Heaney [73], who reported higher protein contents in Portuguese cabbage grown in lower environmental temperatures compared to those grown in higher temperatures.

Specific activity of the cabbages was similar to the myrosinase activity and protein content, with field grown cabbages generally having higher specific activity than the glasshouse cabbages. Savoy and white cabbage accessions had significantly higher specific activities than other cabbage morphotypes, as indeed both were found to have significantly higher total myrosinase activity (Figure 1). White cabbage has previously been reported to have higher specific activity than red cabbage [28], which is in agreement with the results of this study. However, a study conducted by Singh et al. [38] showed red cabbage with a higher specific activity than white and savoy cabbage. This might have been due to the differences in varieties studied or protein content of the cabbages, which was not reported in their study.

3.3. Effect of Cabbage Morphotype and Accession on GSL Profile and Concentration of Field Grown Cabbages

GSL profiles across cabbage accessions are presented in Figure 1; the statistical output of significant differences within and between cabbage morphotypes is documented in Supplementary Table S2. In total, nine different GSLs were identified across all accessions tested (Table 2): seven aliphatic GSLs, namely sinigrin (SIN), gluconapin (GPN) and epi/progoitrin (PROG), glucoibeverin (GIBVN), glucoerucin (GER), glucoiberin (GIBN) and glucoraphanin (GRPN), and two indole GSLs, glucobrassicin (GBSN) and 4-hydroxyglucobrassicin (4-HOH). PROG, GIBN and GRPN were the most abundant GSLs across all accessions studied, with 4-HOH, GIBVN and GER being the least abundant. 4-HOH was present in negligible amounts (<1.0 µmol/g DW) in all accessions, contributing not more than 1% to the total GSL content of the cabbages. When considering the ratio of total aliphatic to indole GSL concentrations in the accessions, over 60% of total GSL concentration was made up of aliphatic GSLs, with less than 30% from indole GSLs, with the exception of savoy SC-PW accession, where indole GSL comprised 36% of total GSL concentration (Supplementary Table S2).

GSL profiles and concentrations varied across cabbage accessions and differed significantly ($p < 0.05$) in some cases between and within cabbage morphotypes and accessions. Only five of the nine individual GSLs identified in the cabbages studied were found in black kale accessions:—GIBN, GRPN, GBSN, 4-HOH and GER—the last of which was present in BK-CNDTT alone. GRPN was the major GSL present in black kale accessions, consisting of over 50% on average of the total GSL content of black kale. The proportion of GRPN is similar to those previously reported by Kushad et al. [74] but much higher than those reported by Cartea et al. [40]. Previous studies detected SIN and PROG in kale and reported SIN as the main GSL in kale varieties [32,40,74]; however, SIN and PROG were not detected in the current study. There was a significant difference in total and individual GSL concentrations within black kale accessions, except for 4-HOH, which did not differ significantly ($p = 0.401$). BK-CPNT had the highest total GSL content (47.5 µmol/g DW).

GIBVN and GER were not identified in any of the wild and tronchuda cabbage accessions studied, while GIBN and GRPN were identified in all accessions except for WD-8707 accession. The concentration of individual GSLs differed significantly ($p < 0.0001$) across all wild and tronchuda cabbages. PROG and GPN were the most abundant GSLs in WD-8707 and WD-8714, while PROG and GRPN were the most abundant in WD-GRU. In tronchuda cabbages, SIN, GIBN and GBSN were at the highest concentrations, with SIN comprising up to 42% in TC-T.

A previous study [40] on GSL profile and concentrations in tronchuda cabbage identified 14 GSLs, compared to seven found in this study. However, GER was not identified in both studies, and proportions of the individual GSLs identified in both studies were similar.

The total GSL content of wild and tronchuda accessions differed significantly ($p < 0.01$ and $p < 0.0001$, respectively) between accessions within each cabbage morphotype. The most abundant GSLs in savoy cabbages were GIBN, SIN and GBSN, with GIBN concentrations as high as 61.3 µmol/g DW (57% of the total GSLs) in SC-SDG. GER was not identified in savoy accessions, and GPN was present in very low amounts in SC-SDG only. Similar proportions of savoy GSLs were reported by Ciska et al. [41] and Hanschen and Schreiner [43], but in both studies more individual GSLs were identified in the savoy varieties investigated than those reported in this study. For example, both studies identified GER in savoy cabbages, although present in trace amounts in the Ciska et al. [41] study. The total GSL content of savoy cabbages ranged from 47.6 µmol/g DW to 108.5 µmol/g DW. SC-SDG accession had significantly higher ($p < 0.0001$) total GSLs than SC-HSC and SC-PW, with SC-HSC having significantly lower total GSLs than the other two accessions.

In red and white cabbages, PROG, GIBN and GRPN were the most abundant GSLs. GBSN was also abundant in WC-CRB and RC-RL accessions, while GER was not identified in either accession. The concentrations of GRPN, GIBVN and GER did not differ significantly between red cabbage accessions. WC-CRB had significantly higher amounts of SIN, GIBN, GBSN and total GSL compared to WC-FEM, but differences in PROG and GRPN content were not significant. The total GSL content of RC-RL was significantly ($p < 0.0001$) higher than the other two red cabbage accessions. The results obtained agree with those previously reported [22,41,43]. However, a few studies disagree with the findings of this study; a previous study conducted by Park et al. [75] quantifying red cabbage GSL reported SIN absent in red cabbage, while Zabaras et al. [76] found GPN as the most abundant GSL in red cabbage.

Individual GSLs and total average GSL concentrations differed significantly ($p < 0.0001$) across all accessions, irrespective of cabbage morphotype. Total average GSL concentrations of accessions studied ranged from 18.9 μmol/g DW (BK-CNDTT) to 163.1 μmol/g DW (WD-8714). These differences were due to variations in GSL profiles and concentrations of individual GSLs. Wild cabbages generally had higher total GSL concentrations (Figure 2b) than other cabbage morphotypes, and these high concentrations were driven by significantly higher amounts of PROG in wild cabbages. Lower concentrations of total GSL observed in black kale accessions (18.9 μmol/g DW to 47.5 μmol/g DW) were due to lower numbers and concentrations of individual GSLs compared to the other cabbage morphotypes studied (Figure 2a). The variability in GSL concentrations between and within cabbage morphotypes and accessions is in agreement with previous reports that GSL profiles and concentrations vary between *Brassica* species and varieties [5,29,39,40,43,52,77]. The difference in GSL profiles of *Brassica* vegetables has been linked to genetic factors, while interactions between environmental and genetic factors are largely responsible for differences in GSL concentrations [8]. In general, concentrations of individual and total GSL of the gene bank cabbages reported in this study are much higher than those reported for commercial and gene bank cabbage varieties/accessions in the literature [29,40,41,43,74]. One reason for this may be due to the different varieties/accessions studied, implying that gene banks may indeed be a useful source from which to select accessions with higher GSL concentrations.

Differences in postharvest handling/time could have also contributed to the higher abundance of GSLs observed in the current study. Most varieties used in the literature were obtained from the supermarket and would have gone through a standard commercial supply chain upon harvest, unlike the samples used in this study, which were transferred to the laboratory immediately after harvest. The absence of commercial postharvest storage and handling processes in the current study could account for the differences observed between the samples and those reported in the literature. Total GSL abundance has been shown to decrease in *Brassica* vegetables stored for 7 days at 4–8 °C [78]. Lastly, differences in the conditions under which the plants were grown and/or harvested could also be responsible for the variations in GSL concentrations observed. This suggests that it is not only important that the right accession/variety is selected, but it must also be grown under optimal conditions and given as short a supply chain as possible to achieve optimum GSL abundance in the plants. The higher GSL concentrations in the present study can enhance the potential health benefits that may be derived from their consumption.

The differences in GSL profiles and concentrations of the accessions studied can potentially influence the sensory and health properties of the cabbages. For example, the absence of SIN and PROG in black kale accessions and higher concentrations of PROG reported in wild cabbage accessions may potentially influence the sensory characteristics of these cabbages, given SIN and PROG have been linked with bitter taste in *Brassica* vegetables [22,79]. On the other hand, higher amounts of GRPN (the precursor GSL for SFP formation linked to several health promoting properties of *Brassicas*) in kale, red and white cabbages could enhance the potential health benefits derived from their consumption [80]. The differences in cabbage accessions, growing conditions and geographical location, as well as environmental factors during cabbage cultivation, all play a vital role in GSL profile and concentration and therefore make comparing results between different studies difficult.

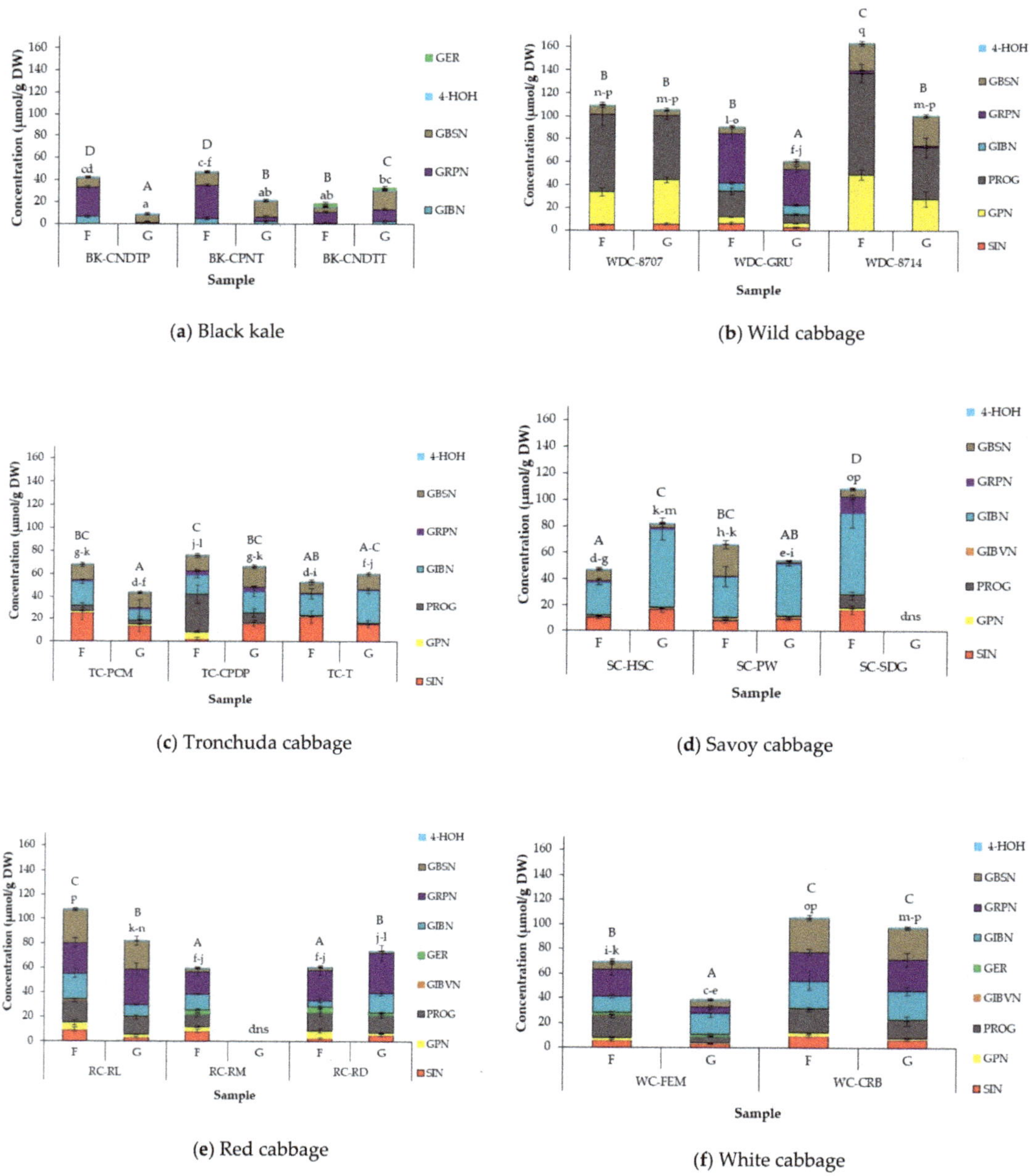

Figure 2. Glucosinolate concentrations (µmol/g DW) in different accessions of (**a**) Black kale; (**b**) Wild cabbage; (**c**) Tronchuda cabbage; (**d**) Savoy cabbage; (**e**) Red cabbage; and (**f**) White cabbage grown in the field and glasshouse. Error bars represent standard deviation from mean values. Letters above bars refer to differences in total GSL concentration. Letters "A-D": bars not sharing a common uppercase letter differ significantly ($p < 0.05$) between accession and growing conditions within a cabbage morphotype (i.e., within each separate graph). Letters "a-q": bars not sharing a common lowercase letter differ significantly ($p < 0.0001$) between cabbage morphotypes, accessions, and growing conditions (i.e., between the separate cabbage morphotype graphs). Abbreviations: F = Field, G = glasshouse; dns = did not survive. For abbreviations of accessions and compounds see Table 1 (cabbage accessions) and Table 2 (GSLs).

3.4. Effect of Growing Conditions on GSL Concentrations in Cabbage Accessions

The effect of growing conditions on GSL concentration is presented in Figure 1, with significant differences within and between cabbage morphotypes presented in Supplementary Table S2. While the GSL profile of cabbage accessions studied did not differ between growing conditions, there was a difference in GSL abundance between glasshouse and field grown cabbages. Total GSL concentrations in field grown samples ranged from 18.9 μmol/g DW (BK-CNDTT) to 163.1 μmol/g DW (WD-8714) and glasshouse samples from 8.81 μmol/g DW (BK-CNDTP) to 105.5 μmol/g DW (WD-8707). WD-8714 had significantly ($p < 0.0001$) higher concentrations of total GSLs compared to all other accessions, and this was largely due to the abundance of PROG and GPN, making up 83% and 69% of total GSLs in field and glasshouse samples, respectively.

Cabbages grown in the field had higher total GSL concentrations than glasshouse samples across most accessions studied, with a few exceptions (BK-CNDTT, TC-T, SC-HSC, and RC-RD), where total GSL concentrations were higher in glasshouse samples. These differences were significant in some but not all cases. Growing conditions significantly affected individual GSL concentrations between and within cabbage morphotypes and accessions. With the exception of black kale accessions, both field and glasshouse cabbages were predominantly abundant in aliphatic GSLs, with averages of 82 and 78%, respectively across all accessions, while indole GSLs comprised only 18 and 22% of total GSLs in field and glasshouse samples, respectively. In black kale accessions, however, growing conditions seemed to influence the ratio of aliphatic to indole GSL present in the samples. All black kale accessions grown in the glasshouse had much higher total indole GSL, with up to seven-fold differences reported in BK-CNDTP samples (Supplementary Table S2). The differences observed are mainly due to differences in the ratio of individual aliphatic to indole GSL present in the samples and not higher concentrations of indole GSLs in the glasshouse samples, as there was no significant difference observed in the concentrations of the most abundant indole GSL, GBSN, present in the samples between growing conditions (except for BK-CNDTT).

There was no clear pattern for the abundance of individual GSLs, as some GSLs were significantly higher in glasshouse samples for some accessions, but lower or not significantly different in others. PROG and GRPN were either significantly higher in field samples or did not significantly differ from glasshouse samples within accessions, except for RC-RD accession, where GRPN was significantly higher ($p < 1.0001$) when grown in the glasshouse. GRPN abundance in BK-CNDTP and BK-CPNT field grown accessions was up to 90% more than the corresponding glasshouse grown cabbages. GBSN was the most stable GSL across growing conditions, as there was no significant difference ($p = 0.101$) in GBSN between field and glasshouse cabbages.

Growing conditions such as growth temperature and photoperiod have been shown to influence the abundance of GSLs. There are several possible reasons for the differences observed in GSL concentrations in the different growing conditions. The higher total GSL content reported in most field samples could be due to production of higher amounts of GSLs by the plant in response to insect and pest attack on the field when compared to glasshouse samples. GSL compounds are plant metabolites produced by plants for defence against stress and attack from insect and pests [8,81]. In addition, the higher amount of GSLs in field samples could also be due to the lower average temperatures during growth (6 to 24 °C) compared to the higher temperatures in the glasshouse (14 and 43 °C) (Supplementary Table S1). Growth temperatures have been reported to influence GSL concentrations in *Brassica* vegetables. *Brassica* vegetables are generally thought to be cool weather crops, with average growing temperatures between 4–30 °C [82]. The optimum temperature for growth varies between different types of *Brassicas* and going below or above that temperature could affect GSL concentrations. The exact mechanism of GSL biosynthesis under different temperature conditions is unclear because of several interacting factors, such as drought and photoperiod, but it has been reported that plant stress due to high or low growing temperatures may enhance activities of transcription factors such as *MYC2* and *MYB28*, which promote GSL biosynthesis [42,83]. Literature studies have, however, generally reported higher GSLs at higher growing temperatures; Rosa and Rodrigues [27] reported a higher GSL content in young cabbage plants when grown at 30 °C compared to 20 °C. Lower GSL concentrations was reported in kale grown at lower temperatures compared to those grown at higher temperatures [32,33]. In addition, several authors have reported higher GSL concentrations in spring/summer grown cabbages (average temperatures between 25–30 °C) compared to autumn grown plants (temperatures < 20 °C) [29,39–41]. The lower amounts of GSL accumulated in glasshouse plants could also be the result of plant growing conditions. Glasshouse samples were grown in pots with drainage holes to allow excess water to seep out. However, this could have also led to sulphur leaching, leading to sulphur deficiency in the soil, and plants were not fed with sulphur fertilizers. Sulphur is a major precursor for GSL biosynthesis, and its deficiency has been reported to reduce GSL concentrations in *Brassica* plants, especially aliphatic GSLs, as sulphur deficiency limits methionine synthesis (basic substrate for aliphatic GSL biosynthesis) as opposed to tryptophan, a non-sulphur amino acid and precursor for indole GSL biosynthesis [84]. On average, reduced amounts of aliphatic GSLs were accumulated in glasshouse plants compared to field plants, while glasshouse samples accumulated higher amounts of indole GSLs than field samples. Sulphur was reported to influence the aliphatic GSL concentrations in rapeseed more than indole GSL [84]. However, glasshouse plants, which had significantly higher GSL concentrations compared to their field counterparts, may have found the glasshouse conditions more favourable than other accessions, which resulted in enhanced GSL production.

The results of this study show that cabbages differ in their requirements for growth, and it is important to plant cabbage accessions in growing conditions that are best suited for their maximum development, as individual plants respond differently under different environmental conditions. Optimizing agronomy practices and applying limited abiotic stress in a controlled manner could be a way of increasing myrosinase activity and GSL production in some *Brassica* species.

3.5. Effect of of Cabbage Morphotype and Accession on Glucosinolate Hydrolysis Products (GHPs) of Field Grown Cabbages

A total of 22 GHPs were identified and quantified from the cabbage accessions studied, comprising 11 ITCs and 11 nitriles/epithionitriles (Table 3). Concentrations of GHPs are presented in Figure 2, with significant differences between and within cabbage morphotypes and accessions presented in Supplementary Table S3. Results are expressed as sulforaphane equivalents.

The type and concentration of GHPs formed differed between cabbage accessions. Predominant GHPs did not differentiate between accessions within a cabbage morphotype but varied across cabbage morphotypes. There was a significant difference in the concentrations of individual and total GHPs formed within and between cabbage morphotypes and accessions (Figure 3 and Supplementary Table S3). Wild cabbage accessions had the highest levels of GHPs formed (8.79 µmol/g DW—8.6 µmol/g DW; Figure 2b) and tronchuda accessions the lowest (0.95 µmol/g DW—3.27 µmol/g DW; Figure 2c).

GHPs of GRPN and GBRN were the main GHPs detected in black kale accessions, with nitrile concentrations accounting for 74–89% of the total GHPs. BK-CPNT accessions had significantly lower total GHPs than BK-CNDTP. Isomers of CHETB, nitriles of PROG hydrolysis, were the most abundant GHPs formed in wild cabbages, except for WD-GRU, which had higher amounts of GN (PROG ITC) compared to the nitriles formed. This was unexpected, and it is unclear why this happened, because more nitriles than ITCs were formed for other GSLs present in the same sample. A possible explanation for this could be the activity of epithiospecifier modifier proteins (ESMs), enhancing the activity of specific myrosinase isoenzymes, which hydrolyse PROG present in the samples. ESM inhibits the activity of ESP, preventing the formation of nitriles and epithionitriles, and instead promotes ITC formation [15,85,86]. GN have been associated with bitter taste [87] and adverse effects on thyroid metabolism, leading to goitre formation. The reports on goitre formation are limited and based on animal studies, which show that average daily intake is not enough to produce adverse effects in humans [8]. However, to limit the health risks, genetic manipulation and selective breeding methods used to increase GRPN contents by threefold in '*Beneforte*' broccoli [88] could be employed to reduce PROG contents in the wild accessions. The main GHPs of tronchuda accessions were CETP and IBN, nitriles of SIN and GIBN, respectively. Total GHPs of TC-CPDP were significantly higher than TC-T. IBN and IB (GIBN hydrolysis products) were the most abundant GHPs in savoy cabbages, and SFP and SFN (hydrolysis products of GRPN) the most abundant in red and white cabbages.

In savoy, SC-HSC varied significantly from SC-PW and SC-SDG accessions, containing up to 60% more GHPs than the other two accessions. The much lower concentrations of GHPs in SC-PW compared to SC-HSC were unexpected due to similar concentrations of GSLs in both accessions. A similar trend was noticed between WC-CRB and WC-FEM accessions, where much lower GHPs were formed in WC-CRB accession, with significantly higher GSLs than WC-FEM. This might be related to variation in myrosinase and ESP activities within the samples. As previously discussed in Section 3.1, WC-FEM had significantly higher myrosinase activity than WC-CRB (Figure 1), which may explain the higher concentrations of GHPs formed. However, this is not the case in savoy cabbages, as SC-PW had the highest myrosinase activity (see Figure 1). It is hypothesized that myrosinase isoenzymes and ESP of SC-PW accession may be less stable than the other accessions and was, therefore, denatured before permitting full hydrolysis. As previously discussed, ESM activities promoting ITC formation may also be responsible for the higher GHP concentrations observed. For example, although GIBN concentration in WC-FEM was significantly ($p < 0.0001$) lower than that of WC-CRB, the amount of IB, the ITC formed from GIBN, was significantly ($p < 0.0001$) higher in WC-FEM than in WC-CRB. Another possible reason for the variation in GHP concentrations could be due to the type of myrosinase isoenzyme present within the samples. It has been reported that myrosinase isoenzymes differ in the rate at which they hydrolyse individual GSLs, though little is known of their substrate specificity. James and Rossiter [89] found that in the presence of ascorbic acid, two myrosinase isoenzymes identified in *Brassica napus* L. differed in the way they degraded SIN and neoglucobrassicin (NEO), with SIN being degraded more rapidly than NEO by both isoenzymes under the same conditions. While there are limited studies on the conversion ratio of GSLs to GHPs, studies on GHP formation in *Brassica oleracea* [43] and rocket salad [42] have shown that conversion of GSLs to GHPs is not always a linear reaction and GHP concentrations are generally much lower than the precursor GSL concentrations.

Several GHPs were identified in cabbage accessions where their GSLs were not detected: tiny amounts of 3BITC (GPN hydrolysis product) were formed in BK-CNDTT; 4MBN (nitrile of GIBVN) in tronchuda; EVN (GPN nitrile) in savoy cabbages; and ER and ERN (GER GHPs) in red and white cabbages. PEITC and BPN (GHPs of gluconasturtiin), PITC and BAN were also formed in most accessions. This could be due to concentration of the respective GSLs being below the limits of detection of the LC-MS^2 instrument used. A previous study of turnips detected GHPs of glucoberteroin, though the intact GSL was not detected [90]. A recent study on horseradish, wasabi, watercress, and rocket also detected GHPs, where their intact glucosinolates were not identified [91]. The profile of GHPs in this study is in agreement with the study of Hanschen and Schreiner [43]. However, in their study, they found CETP (nitrile from SIN hydrolysis) as the main GHP in savoy, red and white cabbages, which is inconsistent with this study, where GIBN GHPs (IB and IBN) and GRPN GHPs (SFP and SFN) were the main compounds detected. This difference can be attributed to the different varieties/accessions studied.

In general, the relationship between individual GSLs and their corresponding GHPs within an accession was as expected, where the dominant GSL resulted in their corresponding dominant GHPs, which is helpful in confirming the efficiency and accuracy of the GHP extraction method. Overall, nitriles and epithionitriles were the major hydrolysis products formed across all cabbage accessions, as has been reported previously in raw cabbage [62,92]. This is due to the activity of ESP and other nitrile forming proteins present in the samples, which hydrolyse GSLs to epithionitriles and nitriles instead of the more beneficial ITCs [92].

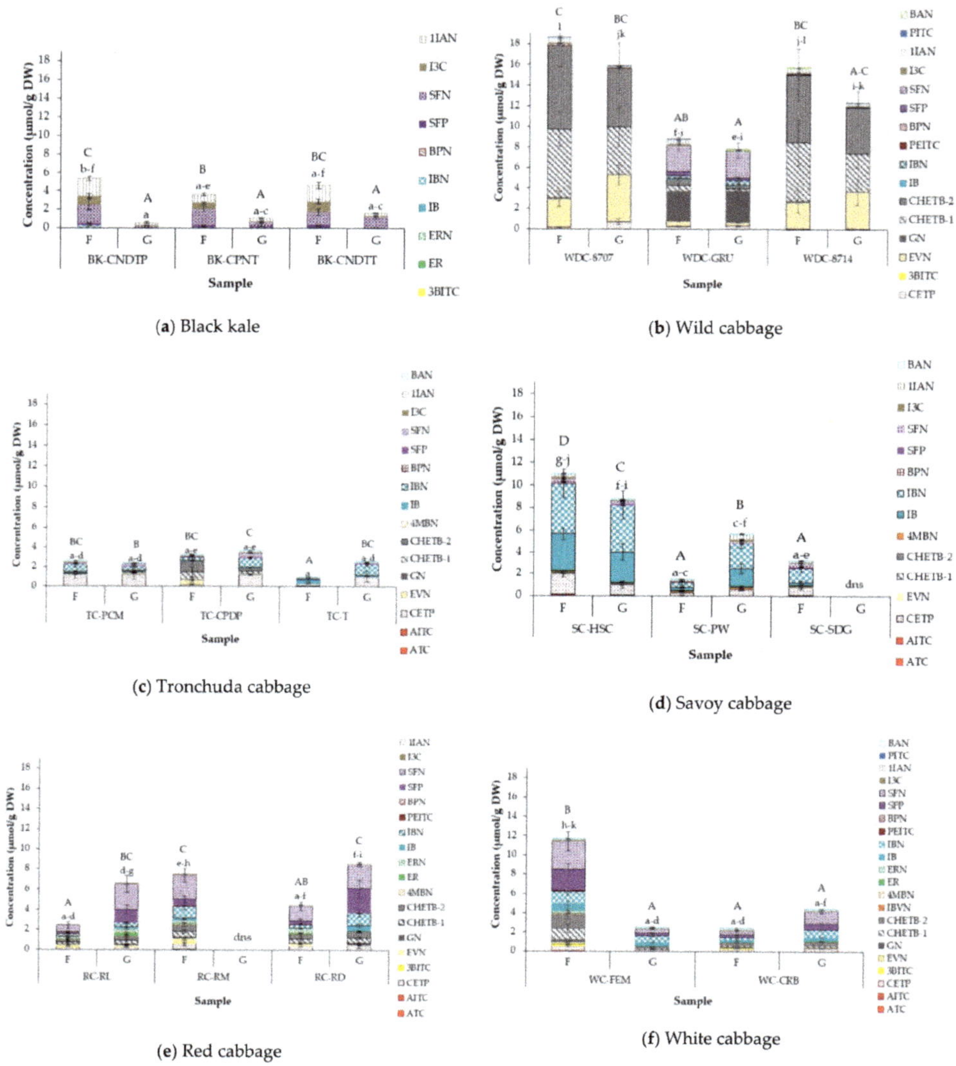

(**a**) Black kale (**b**) Wild cabbage (**c**) Tronchuda cabbage (**d**) Savoy cabbage (**e**) Red cabbage (**f**) White cabbage

Figure 3. Glucosinolate hydrolysis products (GHPs) (µmol/g DW) in different accessions of (**a**) Black kale; (**b**) Wild cabbage; (**c**) Tronchuda cabbage; (**d**) Savoy cabbage; (**e**) Red cabbage; and (**f**) White cabbage grown in the field and glasshouse. Error bars represent standard deviation from mean values. Letters above bars refer to differences in total GHP concentration. Letters "A-D": bars not sharing a common uppercase letter differ significantly ($p < 0.05$) between accessions and growing conditions within a cabbage morphotype (i.e., within each separate graph). Letters "a-l": bars not sharing a common lowercase letter differ significantly ($p < 0.0001$) between cabbage morphotypes, accessions, and growing conditions (i.e., between the separate graphs). Compounds with colour shades similar to one another are GHPs of corresponding GSLs presented in Figure 2. Abbreviations: F = Field; G = glasshouse; dns = did not survive. For abbreviations of accessions and compounds see Table 1 (cabbage accessions) and Table 3 (GHPs).

3.6. Effect of Growing Condition on GHP Concentrations

GHP profile and concentration in the two different growing conditions studied is presented in Figure 3, with the significant differences between growing conditions reported in Supplementary Table S3. The profile of the GHPs detected were similar between growing conditions, with a few exceptions. For example, BPN was identified in black kale field samples but not detected in glasshouse samples. GHP concentrations in field and glasshouse ranged from 0.95 µmol/g DW (TC-T) to 18.6 µmol/g DW (WD-8707) and 0.59 µmol/g DW (BK-CNDTP) to 15.9 µmol/g DW (WD-8707), respectively. Within accessions, total GHP accumulation was significantly higher in field plants than glasshouse, except for wild cabbage accessions, TC-PCM and WC-CRB, where total GHPs were higher in glasshouse samples; however, the differences were not significant, except in WC-CRB, where a significant difference was observed. Generally, total GHP concentrations followed a similar pattern to total GSLs, with a few exceptions. For example, the BK-CNDTT glasshouse sample had significantly lower total GHPs compared to the field sample (Figure 3a), despite the significantly higher total GSL in the glasshouse sample (Figure 2a). Significantly higher myrosinase activity, and possibly ESP activity, in the BK-CNDTT field compared to glasshouse sample may have led to the formation of more GHPs (Figure 1). A similar trend was observed in savoy accessions, where an abundance of GSL under one growing condition did not necessarily result in higher amounts of GHP formed. The results obtained in our study are in agreement with those reported by Jasper et al. [42], where growth temperatures had different effects on the amount of GSL and GHPs formed in rocket salads.

In summary, the results of this study show the importance of having both high myrosinase activity and GSL accumulation in plants, as they have a direct impact on the amount of hydrolysis compounds formed. It is therefore important to ensure that cabbages are cultivated under optimised growing conditions (such as temperature, available sulphur/nitrogen and controlled biotic stress) that favour both high myrosinase and GSL accumulation and not only one or the other.

3.7. Multifactor Analysis (MFA) of GSLs and GHPs Identified in Cabbage Accessions Grown under Two Different Conditions

To investigate the underlying structure of the results, MFA was performed on the GSL and GHP data from the cabbage accessions. Figure 4 shows distribution of the cabbage accessions as well as the scores and loadings of MFA performed on the mean data of GSLs and GHPs. Dimensions 1 and 2 (F1, F2) explained 42% of the variance in the data, but other dimensions did not provide any new information; therefore, only F1 and F2 are presented and discussed. The plot demonstrates that individual GSLs were positively correlated with their corresponding GHPs. From the plot, it is clear that cabbages were mostly distinguished based on morphotype rather than accessions or growing conditions, except for wild cabbage accessions, where there was a clear separation of WD2 (WD-GRU) from WD1 (WD-8707) and WD3 (WD-8714).

Based on the MFA, samples were grouped into three distinct clusters: one cluster comprised of black kale, red cabbage, white cabbage and WD-GRU accessions, another tronchuda and savoy cabbage accessions, and the final cluster WD-8707 and WD-8714 accessions. Black kale, red cabbage, white cabbage and WD-GRU correlated positively with GRPN, GER, 4-HOH and their hydrolysis products. Tronchuda and savoy cabbage samples correlated positively with GIBN, GIBVN, SIN and their hydrolysis products. WD1 and WD2 correlated positively with GPN and PROG and their nitriles, as well as total GSLs and GHPs, but was negatively correlated with black kale, red cabbage, white cabbage and WD-GRU accessions. An additional Pearson correlation demonstrating significant correlations ($p < 0.05$) between various GSLs and GHPs is presented in Supplementary Table S4. GIBN correlated negatively ($r^2 > -0.3$; $p < 0.01$) with PROG and its hydrolysis products, GPN and its hydrolysis products, and PITC. On the contrary, GPN was strongly positively correlated ($r^2 > 0.6$; $p < 0.0001$) with PROG and its hydrolysis products, EVN, PITC, total GSL and total GHPS. Total GSLs were significantly positively correlated ($r^2 = 0.5$;

$p < 0.01$) with total GHPs. Strong significant positive correlations ($r^2 > 0.5$; $p < 0.05$) were observed between individual GSLs and their corresponding GHPs. For example, GRPN was positively correlated with SFP and SFN ($r^2 > 0.5$ and 0.8; $p < 0.01$ and $p < 0.0001$ respectively).

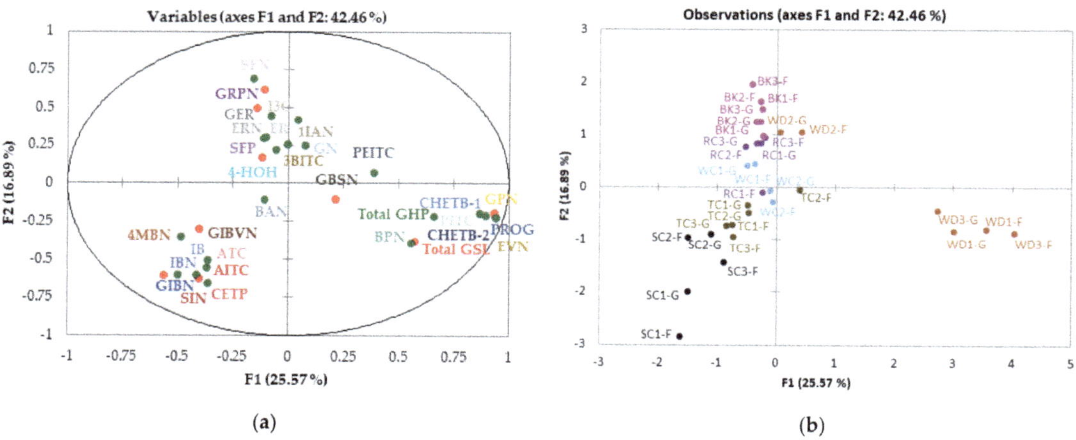

Figure 4. MFA map of glucosinolates and glucosinolate hydrolysis products (**a**) distribution of variables and (**b**) sample distribution. For codes and distribution on plot, refer to Table 1 (cabbage accessions) and Tables 2 and 3 (compounds). Compounds with different shades of the same colour in Figure 3a refer to the GSL and corresponding GHP. Key: F = Field ; G = Glasshouse; • GSL = Glucosinolates; • GHPs = Glucosinolate hydrolysis products; • BK= Black kale; • WD = Wild cabbage; • TC= Tronchuda cabbage; • SC = Savoy cabbage; • RC = Red cabbage; • WC = White cabbage.

It is obvious that the separations observed between samples are mainly driven by differences in GSLs and GHPs most accumulated in the samples: GN, GRPN, GER, 4-HOH and their GHPs in black kale, red cabbage, white cabbage and WD-GRU accessions; GIBN, SIN, GIBVN and their GHPs in tronchuda and savoy cabbage accessions; and lastly, PROG, GPN and their GHPs in WD-8707 and WD-8714 accessions. WD-8707 and WD-8714 had the highest concentration of total GSLs and GHPs, and this was responsible for the positive correlation of these accessions to total GSLs and GHPs observed. It is worth mentioning that PROG and CHETB, which were largely responsible for the high concentrations of total GSLs and GHPs in these accessions, correlated positively with total GSLs and GHPs. The result obtained provides a clear picture of the similarities and differences in GSL and GHP profile and concentrations of the different cabbage morphotypes and accessions studied.

Like any other study, some limitations were encountered in this study. First, the cabbage seeds used in the study were obtained from a gene bank. This means they have not been bred for uniformity in terms of plant characteristics and abundance of phytochemical compounds. Breeding programmes to date have mostly focused on developing disease-resistant and environmentally resilient crops, with less emphasis on the content of phytochemical compounds. This implies that there may be large variations in phytochemical compounds between cabbage heads/plants of the same accession, as has been observed in *Marathon* broccoli heads [93], and this may have influenced the results obtained in the present study. To reduce the effects of possible variation between plant heads, four to five heads were mixed together to obtain a representative sample. However, considering the amounts of heads used during the study, some variations may still have existed within the samples.

Second, the GC-MS method used for GHP analysis was long and required several steps to ensure that all GHPs present in the sample could be identified. However, some GHPs may have been lost or converted into other compounds in the process due to their very volatile and unstable nature. Though care was taken during the analysis to prevent losses, the rigorous analytical method may have led to some losses of the more volatile compounds.

4. Conclusions

In line with the primary hypothesis of the study, the results demonstrated that myrosinase activity as well as profiles and concentrations of GSLs and GHPs were all influenced by growing conditions, cabbage morphotypes and accession. However, in agreement with our secondary hypothesis, the profile and concentration of GSLs and GHPs formed were substantially more influenced by cabbage morphotype than accession. The study showed that planting cabbages in high growth temperatures and stressful conditions resulted in lower myrosinase activity. Myrosinase activity differed between accessions and cabbage morphotypes, although morphotype tended to have the more significant impact. Savoy cabbage accessions had the highest myrosinase activity, while black kale accessions had the lowest myrosinase activity.

The concentration and profile of GSL and GHP compounds accumulated differed between growing conditions and accessions, within and across cabbage morphotypes. While genetic factors had more influence on the GSL profile of the cabbages, differences in the GSL concentration were more affected by environmental factors during growth, which agrees with previous studies [8]. Growing conditions and cabbage accessions seem to have different effects on GSL and GHP formation, with higher GSL concentrations observed within a growing condition or accession not always resulting in a corresponding greater accumulation of GHPs and vice versa. Results obtained from the study showed that a possible reason for the higher GHP concentrations could be higher myrosinase activities in accessions with lower GSLs, as was observed in white cabbage and black kale accessions. However, this this was not the case in all accessions, suggesting there may be other reasons for the differences obtained. The results obtained therefore suggest that it would be incorrect to assume that higher myrosinase activity and/or GSL accumulation would automatically always result in high concentrations of GHPs.

Variations in the GSL and GHP contents imply differences in the potential health-promoting and sensory characteristics of the cabbages studied. For example, the high amounts of SFP present in red and white cabbages could potentially provide more health benefits on consumption when compared to other accessions. Conversely, high concentrations of PROG and GN (compounds linked to bitter taste) in wild accessions may reduce consumer acceptance and liking. However, the contents of GSLs and ITCs in *B. oleracea* vegetables alone does not provide a clear picture of the sensory characteristics of *B. oleracea* vegetables, as other compounds in the plant matrix, such as sugars and sweet tasting amino acids, can influence and modulate the sensory perception of these vegetables, as has been shown in previous studies on *Brassicas* and other crops such as lettuce [22,94–96].

Field grown cabbages had much higher GSLs and GHPs than glasshouse plants, with a few exceptions (SC-HSC and RC-RD). However, the biggest differences observed were between cabbage morphotypes, irrespective of the conditions under which they were grown. The result of this study suggests that cabbage morphotype and accession might be more important factors for GSL and GHP profiles of plants than the conditions under which they are grown. All individual GSLs and their corresponding GHPs were identified in the accessions studied, and a correlation between GSLs and GHPs was found. The difference in myrosinase activity and GSL and GHP concentrations could not be linked to morphology of head formation (closed heart or open leaf). The influence of growing conditions on cabbage biochemistry will be an important consideration, as the use of highly protected environments for crop production becomes more prevalent through indoor farming, which will also lead to breeding of cabbages with more compact morphology. Our data indicate that protected conditions need to be optimised, possibly by inclusion of controlled abiotic stress, in order to generate the GSL abundance that is observed in field grown crops.

Aliphatic GSLs, nitriles and epithionitriles were the most abundant compounds identified. The results suggest that consumption of raw cabbage may provide limited health benefits, as more nitriles and epithionitriles are formed than the more beneficial ITCs. It is therefore recommended to process the cabbages in ways that ensure hydrolysis of GSL to ITCs rather than nitriles. Despite the high amounts of nitriles and epithionitriles formed overall, high amounts of health beneficial SFP were detected in some red and white cabbage accessions. The result suggests that some gene bank accessions can be a good source of beneficial compounds and could be used in breeding programmes to introgress areas of the genome that regulate these compounds from the gene bank accessions into elite commercial cultivars. This can also be helpful for selection of more beneficial accessions for commercial cultivation and production. Given that accessions with lower GSL concentrations and higher myrosinase resulted in high GHP concentrations for some of the accessions studied, breeding programmes should not on only focus on selection of accessions with high GSL concentration but should also consider accessions that have high myrosinase activity and ESM, if maximum conversion of GSLs to ITCs is to be achieved.

Supplementary Materials: The following are available online at https://www.mdpi.com/article/10.3390/foods10122903/s1, Figure S1: Cross-section of planted cabbage morphotypes (**a**) Black kale (**b**) Wild cabbage (**c**) Tronchuda cabbage (**d**) Savoy cabbage (Field grown) (**e**) Savoy cabbage (Glasshouse grown) (**f**) Red cabbage (Field grown) (**g**) Red cabbage (Glasshouse grown) (**h**) White cabbage (Field grown) (**i**) White cabbage (Glasshouse grown). Figure S2: Cross-section of cabbages grown under (**a**) Controlled environment and (**b**) Glasshouse. Figure S3: Examples of GC-MS chromatograms for field and glasshouse grown samples for each morphotype of cabbage studied (**a**) Black kale; (**b**) Wild cabbage; (**c**) Tronchuda cabbage; (**d**) Savoy cabbage; (**e**) Red cabbage; and (**f**) White cabbage. Table S1: Climatic data of field and glasshouse cabbages. Table S2: Glucosinolate concentration in cabbages grown under different conditions (mg/g DW). Table S3: Glucosinolate hydrolysis products concentration in cabbages grown under different conditions (μg/g DW sulforaphane equivalent). Table S4: Pearson correlation matrix table showing correlations between glucosinolates and glucosinolate hydrolysis products identified in cabbage grown under two different conditions (a) correlation coefficients (r) and (b) significance of the correlation (p value).

Author Contributions: Conceptualization, O.O.O., C.W. and L.M.; methodology, O.O.O., C.W. and L.M.; software, O.O.O., and L.M.; validation, O.O.O., C.W. and L.M.; formal analysis, O.O.O.; investigation, O.O.O.; resources, O.O.O.; data curation, O.O.O.; writing—original draft preparation, O.O.O.; writing—review and editing, O.O.O., C.W. and L.M.; visualization, O.O.O., C.W. and L.M.; supervision, C.W. and L.M.; project administration, O.O.O., C.W. and L.M.; funding acquisition, O.O.O. All authors have read and agreed to the published version of the manuscript.

Funding: This research was funded by the Commonwealth Scholarship commission (CSC), UK, as part of the doctoral research of the first author (O.O.O.), scholar ID: NGCS-2013-363.

Institutional Review Board Statement: Not applicable.

Informed Consent Statement: Not applicable.

Data Availability Statement: The data presented in this study are available on request from the corresponding author.

Acknowledgments: We would like to specially thank Warwick Genetic Bank for providing the cabbage seeds used for the study, Chelsea Snell for her advice on the cabbage growing conditions and Valerie A. Jasper, Tobias James Lane and Matthew J. Richardson of the plant growth unit, University of Reading, for their help with growing the cabbages. A big thank you to Denise Macdonald, Bindukala Radha, Chris Bussey, Josh Stapleford and Charwin Piyapinyo for their help with sample preparation. Our thanks go to Sameer Khalil Ghawi and Olukayode Okunade for support and guidance with myrosinase extraction and assay; Luke Bell, Nicholas Michael, Stella Lignou, Hanis Nadia Yahya and Rashed Alarfaj for support and guidance with glucosinolate extraction and LC-MS analysis; and finally, Salah Abukhabta and Stephen Elmore for help and guidance with glucosinolate hydrolysis product extraction and GC-MS analysis, respectively.

Conflicts of Interest: The authors declare no conflict of interest. The funders had no role in the design of the study; in the collection, analyses, or interpretation of data; in the writing of the manuscript, or in the decision to publish the results.

References

1. Plants of the World Online. Available online: http://powo.science.kew.org/taxon/urn:lsid:ipni.org:names:279435-1 (accessed on 24 May 2021).
2. Herr, I.; Buchler, M.W. Dietary constituents of broccoli and other cruciferous vegetables: Implications for prevention and therapy of cancer. *Cancer Treat. Rev.* **2010**, *36*, 377–383. [CrossRef] [PubMed]
3. Verkerk, R.; Dekker, M.; Jongen, W.M.F. Post-harvest increase of indolyl glucosinolates in response to chopping and storage of *Brassica* vegetables. *J. Sci. Food Agric.* **2001**, *81*, 953–958. [CrossRef]
4. Guerrero-Beltran, C.E.; Calderon-Oliver, M.; Pedraza-Chaverri, J.; Chirino, Y.I. Protective effect of sulforaphane against oxidative stress: Recent advances. *Exp. Toxicol. Pathol.* **2012**, *64*, 503–508. [CrossRef] [PubMed]
5. Mithen, R.F.; Dekker, M.; Verkerk, R.; Rabot, S.; Johnson, I.T. The nutritional significance, biosynthesis and bioavailability of glucosinolates in human foods. *J. Sci. Food Agric.* **2000**, *80*, 967–984. [CrossRef]
6. Mithen, R. Glucosinolates—Biochemistry, genetics and biological activity. *Plant Growth Regul.* **2001**, *34*, 91–103. [CrossRef]
7. Redovnikovi, I.R.; Gliveti, T.; Delonga, K.; Vorkapi-Fura, J. Glucosinolates and their potential role in plant. *Period. Biol.* **2008**, *110*, 297–309.
8. Bjorkman, M.; Klingen, I.; Birch, A.N.E.; Bones, A.M.; Bruce, T.J.A.; Johansen, T.J.; Meadow, R.; Molmann, J.; Seljasen, R.; Smart, L.E.; et al. Phytochemicals of Brassicaceae in plant protection and human health—Influences of climate, environment and agronomic practice. *Phytochemistry* **2011**, *72*, 538–556. [CrossRef] [PubMed]
9. Wittstock, U.; Halkier, B.A. Glucosinolate research in the *Arabidopsis* era. *Trends Plant Sci.* **2002**, *7*, 263–270. [CrossRef]
10. Halkier, B.A.; Gershenzon, J. Biology and Biochemistry of Glucosinolates. *Annu. Rev. Plant Biol.* **2006**, *57*, 303–333. [CrossRef]
11. Blažević, I.; Montaut, S.; Burčul, F.; Olsen, C.; Burow, M.; Rollin, P.; Agerbirk, N. Glucosinolate structural diversity, identification, chemical synthesis and metabolism in plants. *Phytochemistry* **2020**, *169*, 112100. [CrossRef]
12. Bones, A.M.; Rossiter, J.T. The myrosinase-glucosinolate system, its organisation and biochemistry. *Physiol. Plant.* **1996**, *97*, 194–208. [CrossRef]
13. Ludikhuyze, L.; Rodrigo, L.; Hendrickx, M. The activity of myrosinase from broccoli (*Brassica oleracea* L. cv. *italica*): Influence of intrinsic and extrinsic factors. *J. Food Prot.* **2000**, *63*, 400–403. [CrossRef]
14. Morimitsu, Y.; Hayashi, K.; Nakagawa, Y.; Fujii, H.; Horio, F.; Uchida, K.; Osawa, T. Antiplatelet and anticancer isothiocyanates in Japanese domestic horseradish, Wasabi. *Mech. Ageing Dev.* **2000**, *116*, 125–134. [CrossRef]
15. Hanschen, F.S.; Lamy, E.; Schreiner, M.; Rohn, S. Reactivity and stability of glucosinolates and their breakdown products in foods. *Angew. Chem. Int. Ed.* **2014**, *53*, 11430–11450. [CrossRef] [PubMed]
16. Tsai, S.-C.; Huang, W.-W.; Huang, W.-C.; Lu, C.-C.; Chiang, J.-H.; Peng, S.-F.; Chung, J.-G.; Lin, Y.-H.; Hsu, Y.-M.; Amagaya, S. ERK-modulated intrinsic signaling and G2/M phase arrest contribute to the induction of apoptotic death by allyl isothiocyanate in MDA-MB-468 human breast adenocarcinoma cells. *Int. J. Oncol.* **2012**, *41*, 2065–2072. [CrossRef] [PubMed]
17. Leoni, O.; Iori, R.; Palmieri, S.; Esposito, E.; Menegatti, E.; Cortesi, R.; Nastruzzi, C. Myrosinase-generated isothiocyanate from glucosinolates: Isolation, characterization and in vitro antiproliferative studies. *Bioorg. Med. Chem.* **1997**, *5*, 1799–1806. [CrossRef]
18. Tripathi, K.; Hussein, U.K.; Anupalli, R.; Barnett, R.; Bachaboina, L.; Scalici, J.; Rocconi, R.P.; Owen, L.B.; Piazza, G.A.; Palle, K. Allyl isothiocyanate induces replication-associated DNA damage response in NSCLC cells and sensitizes to ionizing radiation. *Oncotarget* **2015**, *6*, 5237. [CrossRef]
19. Lambrix, V.; Reichelt, M.; Mitchell-Olds, T.; Kliebenstein, D.J.; Gershenzon, J. The Arabidopsis Epithiospecifier Protein Promotes the Hydrolysis of Glucosinolates to Nitriles and Influences Trichoplusia ni Herbivory. *Plant Cell* **2001**, *13*, 2793–2807. [CrossRef]
20. Baik, H.Y.; Juvik, J.; Jeffery, E.H.; Wallig, M.A.; Kushad, M.; Klein, B.P. Relating glucosinolate content and flavor of broccoli cultivars. *J. Food Sci.* **2003**, *68*, 1043–1050. [CrossRef]
21. Cox, D.N.; Melo, L.; Zabaras, D.; Delahunty, C.M. Acceptance of health-promoting Brassica vegetables: The influence of taste perception, information and attitudes. *Public Health Nutr.* **2012**, *15*, 1474–1482. [CrossRef]
22. Beck, T.K.; Jensen, S.; Bjoern, G.K.; Kidmose, U. The Masking Effect of Sucrose on Perception of Bitter Compounds in Brassica Vegetables. *J. Sens. Stud.* **2014**, *29*, 190–200. [CrossRef]
23. Doorn, H.E.v.; Kruk, G.C.v.d.; Holst, G.-J.v.; Raaijmakers-Ruijs, N.C.; Postma, E.; Groeneweg, B.; Jongen, W.H.F. The Glucosinolates Sinigrin and Progoitrin are important determination for taste. *J. Sci. Food Agric.* **1998**, *78*, 30–38. [CrossRef]
24. Wei, J.; Miao, H.; Wang, Q. Effect of glucose on glucosinolates, antioxidants and metabolic enzymes in *Brassica* sprouts. *Sci. Hortic.* **2011**, *129*, 535–540. [CrossRef]
25. Charron, C.S.; Saxton, A.M.; Sams, C.E. Relationship of climate and genotype to seasonal variation in the glucosinolate–myrosinase system. II. Myrosinase activity in ten cultivars of *Brassica oleracea* grown in fall and spring seasons. *J. Sci. Food Agric.* **2005**, *85*, 682–690. [CrossRef]
26. Charron, C.S.; Sams, C.E. Glucosinolate content and myrosinase activity in rapid-cycling *Brassica oleracea* grown in a controlled environment. *J. Am. Soc. Hort. Sci.* **2004**, *129*, 321–330. [CrossRef]
27. Rosa, E.A.; Rodrigues, P.M. The effect of light and temperature on glucosinolate concentration in the leaves and roots of cabbage seedlings. *J. Sci. Food Agric.* **1998**, *78*, 208–212. [CrossRef]
28. Yen, G.-C.; Wei, Q.-K. Myrosinase activity and total glucosinolate content of cruciferous vegetables, and some properties of cabbage myrosinase in Taiwan. *J. Sci. Food Agric.* **1993**, *61*, 471–475. [CrossRef]

29. Penas, E.; Frias, J.; Martinez-Villaluenga, C.; Vidal-Valverde, C. Bioactive compounds, myrosinase activity, and antioxidant capacity of white cabbages grown in different locations of Spain. *J. Agric. Food Chem.* **2011**, *59*, 3772–3779. [CrossRef]
30. Farnham, M.W.; Wilson, P.E.; Stephenson, K.K.; Fahey, J.W. Genetic and environmental effects on glucosinolate content and chemoprotective potency of broccoli. *Plant Breed.* **2004**, *123*, 60–65. [CrossRef]
31. Verkerk, R.; Schreiner, M.; Krumbein, A.; Ciska, E.; Holst, B.; Rowland, I.; Schrijver, R.D.; Hansen, M.; Gerhaeuser, G.; Mithen, R.; et al. Glucosinolates in *Brassica* vegetables: The influence of the food supply chain on intake, bioavailability and human health. *Mol. Nutr. Food Res.* **2009**, *53*, S219. [CrossRef] [PubMed]
32. Velasco, P.; Cartea, M.E.; González, C.; Vilar, M.; Ordás, A. Factors affecting the glucosinolate content of kale (*Brassica oleracea acephala* group). *J. Agric. Food Chem.* **2007**, *55*, 955–962. [CrossRef]
33. Steindal, A.L.H.; Rødven, R.; Hansen, E.; Mølmann, J. Effects of photoperiod, growth temperature and cold acclimatisation on glucosinolates, sugars and fatty acids in kale. *Food Chem.* **2015**, *174*, 44–51. [CrossRef] [PubMed]
34. Rungapamestry, V.; Duncan, A.J.; Fuller, Z.; Ratcliffe, B. Changes in glucosinolate concentrations, myrosinase activity, and production of metabolites of glucosinolates in cabbage (*Brassica oleracea* var. *capitata*) cooked for different durations. *J. Agric. Food Chem.* **2006**, *54*, 7628–7634. [CrossRef] [PubMed]
35. Rungapamestry, V.; Duncan, A.J.; Fuller, Z.; Ratcliffe, B. Influence of blanching and freezing broccoli (*Brassica oleracea* var. *italica*) prior to storage and cooking on glucosinolate concentrations and myrosinase activity. *Eur. Food Res. Technol.* **2008**, *227*, 37.
36. Ghawi, S.K.; Methven, L.; Rastall, R.A.; Niranjan, K. Thermal and high hydrostatic pressure inactivation of myrosinase from green cabbage: A kinetic study. *Food Chem.* **2012**, *131*, 1240–1247. [CrossRef]
37. Okunade, O.A.; Methven, L.; Niranjan, K. A comparison of myrosinase activity and stability in fresh broccoli (*B. oleracea* var. *italica*) and Brown Mustard (B. *juncea*) Seeds. *Turk. J. Agric.-Food Sci. Technol.* **2020**, *8*, 64. [CrossRef]
38. Singh, J.; Rai, M.; Upadhyay, A.K.; Prasad, K. Sinigrin (2-propenyl glucosinolate) content and myrosinase activity in *Brassica* vegetables. *Int. J. Veg. Sci.* **2007**, *13*, 21–31. [CrossRef]
39. Charron, C.S.; Saxton, A.M.; Sams, C.E. Relationship of climate and genotype to seasonal variation in the glucosinolate-myrosinase system. I. Glucosinolate content in ten cultivars of *Brassica oleracea* grown in fall and spring seasons. *J. Sci. Food Agric.* **2005**, *85*, 671–681. [CrossRef]
40. Cartea, M.E.; Velasco, P.; Obregon, S.; Padilla, G.; de Haro, A. Seasonal variation in glucosinolate content in *Brassica oleracea* crops grown in northwestern Spain. *Phytochemistry* **2008**, *69*, 403–410. [CrossRef]
41. Ciska, E.; Martyniak-Przybyszewska, B.; Kozlowska, H. Content of glucosinolates in cruciferous vegetables grown at the same site for two years under different climatic conditions. *J. Agric. Food Chem.* **2000**, *48*, 2862–2867. [CrossRef]
42. Jasper, J.; Wagstaff, C.; Bell, L. Growth temperature influences postharvest glucosinolate concentrations and hydrolysis product formation in first and second cuts of rocket salad. *Postharvest Biol. Technol.* **2020**, *163*, 111157. [CrossRef] [PubMed]
43. Hanschen, F.S.; Schreiner, M. Isothiocyanates, nitriles, and epithionitriles from glucosinolates are affected by genotype and developmental stage in *Brassica oleracea* varieties. *Front. Plant Sci.* **2017**, *8*, 1095. [CrossRef] [PubMed]
44. Daxenbichler, M.E.; Van Etten, C.H.; Spencer, G.F. Glucosinolates and derived products in cruciferous vegetables. Identification of organic nitriles from cabbage. *J. Agric. Food Chem.* **1977**, *25*, 121–124. [CrossRef]
45. Van Eylen, D.; Bellostas, N.; Strobel, B.W.; Oey, I.; Hendrickx, M.; Van Loey, A.; Sørensen, H.; Sørensen, J.C. Influence of pressure/temperature treatments on glucosinolate conversion in broccoli (*Brassica oleraceae* L. cv *Italica*) heads. *Food Chem.* **2009**, *112*, 646–653. [CrossRef]
46. Church, A.; Mitchell, R.; Ravenscroft, N.; Stapleton, L.M. 'Growing your own': A multi-level modelling approach to understanding personal food growing trends and motivations in Europe. *Ecol. Econ.* **2015**, *110*, 71–80. [CrossRef]
47. Tolstrup, J. The Rise of the Indoor Gardening Movement. Available online: https://blog.backtotheroots.com/2018/03/14/rise-indoor-gardening-movement/ (accessed on 25 May 2021).
48. Ghawi, S.K.; Methven, L.; Niranjan, K. The potential to intensify sulforaphane formation in cooked broccoli (*Brassica oleracea* var. *italica*) using mustard seeds (*Sinapis alba*). *Food Chem.* **2013**, *138*, 1734–1741. [CrossRef]
49. Gatfield, I.L.; Sand, T. A coupled enzymatic procedure for the determination of myrosinase activity. *Lebensm.-Wiss. Technol.* **1983**, *16*, 73–75.
50. Wilkinson, A.P.; Rhodes, M.J.C.; Fenwick, R.G. Myrosinase Activity of Cruciferous Vegetables. *J. Sci. Food Agric.* **1984**, *35*, 543–552. [CrossRef]
51. Bradford, M.M. A rapid and sensitive method for the quantitation of microgram quantities of protein utilizing the principle of protein-dye binding. *Anal. Biochem.* **1976**, *72*, 248–254. [CrossRef]
52. Bell, L.; Oruna-Concha, M.J.; Wagstaff, C. Identification and quantification of glucosinolate and flavonol compounds in rocket salad (*Eruca sativa*, *Eruca vesicaria* and *Diplotaxis tenuifolia*) by LC–MS: Highlighting the potential for improving nutritional value of rocket crops. *Food Chem.* **2015**, *172*, 852–861. [CrossRef]
53. Clarke, D.B. Glucosinolates, structures and analysis in food. *Anal. Methods* **2010**, *2*, 310. [CrossRef]
54. Rochfort, S.J.; Trenerry, V.C.; Imsic, M.; Panozzo, J.; Jones, R. Class targeted metabolomics: ESI ion trap screening methods for glucosinolates based on MSn fragmentation. *Phytochemistry* **2008**, *69*, 1671–1679. [CrossRef] [PubMed]
55. Lelario, F.; Bianco, G.; Bufo, S.A.; Cataldi, T.R.I. Establishing the occurrence of major and minor glucosinolates in Brassicaceae by LC–ESI-hybrid linear ion-trap and Fourier-transform ion cyclotron resonance mass spectrometry. *Phytochemistry* **2012**, *73*, 74–83. [CrossRef]

56. Bennett, R.N.; Mellon, F.A.; Kroon, P.A. Screening crucifer seeds as sources of specific intact glucosinolates using ion-pair high-performance liquid chromatography negative ion electrospray mass spectrometry. *J. Agric. Food Chem.* **2004**, *52*, 428–438. [CrossRef]
57. Bell, L.; Yahya, H.N.; Oloyede, O.O.; Methven, L.; Wagstaff, C. Changes in rocket salad phytochemicals within the commercial supply chain: Glucosinolates, isothiocyanates, amino acids and bacterial load increase significantly after processing. *Food Chem.* **2017**, *221*, 521–534. [CrossRef]
58. Al-Gendy, A.A.; Lockwood, G.B. GC-MS analysis of volatile hydrolysis products from glucosinolates in Farsetia aegyptia var. ovalis. *Flavour Fragr. J.* **2003**, *18*, 148–152. [CrossRef]
59. Arora, R.; Sharma, D.; Kumar, R.; Singh, B.; Vig, A.P.; Arora, S. Evaluating extraction conditions of glucosinolate hydrolytic products from seeds of *Eruca sativa* (Mill.) Thell. using GC-MS. *J. Food Sci.* **2014**, *79*, C1964–C1969. [CrossRef]
60. Hong, E.; Kim, G.-H. GC-MS analysis of the extracts from Korean cabbage (*Brassica campestris* L. ssp. *pekinensis*) and its seed. *Prev. Nutr. Food Sci.* **2013**, *18*, 218–221. [CrossRef]
61. Spencer, G.F.; Daxenbichler, M.E. Gas chromatography-mass spectrometry of nitriles, isothiocyanates and oxazolidinethiones derived from cruciferous glucosinolates. *J. Sci. Food Agric.* **1980**, *31*, 359–367. [CrossRef]
62. Hanschen, F.S.; Klopsch, R.; Oliviero, T.; Schreiner, M.; Verkerk, R.; Dekker, M. Optimizing isothiocyanate formation during enzymatic glucosinolate breakdown by adjusting pH value, temperature and dilution in *Brassica* vegetables and *Arabidopsis thaliana*. *Sci. Rep.* **2017**, *7*, 40807. [CrossRef]
63. de Pinho, P.G.; Valentão, P.; Gonçalves, R.F.; Sousa, C.; Andrade, P.B. Volatile composition of *Brassica oleracea* L. var. *costata* DC leaves using solid-phase microextraction and gas chromatography/ion trap mass spectrometry. *Rapid Commun. Mass Spectrom.* **2009**, *23*, 2292–2300. [CrossRef] [PubMed]
64. Vaughn, S.F.; Palmquist, D.E.; Duval, S.M.; Berhow, M.A. Herbicidal activity of glucosinolate-containing seedmeals. *Weed Sci.* **2017**, *54*, 743–748. [CrossRef]
65. Travers-Martin, N.; Kuhlmann, F.; Muller, C. Revised determination of free and complexed myrosinase activities in plant extracts. *Plant Physiol. Biochem.* **2008**, *46*, 506–516. [CrossRef] [PubMed]
66. Grace, J. 3. Plant response to wind. *Agric. Ecosyst. Environ.* **1988**, *22-23*, 71–88. [CrossRef]
67. Hirai, M.Y.; Yano, M.; Goodenowe, D.B.; Kanaya, S.; Kimura, T.; Awazuhara, M.; Arita, M.; Fujiwara, T.; Saito, K. Integration of transcriptomics and metabolomics for understanding of global responses to nutritional stresses in *Arabidopsis thaliana*. *Proc. Natl. Acad. Sci. USA* **2004**, *101*, 10205–10210. [CrossRef]
68. Yuan, G.; Wang, X.; Guo, R.; Wang, Q. Effect of salt stress on phenolic compounds, glucosinolates, myrosinase and antioxidant activity in radish sprouts. *Food Chem.* **2010**, *121*, 1014–1019. [CrossRef]
69. Rodríguez-Hernández, M.D.C.; Moreno, D.A.; Carvajal, M.; Martínez-Ballesta, M.D.C. Genotype influences sulfur metabolism in broccoli (*Brassica oleracea* L.) under elevated CO2 and NaCl stress. *Plant Cell Physiol.* **2014**, *55*, 2047–2059. [CrossRef] [PubMed]
70. Pang, Q.; Guo, J.; Chen, S.; Chen, Y.; Zhang, L.; Fei, M.; Jin, S.; Li, M.; Wang, Y.; Yan, X. Effect of salt treatment on the glucosinolate-myrosinase system in Thellungiella salsuginea. *Plant Soil* **2012**, *355*, 363–374. [CrossRef]
71. Kissen, R.; Eberl, F.; Winge, P.; Uleberg, E.; Martinussen, I.; Bones, A.M. Effect of growth temperature on glucosinolate profiles in *Arabidopsis thaliana* accessions. *Phytochemistry* **2016**, *130*, 106–118. [CrossRef]
72. Koh, J.; Chen, G.; Yoo, M.-J.; Zhu, N.; Dufresne, D.; Erickson, J.E.; Shao, H.; Chen, S. Comparative proteomic analysis of *Brassica napus* in response to drought stress. *J. Proteome Res.* **2015**, *14*, 3068–3081. [CrossRef]
73. Rosa, E.; Heaney, R. Seasonal variation in protein, mineral and glucosinolate composition of Portuguese cabbages and kale. *Anim. Feed Sci. Technol.* **1996**, *57*, 111–127. [CrossRef]
74. Kushad, M.M.; Brown, A.F.; Kurilich, A.C.; Juvik, J.A.; Klein, B.P.; Wallig, M.A.; Jeffery, E.H. Variation of glucosinolates in vegetable crops of *Brassica oleracea*. *J. Agric. Food Chem.* **1999**, *47*, 1541–1548. [CrossRef]
75. Park, S.; Arasu, M.V.; Lee, M.-K.; Chun, J.-H.; Seo, J.M.; Al-Dhabi, N.A.; Kim, S.-J. Analysis and metabolite profiling of glucosinolates, anthocyanins and free amino acids in inbred lines of green and red cabbage (*Brassica oleracea* L.). *LWT-Food Sci. Technol.* **2014**, *58*, 203–213. [CrossRef]
76. Zabaras, D.; Roohani, M.; Krishnamurthy, R.; Cochet, M.; Delahunty, C.M. Characterisation of taste-active extracts from raw Brassica oleracea vegetables. *Food Funct.* **2013**, *4*, 592–601. [CrossRef]
77. Fahey, J.W.; Zalcmann, A.T.; Talalay, P. The chemical diversity and distribution of glucosinolates and isothiocyanates among plants. *Phytochemistry* **2001**, *56*, 5–51. [CrossRef]
78. Song, L.; Thornalley, P.J. Effect of storage, processing and cooking on glucosinolate content of *Brassica* vegetables. *Food Chem. Toxicol.* **2007**, *45*, 216–224. [CrossRef]
79. Drewnowski, A.; Gomez-Carneros, C. Bitter taste, phytonutrients, and the consumer: A review. *Am. J. Clin. Nutr.* **2000**, *72*, 1424–1435. [CrossRef]
80. Vaughn, S.F.; Berhow, M.A. Glucosinolate hydrolysis products from various plant sources: PH effects, isolation, and purification. *Ind. Crop. Prod.* **2005**, *21*, 193–202. [CrossRef]
81. Rohr, F.; Ulrichs, C.; Mucha-Pelzer, T.; Mewis, I. Variability of aliphatic glucosinolates in Arabidopsis and their influence on insect resistance. *Commun. Agric. Appl. Biol. Sci.* **2006**, *71*, 507–515. [PubMed]
82. Wurr, D.C.E.; Fellows, J.R.; Phelps, K. Investigating trends in vegetable crop response to increasing temperature associated with climate change. *Sci. Hortic.* **1996**, *66*, 255–263. [CrossRef]

83. Gigolashvili, T.; Berger, B.; Flügge, U.-I. Specific and coordinated control of indolic and aliphatic glucosinolate biosynthesis by R2R3-MYB transcription factors in Arabidopsis thaliana. *Phytochem. Rev.* **2009**, *8*, 3–13. [CrossRef]
84. Zhao, F.; Evans, E.J.; Bilsborrow, P.E.; Syers, J.K. Influence of nitrogen and sulphur on the glucosinolate profile of rapeseed (*Brassica napus* L.). *J. Sci. Food Agric.* **1994**, *64*, 295–304. [CrossRef]
85. Burow, M.; Wittstock, U. Regulation and function of specifier proteins in plants. *Phytochem. Rev.* **2009**, *8*, 87–99. [CrossRef]
86. Angelino, D.; Jeffery, E. Glucosinolate hydrolysis and bioavailability of resulting isothiocyanates: Focus on glucoraphanin. *J. Funct. Foods* **2014**, *7*, 67–76. [CrossRef]
87. Fenwick, G.R.; Griffiths, N.M.; Heaney, R.K. Bitterness in brussels sprouts (*Brassica oleracea* L. var.gemmifera): The role of glucosinolates and their breakdown products. *J. Sci. Food Agric.* **1983**, *34*, 73–80. [CrossRef]
88. Traka, M.H.; Saha, S.; Huseby, S.; Kopriva, S.; Walley, P.G.; Barker, G.C.; Moore, J.; Mero, G.; van den Bosch, F.; Constant, H.; et al. Genetic regulation of glucoraphanin accumulation in Beneforté broccoli. *New Phytol.* **2013**, *198*, 1085–1095. [CrossRef]
89. James, D.C.; Rossiter, J.T. Development and characteristics of myrosinase in Brassica napus during early seedling growth. *Physiol. Plant.* **1991**, *82*, 163–170. [CrossRef]
90. Klopsch, R.; Witzel, K.; Börner, A.; Schreiner, M.; Hanschen, F.S. Metabolic profiling of glucosinolates and their hydrolysis products in a germplasm collection of *Brassica rapa* turnips. *Food Res. Int.* **2017**, *100*, 392–403. [CrossRef]
91. Bell, L.; Kitsopanou, E.; Oloyede, O.O.; Lignou, S. Important odorants of four *Brassicaceae* species, and discrepancies between glucosinolate profiles and observed hydrolysis products. *Foods* **2021**, *10*, 1055. [CrossRef] [PubMed]
92. Matusheski, N.V.; Swarup, R.; Juvik, J.A.; Mithen, R.; Bennett, M.; Jeffery, E.H. Epithiospecifier protein from broccoli (*Brassica oleracea* L. ssp. *italica*) inhibits formation of the anticancer agent sulforaphane. *J. Agric. Food Chem.* **2006**, *54*, 2069–2076. [CrossRef] [PubMed]
93. Winkler, S.; Faragher, J.; Franz, P.; Imsic, M.; Jones, R. Glucoraphanin and flavonoid levels remain stable during simulated transport and marketing of broccoli (*Brassica oleracea* var. *italica*) heads. *Postharvest Biol. Technol.* **2007**, *43*, 89–94. [CrossRef]
94. Jones, R.B.; Faragher, J.D.; Winkler, S. A review of the influence of postharvest treatments on quality and glucosinolate content in broccoli (Brassica oleracea var. italica) heads. *Postharvest Biol. Technol.* **2006**, *41*, 1–8. [CrossRef]
95. van Doorn, J.E. *Development of Vegetables with Improved Consumer Quality: A Case Study in Brussels Sprouts*; University of Wageningen: Wageningen, The Netherlands, 1999.
96. Chadwick, M.; Gawthrop, F.; Michelmore, R.W.; Wagstaff, C.; Methven, L. Perception of bitterness, sweetness and liking of different genotypes of lettuce. *Food Chem.* **2016**, *197*, 66–74. [CrossRef] [PubMed]

foods

MDPI

Article

Important Odorants of Four Brassicaceae Species, and Discrepancies between Glucosinolate Profiles and Observed Hydrolysis Products

Luke Bell [1], Eva Kitsopanou [2], Omobolanle O. Oloyede [2] and Stella Lignou [2,*]

1 School of Agriculture, Policy and Development, University of Reading, Whiteknights, Reading RG6 6AR, UK; luke.bell@reading.ac.uk

2 Sensory Science Centre, Department of Food and Nutritional Sciences, Harry Nursten Building, University of Reading, Whiteknights, Reading RG6 6DZ, UK; evakitsopanou@gmail.com (E.K.); bola.oloyede@reading.ac.uk (O.O.O.)

* Correspondence: s.lignou@reading.ac.uk; Tel.: +44-(0)118-378-8717

Abstract: It is widely accepted that the distinctive aroma and flavour traits of Brassicaceae crops are produced by glucosinolate (GSL) hydrolysis products (GHPs) with other non-GSL derived compounds also reported to contribute significantly to their aromas. This study investigated the flavour profile and glucosinolate content of four Brassicaceae species (salad rocket, horseradish, wasabi, and watercress). Solid-phase microextraction followed by gas chromatography-mass spectrometry and gas chromatography-olfactometry were used to determine the volatile compounds and odorants present in the four species. Liquid chromatography-mass spectrometry was used to determine the glucosinolate composition, respectively. A total of 113 compounds and 107 odour-active components were identified in the headspace of the four species. Of the compounds identified, 19 are newly reported for 'salad' rocket, 26 for watercress, 30 for wasabi, and 38 for horseradish, marking a significant step forward in understanding and characterising aroma generation in these species. There were several non-glucosinolate derived compounds contributing to the 'pungent' aroma profile of the species, indicating that the glucosinolate-derived compounds are not the only source of these sensations in Brassicaceae species. Several discrepancies between observed glucosinolates and hydrolysis products were observed, and we discuss the implications of this for future studies.

Keywords: volatile compounds; odorants; glucosinolate; Brassicaceae; 'salad' rocket; wasabi; horseradish; watercress

Citation: Bell, L.; Kitsopanou, E.; Oloyede, O.O.; Lignou, S. Important Odorants of Four Brassicaceae Species, and Discrepancies between Glucosinolate Profiles and Observed Hydrolysis Products. *Foods* **2021**, *10*, 1055. https://doi.org/10.3390/foods10051055

Academic Editors: Franziska S. Hanschen and Sascha Rohn

Received: 8 April 2021
Accepted: 7 May 2021
Published: 11 May 2021

1. Introduction

Crops of the Brassicaceae family are grown all over the world, and they form an important part of many different cuisines and cultures [1]. Some species are noted for their distinctive, and often very strong, tastes and flavours. *Armoracia rusticana* (horseradish), *Eruca sativa* ('salad' rocket), *Eutrema japonicum* (wasabi), and *Nasturtium officinale* (watercress) are four such examples, which are noted for their pungent, peppery, and aromatic organoleptic properties [2–4].

Horseradish and wasabi produce large roots that are grated and used as a condiment in many cultures across the world, most notably in Eastern Europe and the United Kingdom (horseradish) and Japan (wasabi; [5]). Horseradish is a vegetative perennial that grows widely in temperate regions [6], whereas wasabi can only be grown in a very few locations, owing to its sensitivity to temperature change and root oxygen availability [7]. Wasabi is traditionally cultivated in damp river valleys of Japan, although commercial operations have been established elsewhere, such as in the UK.

Salad rocket originates from the Middle East and has spread throughout the Mediterranean basin [8]. It has become naturalised on every inhabited continent and is considered an invasive weed in some regions. Watercress has similarly become naturalised (for example in North America) and grows in shallow rivers and streams. It can be cultivated commercially on large scales using artificial growing 'pools' flooded with stream water. Its leaves and shoots are a popular salad and sandwich garnish, and they can also be made into soups [9].

It is widely accepted that the distinctive aroma and flavour traits are produced by glucosinolate (GSL) hydrolysis products (GHPs [10]). GSLs are sulphur-containing secondary metabolites produced by Brassicales plants in response to biotic and abiotic stress [11]. Myrosinase enzymes are responsible for the hydrolysis of GSLs in water to form a plethora of GHPs, the conformation of which can be determined by the presence of enzyme cofactors, pH level, metallic ion concentration, and the precursor GSLs side-chain structure [8]. These products include: isothiocyanates (ITCs), nitriles, epithionitriles, indoles, oxazolidine-2-thiones, and other diverse products that result from tautomeric rearrangements (Figure 1).

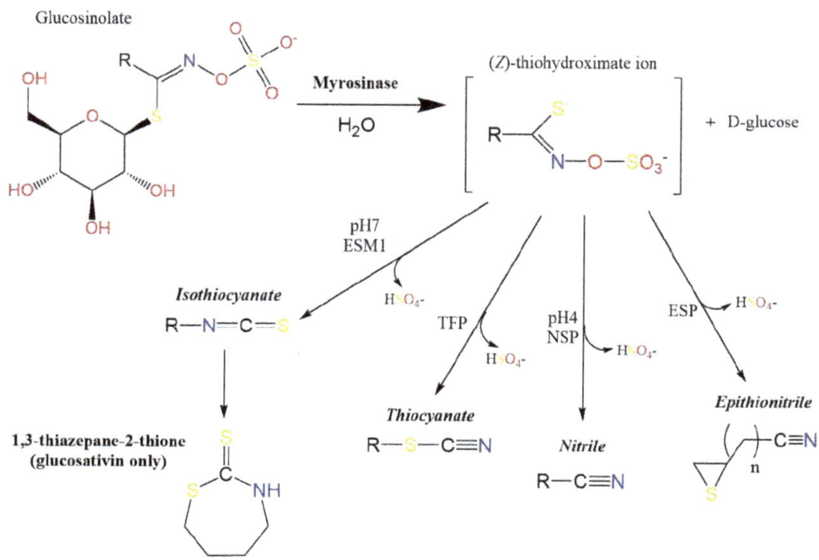

Figure 1. The glucosinolate–myrosinase reaction and hydrolysis products. Abbreviations: ESM1, epithiospecifier modifier protein 1; TFP, thiocyanate forming protein; NSP, nitrile specifier forming protein; ESP, epithiospecifier protein.

Previous work has reported olfactometry data for each of these crops; however, data are generally very scarce. Only five studies of *E. sativa* volatile compounds have been published in the last twenty years [3,12–15]. Very little information regarding wasabi root volatile composition and aroma is available outside of Japanese language journals [7,16]. Similarly, very little information is available for watercress, with only four papers published in the last 40 years [17–20]. Horseradish is the most well characterised of these four species, but still, only six studies of note have been published in the last 50 years [6,21–25].

There is also an 'elephant in the room' regarding previous reports of GHPs present in aroma profiles of these Brassicaceae vegetables. Many of the reported GHP compounds are derived from GSL precursors that are not regularly reported as part of the profile for each respective species. In *E. sativa*, for example, GHPs such as methyl ITC, 3-butenyl ITC, 1-isothiocyanato-4-methylpentane, and 5-(methylsulfanyl)pentanenitrile have all been previously reported [13]. The GSL precursors to these compounds (glucocapparin, GCP; gluconapin, GNP; 4-methylpentyl GSL, 4MP; and glucoberteroin, GBT; respectively) have never been reliably or consistently reported as being part of the GSL profile in this species [26,27]. This begs the question whether these identifications are correct or if the reports of GSL components in these species are incomplete. This may be due to a lack of sensitivity in reported mass spectrometry methods or because there is a lack of analytical standards to confirm compound identities. Another possibility is that there is a post-hydrolysis modification of GHPs, either by enzymatic means or through reactions with other phytochemical components, or as part of thermolytic reactions during gas chromatography. Very few examples of such modifications have been reported within the literature [28], but this may explain the presence of some GHPs within profiles of species where the GSL precursor is absent.

Other non-GSL derived compounds have also been reported to contribute significantly to the aromas of Brassicaceae crops. 2-Isopropyl-3-methoxypyrazine has been found to produce a strong pea, or green, pepper-like aroma in horseradish, for example ([6] Figure 2). In rocket, 'green-leaf' volatiles such as 3-hexenal and 1-penten-3-one have also been highlighted as having high odour potency [12]. The role of these compounds in aroma generation in GSL-containing crops is often not fully appreciated, and there are many diverse compounds with equally high odour intensities to GHPs present within the volatile bouquet.

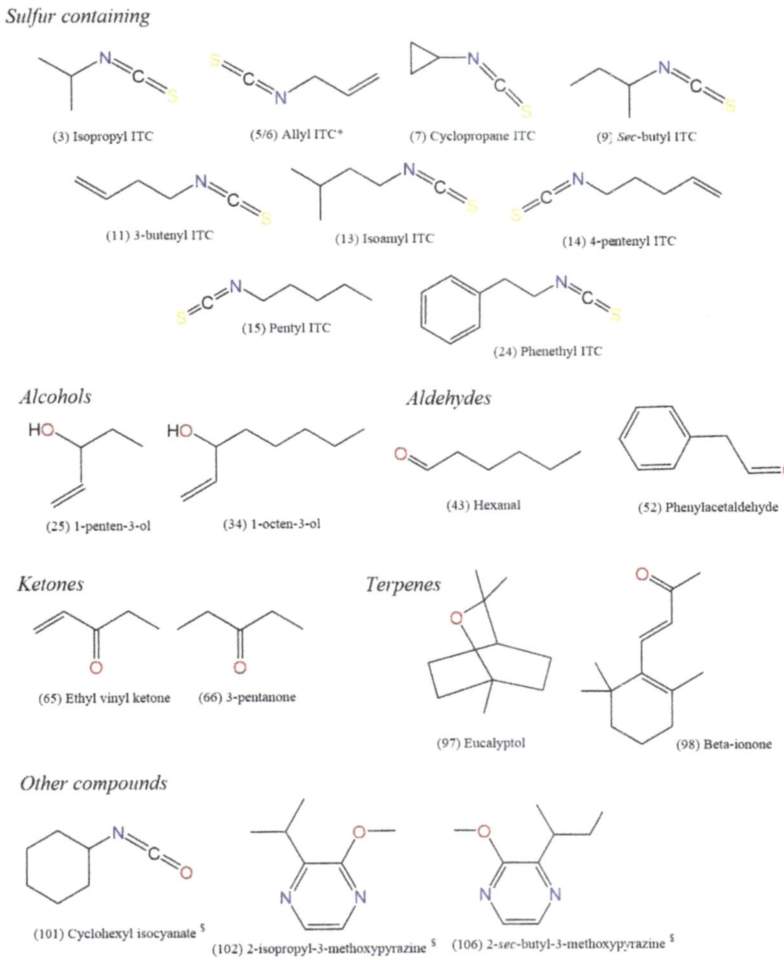

Figure 2. Chemical structures of volatile compounds found in horseradish, rocket, wasabi, and watercress samples (numbers in parentheses refer to compound codes in Table 2); * two separate peaks present for this compound; $ tentatively identified.

The aims of this study were to (i) identify and describe in detail the key odorants of four Brassicaceae species (salad rocket, horseradish, wasabi, and watercress) by gas chromatography-olfactometry (GC-O) and gas chromatography-mass spectrometry (GC-MS), and (ii) associate the observed GHPs with their respective GSL profiles by liquid chromatography-mass spectrometry (UPLC-MS/MS). Our goal was to improve upon existing compound characterisation and odour descriptors for compounds in salad rocket, horseradish, wasabi, and watercress. The contribution of pungency by GHPs to aroma profiles is well studied; however, other aroma traits generated by non-GHPs are not well described for these species, and they likely create distinctive and subtle sensory characteristics. Additionally, detailed GSL compositions and MS/MS spectra for rocket, wasabi, and watercress is presented, highlighting discrepancies with observed hydrolysis products.

2. Materials and Methods

2.1. Samples

Individual horseradish and wasabi roots were purchased from Morrison's supermarket (Reading, UK) and The Wasabi Company (Dorchester, UK) respectively. Watercress was purchased as whole bags of leaves from ASDA supermarket (Reading, UK). Salad rocket was grown in controlled environment conditions at the University of Reading using seeds donated by Elsoms Seeds Ltd. (Spalding, UK) and designated RS4 and RS8. Seeds of each cultivar were sown into module trays containing peat-based seedling compost and germinated at 30 °C (daytime; 25 °C night). Lighting conditions were set to a long-day cycle (16 h light, 8 h dark). Light intensity was set to 380 μmol m^{-2} s^{-1}. Humidity was ambient. Plants were considered mature upon the development of 10 to 15 leaves. Harvested leaves were taken intact and placed inside a Zip-loc bag.

Roots and leaves were placed in a fridge upon either purchase or harvest (4 °C) until further analysis was performed.

2.2. Chemicals

For headspace solid-phase-microextraction (HS-SPME), calcium chloride and the alkane standards C_6-C_{25} (100 μg/mL) in diethyl ether were obtained from Sigma-Aldrich (now Merck; Poole, UK). For ultra-high performance liquid chromatography mass spectrometry (UPLC-MS), authentic compounds of glucoiberin (GIB; 99.61%, HPLC), progoitrin (PRO; 99.07%, HPLC), sinigrin (SIN; 99%, HPLC), glucoraphanin (GRA; 99.86%, HPLC), glucoalyssin (GAL, 98.8%, HPLC), gluconapin (GNP, 98.66%, HPLC), 4-hydroxyglucobrassicin (4HGB; 96.19%, HPLC), glucobrassicanapin (GBN; 99.22%, HPLC), glucotropaeolin (GTP; 99.61%, HPLC), glucoerucin (GER; 99.68%, HPLC), glucobrassicin (GBR; 99.38%, HPLC), and gluconasturtiin (GNT; 98.38%, HPLC) were purchased from PhytoPlan (Heidelberg, Germany). Methanol (HPLC grade), formic acid (LC-MS grade), and acetonitrile (LC-MS grade) were purchased from VWR (Leicestershire, UK).

2.3. Volatile Compounds

2.3.1. Headspace Solid Phase Microextraction (SPME)

Samples of respective leaf and root tissues were homogenised by means of a commercial blender for 30 s, and 2 g of each was weighed into a SPME vial of 15 mL fitted with a screw cap. Samples were left aside for 10 min for the enzymatic hydrolysis of GSLs to take place. After exactly 10 min, 2 mL of saturated $CaCl_2$ was added in order to cease the enzymatic reactions. After equilibration at 40 °C for 10 min, a 50/30 μm DVB/CAR/PDMS fibre was exposed to the headspace above the sample for 20 min. Three biological replicates were prepared for GC-MS analysis, and two replicates for each of the three assessors were prepared for the GC-O analysis.

2.3.2. GC-MS Analysis of SPME Extracts

After extraction, the SPME device was inserted into the injection port of an Agilent 7890A gas chromatography system coupled to an Agilent 5975C detection system equipped with an automated injection system (CTC-CombiPAL). A capillary column HP-5MS (30 m × 0.25 mm × 0.25 µm film thickness) (Agilent, Santa Clara, CA, USA) coated with (5% Phenyl Methyl Silox) was used for the chromatographic separation of volatile compounds. The oven temperature program used was 2 min at 40 °C isothermal and an increase of 4 °C/min to 250 °C. Helium was used at 3 mL/min as carrier gas. The sample injection mode was splitless. Mass spectra were measured in electron ionisation mode with an ionisation energy of 70 eV, the scan range from 20 to 280 *m/z* and the scan rate of 5.3 scans/s. The data were controlled and stored by the HP G1034 Chemstation system. Identities were confirmed by running the samples on a Stabilwax-DA (30 m × 0.25 mm × 0.25 µm film thickness) polar column from Restek (Bellefonte, PA, USA). Volatile compounds were identified or tentatively identified by comparison of each mass spectrum with spectra from authentic compounds analysed in our laboratory, or from the NIST mass spectral database (NIST/EPA/NIH Mass Spectral database, 2014), or spectra published elsewhere (see Supplementary Data S1 for GC-MS chromatograms and compound fragmentation spectra). A spectral quality value >80 was used alongside linear retention index to support the identification of compounds where no authentic standards were available. LRI was calculated for each volatile compound using the retention times of a homologous series of C_6-C_{25} *n*-alkanes and by comparing the LRI with those of authentic compounds analysed under similar conditions. The compound peak areas were normalised and converted to the relevant abundance of each component as a percentage of the total peak area.

2.3.3. GC-O Analysis of SPME Extracts

After extraction, the SPME device was inserted into the injection port of an Agilent 7890B Series ODO 2 (SGE) GC-O system equipped with a non-polar HP-5MS column (30 m × 0.25 mm × 0.25 µm film thickness). The outlet was split between a flame ionisation detector and a sniffing port. The contents of the SPME fibre were desorbed for 3 min in a split/splitless injection port, in splitless mode, onto five small loops of the column in a coil, which were cooled in solid carbon dioxide and contained within a 250 mL beaker. The injector and detector temperatures were maintained at 280 °C and 250 °C, respectively. During desorption, the oven was held at 40 °C. After desorption, the solid carbon dioxide was removed from the oven. The oven was maintained at 40 °C for a further 2 min and then, the temperature was raised at 4 °C/min to 200 °C and at 8 °C/min to 300 °C. Helium was the carrier gas, and the flow rate was 2.0 mL/min. Three assessors were used for the detection and verbal description of the odour active components of the SPME extracts. Each assessor participated in three training sessions for each sample species prior to scoring sessions. Each assessor evaluated by sniffing each sample in duplicate and documented the odour description, retention time, and odour intensity (OI) on a seven-point scale (2–8), where <3 = weak, 5 = medium, and 7 = strong. Only those odours that were detected by all three assessors were recorded in the results. *n*-Alkanes C_6-C_{25} were analysed under the same conditions to obtain LRI values for comparison with the GC-MS data.

2.4. Non-Volatile Compounds

2.4.1. Glucosinolate (GSL) Extraction

GSL extraction was performed as per the protocol presented by [29] with modifications. Briefly, 40 mg of dried leaf powder was placed into Eppendorf tubes and put into a heat block (80 °C for ten minutes). Afterwards, 1 mL of preheated methanol water (70% *v/v*) was added to dried powder, vortexed vigorously, and placed in a water bath (75 °C) for 20 min. Samples were cooled and centrifuged at full speed for five minutes at room temperature (≈22 °C); the supernatant was collected and filtered (0.22 µm PVDF Acrodisc syringe filters; VWR, Lutterworth, UK). Crude extracts were dried using a centrifugal evaporator and re-suspended in 1 mL of LC-MS-grade H_2O and stored at −80 °C until analysis. Immediately before analysis by UPLC-MS, samples were diluted five-fold with LC-MS-grade H_2O.

2.4.2. UPLC-MS Analysis of GSL Extracts

UPLC-MS was performed on a Shimadzu Nexera X2 series UHPLC, coupled with an 8050 triple quadrupole mass spectrometer system (Shimadzu UK Ltd., Milton Keynes, UK). Separation of standards and samples was achieved using a Waters BEH C_{18} Acquity column (100 × 2.1 mm, 1.7 µm; Waters Corp., Wilmslow, UK) with an Acquity in-line filter. Mobile phases consisted of 0.1% formic acid in LC-MS grade H_2O (A), and 0.1% formic acid in LC-MS grade acetonitrile (B) and GSLs were separated during a five minute run with the following gradient timetable: (i) 0–50 s (A-B, 98:2, *v/v*), (ii) 50 s–3 min (A-B, 70:30, *v/v*), (iii) 3–3 min 10 s (A-B, 5:95, *v/v*), (iv) 3 min 10 s–4 min (A-B, 5:95, *v/v*), (v) 4–4 min 10 s (A-B, 98:2, *v/v*), (vi) 4 min 10 s–5 min (A-B, 98:2, *v/v*). The flow rate was 0.4 mL per min and the column oven temperature was 35 °C.

Two MS methods were used for the identification and quantification of GSLs. First, a Product Ion Scan (PIS) method was established to identify GSLs based on known primary ion masses ([M-H]-) characteristic fragment ions (357, 258, and 97 *m/z*; Table 1). Then, MS/MS spectra were compared to authentic standards and available literature sources [30–38]. Pentyl GSL (PEN), isobutyl GSL (ISO), glucoputranjivin (GPJ), and butyl GSL (BUT) were tentatively identified due to the possible presence of isomers [38], and/or no reliable reference MS spectra could be found in the literature. Total ion chromatograms of glucosinolates identified were included in the Supplementary Data (S2).

MS/MS settings for the PIS method were as follows: samples were analysed in the negative ion mode with a scan range of 70–820 *m/z*. A collision energy of 25 eV and a scan speed of 30,000 u per s^{-1}. For the quantification of GSLs, a Multiple Reaction Monitoring (MRM) method was established. Based on the fragmentation observed in the PIS method, confirmation and quantification transitions were established (Table 1). Dwell times for each precursor and product ion were set to 5 s.

Authentic GSL compounds were run as external standards. Limits of detection (LOD) and limits of quantification (LOQ) were established for each and are presented in Table 1. As standard compounds are not available for all GSLs, SIN was used to semi-quantify glucorucolamine (GRM), glucoputranjivin (GPJ), diglucothiobeinin (DGTB), glucoberteroin (GBT), glucocochlearin (GCL), glucosativin (GSV), dimeric 4-mercaptobutyl GSL (DMB), glucobarbarin (GBB), and tentatively identified GSL compounds. Similarly, GBR was used to semi-quantify the indolic GSLs 4-methoxyglucobrassicin (4MGB) and neoglucobrassicin (NGB).

3. Results and Discussion

3.1. Volatile Compounds

The volatile compounds identified in the headspace of the four Brassicales species are listed in Table 2, detailing their PubChem compound identification (PubChem CID) as well as their linear retention indices (LRI) in a polar and non-polar column. Semiquantitative characterisation results are also shown in Table 2 as relative area. ITCs and alcohols were the chemical classes of compounds dominating the volatile profile of the samples with other compounds such as aldehydes, esters, and terpenes also present.

GC-olfactometry analysis of the samples yielded a total of 107 odorants across the four species, which are presented in Table 3. Qualitative differences were observed between the samples with horseradish and watercress yielding a total of 52 and 51 odorants, respectively. Green/grassy, radish, sulphury, and horseradish were some of the terms that were mostly used by the assessors to describe the odours. Additionally, a total of 46 odorants of unknown identity were detected within the headspace of the four Brassicales analysed that may contribute to the odour profiles of these crops. These compounds matched no corresponding peaks and LRI values within the GC-MS data. This suggests that the compounds responsible for generating the perceived aromas were present at levels below the detection threshold of the instrumentation used. The number and diversity of unidentified compounds and aromas is indicative of the fact that characterisation of the species' volatile profiles is far from complete, and it is likely that many more will be discovered in future studies.

Table 1. Glucosinolate compounds identified in salad rocket, wasabi, and horseradish by UPLC-MS/MS.

Glucosinolate	Common Name	Abbreviation	Rt	LOD (nmol g^{-1})	LOQ (nmol g^{-1})	R^2	Precursor Ion ([M-H]-)	Quantification Transition	Confirmation Transition	MS/MS of [M-H]- Ion m/z (Relative Intensity)	References
3-(methylsulfinyl)propyl	glucoiberin *	GIB	1.061	0.053	0.161	0.969	422	422 > 357	422 > 97	**357**(57), **97**(100), 95(61)	[30,31]
pentyl [a,$]	-	PEN	1.08	-	-	-	388	388 > 75	388 > 97	273(15), 258(13), 194(14), 192(13), 97(100), 89(24), 79(18),75(71), 74(11)	-
(*R*)-2-hydroxy-3-butenyl	progoitrin *	PRO	1.336	0.066	0.199	0.933	388	388 > 74	388 > 97	**97**(100), 74(75)	[32]
allyl	sinigrin *	SIN	1.523	0.105	0.319	0.982	358	358 > 258	358 > 97	**258**(10), **97**(100), **74**(25)	[30,32,33]
isobutyl [a,$]	-	ISO	1.56	-	-	-	374	374 > 240	374 > 115	277(11), 274(20), 258(86), 257(62), 240(95), 115(100), 97(26), 95(38)	-
4-(methylsulfinyl)butyl	glucoraphanin *	GRA	1.598	0.059	0.179	0.989	436	436 > 371	436 > 97	**371**(80), **97**(100)	[30,32]
4-(cystein-*S*-yl)butyl	glucorucolamine [a,$]	GRM	1.654	-	-	-	494	494 > 406	494 > 217	413(32), 406(66), 404(38), 295(32), 291(17), 275(38), 250(66), 217(100), 209(33), 195(20), 171(38), 145(38), 129(17), 114(17), 112(40), 97(47), 96(32), 75(16)	-
5-(methylsulfinyl)pentyl	glucoalyssin *	GAL	2.421	0.064	0.194	0.971	450	450 > 206	450 > 97	**386**(42), **275**(49), 263(21), **208**(21), 206(100), **190**(71), **97**(51), 75(49)	[31,32,34]
1-methylethyl	glucoputranjivin [a,$]	GPJ	2.447	-	-	-	360	360 > 75	360 > 97	359(24), 119(13), 97(100), 94(34), 75(54)	-
3-butenyl	gluconapin *	GNP	2.489	0.151	0.459	0.953	372	372 > 258	372 > 97	**258**(6), **97**(100)	[30,31]
4-(β-D-glucopyrano syldisulfanyl)butyl	diglucothiobeinin [a,$]	DGTB	2.501	-	-	-	600	600 > 290	600 > 97	290(7), **97**(100)	[32]
5-(methylthio)pentyl	glucoberteroin [a,$]	GBT	2.593	-	-	-	434	434 > 95	434 > 97	146(60), **97**(73), 95(100)	[31,35]
4-hydroxy-3-indolylmethyl	4-hydroxygluco brassicin *	4HGB	2.619	0.15	0.454	0.943	463	463 > 285	463 > 97	**285**(17), **97**(100)	[30,32]
1-methylpropyl	glucocochlearin [a,$]	GCL	2.682	-	-	-	374	374 > 75	374 > 97	**293**(10), **275**(19), 98(58), 97(100), 95(31), 84(10), **75**(81)	[33]
4-mercaptobutyl	glucosativin [a,$]	GSV	2.684	-	-	-	406	406 > 74	406 > 97	**259**(12), 97(100), 76(19), 74(23)	[32]
7-(methylsulfinyl)heptyl [a,$]	-	7MSH	2.737	-	-	-	478	478 > 413	478 > 97	413(46), 259(12), 219(11), 98(68), **97**(100), 75(12)	[36]
4-pentenyl	glucobrassicanapin [a]	GBN	2.819	0.081	0.245	0.945	386	386 > 75	386 > 96	**97**(100), **96**(44), **75**(50)	[33]
dimeric 4-mercaptobutyl [a,$]	-	DMB	2.839	-	-	-	811	405 > 80	405 > 97	208(11), 97(100), 81(15), 80(16), 75(15)	[32]

Table 1. *Cont.*

Glucosinolate	Common Name	Abbreviation	Rt	LOD (nmol g^{-1})	LOQ (nmol g^{-1})	R^2	Precursor Ion ([M-H]-)	Quantification Transition	Confirmation Transition	MS/MS of [M-H]- Ion m/z (Relative Intensity)	References
2(*S*)-hydroxy-2-phenylethyl	glucobarbarin [a,$]	GBB	2.847	-	-	-	438	438 > 98	438 > 96	437(16), **332**(20), **274**(17), 195(11), 137(29), **135**(18), 98(48), **96**(100), **74**(30)	[37]
benzyl	glucotropaeolin *	GTP	2.879	0.129	0.39	0.962	408	408 > 259	408 > 97	**259**(10), 97(100)	[30,31]
4-(methylthio)butyl	glucoerucin *	GER	2.919	0.04	0.121	0.948	420	420 > 74	420 > 96	**258**(16), **241**(17), **178**(15), **96**(100), 75(13), 74(27)	[30,35]
indolyl-3-methyl	glucobrassicin *	GBC	3.102	0.096	0.29	0.955	447	447 > 259	447 > 97	**259**(10), 97(100)	[31,32]
4-methoxyindolyl-3-methyl	4-methoxyglucobrassicin [b,$]	4MGB	3.173	-	-	-	477	477 > 75	477 > 97	**259**(12), **258**(13), 127(10), 119(13), 98(53), 97(100), 84(15), **75**(32), 74(15)	[30,31,33,34]
2-phenethyl	gluconasturtiin *	GNT	3.419	0.062	0.189	0.968	422	422 > 259	422 > 97	**259**(10), 97(100)	[32,33]
1-methoxyindolyl-3-methyl	neoglucobrassicin [b,$]	NGB	3.526	-	-	-	477	477 > 75	477 > 97	**144**(13), 97(100), 84(16), 82(11), 75(22)	[32]
4-methylpentyl [a,$]	-	4MP	3.695	-	-	-	402	402 > 259	402 > 97	**275**(11), **259**(24), 195(16), 179(10), 159(20), 97(100), 85(12)	-
hexyl [a,$]	-	HEX	3.726	-	-	-	402	402 > 119	402 > 97	401(22), **274**(15), **241**(59), **226**(32), 204(14), **198**(15), 197(14), 168(14), 161(56), **160**(32), **138**(32), 121(32), 119(64), 116(15), 114(14), 98(46), 97(100), 96(65), 85(15), 79(15)	[31,32]
7-(methylthio)heptyl [a,$]	-	7MTH	4.109	-	-	-	462	462 > 75	462 > 97	283(11), 275(11), 274(11), 220(17), 97(100), 75(32)	[36]
butyl [a,$]	-	BUT	4.176	-	-	-	375	375 > 256	375 > 180	328(48), 316(41), 307(41), 260(41), 256(58), 235(41), 207(41), 195(50), 185(20), 180(100), 143(20), 120(63), 97(41)	-

Ions in bold agree with previous studies; ions underlined indicate characteristic ions associated with glucosinolates. * Authentic standard; [a] quantified using sinigrin; [b] quantified using glucobrassicin; [$] tentative identification.

Table 2. Volatile compounds identified in the headspace of four Brassicales species analysed by HS-SPME GC-MS.

Compound Number	Compound	PubChem ID	LRI [a]		ID [b]	Peak Areas (% of total)				References
			HP-5MS	Stabilwax		Salad Rocket	Wasabi	Horseradish	Watercress	
	Sulphur-containing compounds									
1	carbon disulphide	6348	<600	738	B	0.43 *	0.04 *	0.02	0.11 *	[21]
2	methyl thiocyanate	11168	711	1266	B	0.13	nd	nd	0.48	[12]
3	isopropyl ITC	75263	835	1176	B	nd	4.07	0.03	nd	[7,23]
4	allyl thiocyanate	69816	870	1358	B	nd	0.48 *	0.73	nd	[22]
5	allyl ITC ^	5971	881	1371	B	nd	8.61	7.35	nd	[7,14,18,24]
6	allyl ITC ^		890	1392	B	nd	52.11	39.34	nd	
7	cyclopropane ITC	92463	899	1223	B	nd	0.07 *	0.05 *	nd	-
8	cyclopentyl-1-thiaethane	138938	922	1068	B	nd	nd	nd	0.51 *	-
9	*sec*-butyl ITC	78151	933	1265	B	0.46 *	4.95	1.84	nd	[7,23]
10	isobutyl ITC	68960	955	1316	B	nd	1.80	0.45	nd	[7,21]
11	3-butenyl ITC	76922	982	1455	B	0.19	4.24	1.52	nd	[3,7,22]
12	butyl ITC	11613	998	1597	B	0.67	0.12	0.32	nd	[7,15]
13	isoamyl ITC	79086	1059	1431	B	0.10	0.41 *	0.55	0.07	[3,18,22]
14	4-pentenyl ITC	87436	1086	1543	B	nd	12.42	2.26	nd	[7,22]
15	pentyl ITC	69415	1098	1488	B	0.56 *	0.12 *	0.20	nd	[21]
16	1-isothiocyanato-4-methylpentane	519452	1162	1522	B	4.96	0.01 *	0.04	0.18	[3,18,21]
17	cyclohexyl ITC	14289	1177	1650	B	nd	0.01 *	nd	nd	-
18	<unidentified ITC>	-	1193			nd	8.57	nd	nd	-
19	ibervirin	62351	1315	1991	B	0.58	0.29	1.02	nd	[3,7,22]
20	sativin	85704368	1353	>2000	B	0.55	nd	nd	nd	[12]
21	octyl ITC	78161	1372	1760	B	nd	nd	nd	0.16 *	-
22	benzyl ITC	2346	1372	>2000	B	nd	0.03 *	1.14	nd	[12,22]
23	erucin	78160	1440	>2000	B	7.16	0.01	0.02	nd	[7,12,22]
24	phenethyl ITC	16741	1477	>2000	B	nd	0.68	32.47	30.72	[7,12,18,24]
			Total sulphur-containing compounds			15.79	99.07	89.78	32.92	
	Alcohols									
25	1-penten-3-ol	12020	678	1159	A	0.78	nd	0.10 *	8.06	[3,7,18]
26	pentan-1-ol	6276	763	1251	A	0.13 *	0.02 *	0.02 *	0.25 *	-
27	(*E*)-2-penten-1-ol	5364919	763	1312	A	0.17 *	nd	nd	0.31 *	-
28	(*Z*)-2-penten-1-ol	15306	767	1322	A	0.76	nd	0.06	4.78	[3,22]
29	1-propoxy-2-propanol	15286	840		B	nd	nd	nd	0.24 *	-
30	(*E*)-3-hexen-1-ol	5284503	850	1365	A	nd	nd	nd	0.20	[3,19,22]
31	(*Z*)-3-hexen-1-ol	5281167	856	1387	A	40.88	nd	1.88	35.99	[3,18,22]
32	2-hexen-1-ol	5318042	863	1405	A	1.25	nd	nd	0.35	[7,13,18,22]
33	hexan-1-ol	8103	866	1355	A	4.82 *	nd	0.22	2.61	[7,18]
34	1-octen-3-ol	18827	977		A	0.18	0.01 *	0.02 *	0.06 *	[14]

Table 2. *Cont.*

Compound Number	Compound	PubChem ID	LRI [a]		ID [b]	Peak Areas (% of total)				References
			HP-5MS	Stabilwax		Salad Rocket	Wasabi	Horseradish	Watercress	
35	2-ethylhexanol	7720	1025	1501	B	0.12 *	0.01 *	nd	0.12 *	-
36	benzyl alcohol	244	1035		A	nd	nd	0.05 *	nd	-
37	2-phenylethanol	6054	1116	1930	A	0.16 *	nd	0.03 *	2.64 *	-
38	1-nonanol	8914	1167		A	nd	0.03 *	nd	nd	-
39	terpinen-4-ol	11230	1182		A	nd	0.01 *	nd	nd	-
					Total alcohols	49.25	0.09	2.38	55.62	
	Aldehydes									
40	2-pentanal	7895	699		A	nd	0.02 *	nd	nd	-
41	2-pentenal	5364752	753	1136	A	0.31	nd	0.02 *	0.23 *	[3]
42	3-hexenal	643139	796	1178	A	1.72	nd	0.04	0.23	[3,18,22]
43	hexanal	6184	798	1094	A	8.64	<0.01 *	0.37	1.16 *	[12,21]
44	(*E*)-2-hexenal	5281168	852	1224	A	14.30	nd	0.31	2.10	[3,7,18,21]
45	4-heptenal	5283318	901		A	0.08 *	nd	nd	0.51	-
46	heptanal	8130	902		A	0.13	nd	0.02	0.10 *	[13,21]
47	2,4-hexadienal	637564	908		B	0.09	nd	0.01 *	nd	[3,18]
48	benzaldehyde	240	964	1654	A	0.17	0.02 *	nd	nd	[12,22]
49	2,4-heptadienal (isomer 1)	5283321	994		B	0.09	nd	nd	nd	[12]
50	octanal	454	1001		A	nd	0.02 *	nd	nd	-
51	2,4-heptadienal (isomer 2)	5283324	1009	1472	B	0.15	nd	0.25 *	0.10 *	[12]
52	phenylacetaldehyde	998	1047		A	nd	nd	0.02	0.24 *	[22]
53	2-octenal	5283324	1058		A	nd	nd	0.06 *	nd	-
54	nonanal	31289	1101		A	0.41	0.10 *	0.04	0.31 *	[13,21]
55	decanal	8175	1202		A	nd	0.12 *	0.04	0.19 *	[21]
56	vanillin	1183	1403		A	0.02	nd	0.01 *	nd	[13]
					Total aldehydes	26.11	0.28	1.25	5.17	
	Esters									
57	(*Z*)-pent-2-en-1-yl acetate	5363400	910		B	nd	nd	nd	0.07 *	-
58	3-hexenyl acetate	5352557	1003	1284	A	2.54	nd	nd	0.50	[14,18]
59	(*Z*)-3-hexenyl butanoate	5352438	1181	1440	A	0.99	nd	nd	nd	[14]
60	methyl salicylate	4133	1201		A	nd	nd	0.04 *	nd	-
61	ethyl decanoate	8048	1389		A	0.06 *	nd	0.03 *	nd	-
62	methyl dodecanoate	8139	1521		B	0.04	0.01 *	0.02 *	0.08 *	[14]
63	diethyl phthalate	6781	1588		B	0.08 *	nd	nd	nd	-
64	ethyl laurate	7800	1591		B	nd	nd	0.06 *	nd	-
					Total esters	3.71	0.01	0.15	0.65	

Table 2. *Cont.*

Compound Number	Compound	PubChem ID	LRI [a]		ID [b]	Peak Areas (% of total)				References
			HP-5MS	Stabilwax		Salad Rocket	Wasabi	Horseradish	Watercress	
	Ketones									
65	ethyl vinyl ketone	15394	682	1035	A	0.43	nd	0.05 *	nd	[3]
66	3-pentanone	7288	697	993	A	0.66	nd	0.05 *	1.86	[3,19]
67	2,3-octanedione	11449	980		A	nd	nd	0.01	nd	[22]
68	6-methyl-5-hepten-2-one	9862	983		A	0.55	nd	nd	0.23 *	[12]
69	2,2,6-trimethylcyclohexanone	17000	1038		B	nd	nd	nd	0.11 *	-
70	(*E*,*E*)-3,5-octadien-2-one	5352876	1070		B	nd	nd	0.08 *	nd	-
71	3,5-octadien-2-one	5352876	1092		B	nd	nd	0.09 *	nd	-
72	dihydro-2H-thiopyran-3(4H)-one	140474	1160	1856	B	0.58	nd	nd	nd	-
73	geranylacetone	1549778	1451		A	0.10 *	0.02 *	nd	nd	-
					Total ketones	2.32	0.02	0.27	2.21	
	Nitriles									
74	3-butenenitrile	8009	654		B	nd	0.15	0.37	nd	[7,21]
75	5-methylhexanenitrile	29593	943		B	0.27	nd	nd	nd	[13]
76	6-heptenenitrile	4140856	971		B	nd	0.08 *	nd	nd	-
77	thiiraneacetonitrile	148821	1004		B	nd	nd	3.58 *	nd	[39]
78	phenylacetonitrile	8794	1142		B	nd	nd	0.01 *	nd	-
79	5-(methylsulfanyl)pentanenitrile	93320	1200		B	1.13	nd	nd	nd	[12]
80	4-(methylthio)-butanenitrile	100962	1213		B	nd	nd	0.11 *	nd	-
81	benzenepropanenitrile	12581	1243	>2000	B	nd	nd	0.75	1.47	[17,22]
82	<unidentified nitrile>	-	1559		B	nd	0.22	nd	nd	-
					Total nitriles	1.39	0.45	4.83	1.47	
	Hydrocarbons									
83	2,2,4,6,6-pentamethylheptane	26058	991	947	A	nd	nd	nd	0.69 *	-
84	undecane	14257	1096		A	0.23	nd	0.01 *	nd	[15]
85	1-dodecene	8183	1187		A	nd	0.06 *	0.12 *	0.64 *	-
86	dodecane	8182	1196		A	nd	nd	0.01	nd	[22]
87	tridecane	12388	1296		A	nd	nd	0.03	nd	
88	tetradecane	12389	1396		A	0.10	0.01 *	0.06 *	nd	[15]
89	pentadecane	12391	1497		A	nd	nd	0.09 *	nd	-
90	hexadecane	11006	1599		A	nd	nd	0.10 *	nd	-
91	heptadecane	12398	1699		A	nd	nd	0.06 *	nd	-
92	octadecane	11635	1799		A	nd	nd	0.03 *	nd	-
93	nonadecane	12401	1899		A	nd	nd	0.01 *	nd	-

Table 2. *Cont.*

Compound Number	Compound	PubChem ID	LRI [a]		ID [b]	Peak Areas (% of total)				References
			HP-5MS	Stabilwax		Salad Rocket	Wasabi	Horseradish	Watercress	
				Total hydrocarbons		0.39	0.06	0.62	1.33	
	Terpenes									
94	*p*-cymene	7463	1023	1250	A	nd	<0.01 *	0.01 *	0.06	[18]
95	*o*-cymene	10703	1027		A	0.12 *	0.01 *	nd	nd	-
96	d-limonene	440917	1033	1169	A	nd	<0.01 *	0.07 *	nd	-
97	eucalyptol	2758	1037		A	nd	<0.01 *	nd	nd	-
98	β-ionone	638014	1494	1952	A	0.16	nd	nd	0.43	[12,20]
				Total terpenes		0.28	0.01	0.08	0.49	
	Other compounds									
99	allyl isocyanate	15123	645		B	nd	nd	0.02 *	nd	-
100	2-ethylfuran	18554	701		A	0.16	nd	0.02	0.14 *	[3,21]
101	cyclohexyl isocyanate	18502	961		B	nd	0.01 *	nd	nd	-
102	1-isopropyl-3-methoxypyrazine	33166	1093		B	0.25	nd	0.09	nd	[15,25]
103	1,2,3,5-tetramethylbenzene	10695	1124		B	0.06 *	nd	nd	nd	-
104	veratrole	7043	1144		B	nd	nd	0.09 *	nd	-
105	octanoic acid	379	1157		B	0.10	nd	0.06	nd	[14,22]
106	2-*sec*-butyl-3-methoxypyrazine	520098	1172		B	0.15 *	nd	0.43	nd	[25]
107	phenethyl isocyanate	160602	1226	1807	B	nd	nd	0.41 *	0.70 *	-
108	quinoline	7047	1247		B	nd	nd	0.02 *	nd	-
109	caprolactam	7768	1251		B	0.05 *	nd	nd	nd	-
110	4-bromophenol	7808	1283		B	0.02 *	nd	nd	nd	-
111	6-methylquinoline	7059	1331		B	nd	nd	0.01 *	nd	-
112	methyleugenol	7127	1400		B	0.02 *	nd	nd	nd	-
113	benzyl tiglate	250096	1504		B	nd	nd	0.01 *	nd	-
				Total other compounds		0.77	0.01	0.63	0.14	

[a] Linear retention index on a HP-5MS and Stabilwax columns. [b] A, mass spectrum and LRI agree with those of authentic compound; B, mass spectrum agrees with reference spectrum in the NIST/EPA/NIH mass spectra database and LRI agree with those in the literature, tentatively identified. ^ Based on library identification and spectra but two separate peaks were present; nd, not detected; * newly reported for species.

Table 3. Odorants identified by HS-SPME GC-O in the headspace of four Brassicales species.

Compound Number	Compound	LRI [a]	Odour Description		Odour Intensity [c]				References
			This Study	Previous Description(s) [b]	Salad Rocket	Wasabi	Horseradish	Watercress	
114	<unknown>	<600	sulphury	-	nd	3	3	4	-
115	<unknown>	<600	cooked onions	-	nd	nd	nd	4	-
116	<unknown>	<600	buttery	-	3	5	5	5	-
117	<unknown>	<600	sulphury, horseradish, rancid	-	3	6	4	nd	-
118	<unknown>	602	horseradish	-	nd	nd	5	nd	-
119	<unknown>	609	mustard, horseradish	-	nd	6	nd	nd	-
120	<unknown>	615	rotten cabbage	-	nd	nd	5	4	-
121	<unknown>	622	onions	-	nd	6	nd	nd	-
99	allyl isocyanate $	648	musty, burnt plastic †	-	nd	nd	2 *	nd	-
122	<unknown>	655	sulphury, cabbage-like	-	6	nd	nd	nd	-
74	3-butenenitrile $	657	sulphury, green, pungent †	-	nd	3	4	nd	[21]
25	1-penten-3-ol	676	cabbage, sulphury	pungent, horseradish-like	3	nd	nd	7	[40]
65	ethyl vinyl ketone	680	pungent, rotten, green	pungent	5	nd	nd	nd	[41]
66	3-pentanone	687	green, grassy, floral	acetone-like	nd	nd	nd	6	[42]
123	<unknown>	713	sulphury, garlic	-	nd	2	3	nd	-
2	methyl thiocyanate $	719	sulphury, oniony	sulphur	4	nd	nd	4 *	-
124	<unknown>	737	sulphury, rotten onion	-	3	nd	nd	3	-
40	2-pentenal	750	apple, green	green, apple-like	3	nd	3 *	4 *	[43]
125	<unknown>	758	sulphury, cooked onion, pungent	-	nd	nd	nd	6	-
126	<unknown>	775	sulphury, oniony	-	nd	nd	nd	4	-
42	3-hexenal	796	grassy, green, floral	leafy, green	4	nd	3	4	[25]
43	hexanal	799	green, grassy, pungent	green, grass-like, leafy	5 *	nd	4	4 *	[44]
127	<unknown>	811	green, parsley	-	nd	nd	nd	4	-
128	<unknown>	812	mustard, pungent, oniony	-	5	6	3	6	-
3	isopropyl ITC $	834	pungent, grassy notes	pungent	nd	5	nd	nd	[12]
44	(*E*)-2-hexenal	852	green, fresh, apples (weak)	green	3	nd	2	4	[45]
30	(*Z*)-3-hexen-1-ol	856	green, radishy	green	2	nd	nd	4	[46]
32	2-hexen-1-ol	864	green, leafy	green, leafy	3	nd	nd	nd	[47]
129	<unknown>	866	cooked, roasted chicken, chicken soup	-	nd	3	3	5	-
130	<unknown>	870	nutty, spicy	-	nd	nd	3	nd	-

Table 3. *Cont.*

Compound Number	Compound	LRI [a]	Odour Description		Odour Intensity [c]				References
			This Study	Previous Description(s) [b]	Salad Rocket	Wasabi	Horseradish	Watercress	
4	allyl thiocyanate $	879	peppery, horseradish, pungent	strong, pungent, mustard-like	nd	nd	4	nd	[48]
131	<unknown>	892	green, sour apples	-	5	nd	nd	6	-
5	allyl ITC $	895	pungent, horseradish	strong, pungent, mustard-like	nd	nd	4	nd	[48]
6	allyl ITC $	898	garlic, mustard, horseradish, very pungent		nd	7	7	nd	
7	cyclopropane ITC $	900	horseradish, garlic, onion, sulphur †	-	nd	5 *	7 *	nd	[49]
45	4-heptenal	901	grassy, green †	mushroom-like, fatty, fishy, cooked, potato	4 *	nd	nd	3 *	[50]
46	heptanal	902	fatty, green	fatty, green	4	nd	nd	5 *	[13]
132	<unknown>	907	apples, grass	-	nd	nd	nd	4	-
47	2,4-hexadienal $	907	green, rotten	green	3	nd	3 *	nd	[51]
57	(*Z*)-2-pent-2-en-1-yl acetate $	914	sulphury, rotten †	-	nd	nd	nd	3 *	-
8	cyclopentyl-1-thiaethane	918	sulphur, sweaty †	-	nd	nd	nd	3 *	-
133	<unknown>	921	potato	-	nd	5	nd	4	-
9	*sec*-butyl ITC $	934	radish, vegetative	green	nd	5	3	nd	[52]
10	isobutyl ITC $	956	cooked, pungent, sulphury †	-	nd	4	5	nd	[21]
101	cyclohexyl isocyanate $	965	cooked, peppery, potato †	-	nd	5 *	nd	nd	[53]
34	1-octen-3-ol	978	mushroom	mushroom	5	5 *	3 *	5 *	-
11	3-butenyl ITC $	983	green, pungent, aromatic	aromatic, pungent	5	6	4	nd	[54]
68	6-methyl-5-hepten-2-one	984	perfume, floral, citrus	citrus, lemongrass	4	nd	nd	5 *	[55]
12	butyl ITC $	997	peppery, sulphurous, oniony	sulphury	nd	4	2	nd	[21]
77	thiiraneacetonitrile $	1005	sweaty, gas-like †	-	nd	nd	4	nd	[56]
134	<unknown>	1010	grassy	-	nd	nd	nd	3	-
135	<unknown>	1020	earthy, musty, petrol, aromatic	-	nd	nd	nd	5	-
136	<unknown>	1025	bread-like	-	nd	nd	nd	5	-
137	<unknown>	1032	green	-	nd	nd	nd	3	-
96	*d*-limonene	1034	lemon, vegetable	citrus, herbal	nd	nd	3 *	nd	[57]
36	benzyl alcohol	1037	fruity, medicinal, wine	phenolic	nd	nd	4 *	nd	[58]
97	eucalyptol	1037	eucalyptus, mint	eucalyptus	nd	5 *	nd	nd	[42]

Table 3. *Cont.*

Compound Number	Compound	LRI [a]	Odour Description		Odour Intensity [c]				References
			This Study	Previous Description(s) [b]	Salad Rocket	Wasabi	Horseradish	Watercress	
69	2,2,6-trimethylcyclohexanone [$]	1038	floral, green with citrus notes	thujonic	nd	nd	nd	4 *	[59]
52	phenylacetaldehyde	1044	honey-sweet	honey, sweet	nd	nd	4	7 *	[21]
138	<unknown>	1051	oniony	-	5	nd	nd	nd	-
139	<unknown>	1053	soily, earthy	-	6	nd	nd	nd	-
13	isoamyl ITC [$]	1057	pungent, grassy	green	nd	6 *	3	nd	[3]
140	<unknown>	1066	medicinal, floral	-	nd	nd	nd	5	-
141	<unknown>	1076	gas, sulphur, burnt, roasted	-	3	nd	5	4	-
142	<unknown>	1080	roasted, smoky	-	3	nd	nd	3	-
14	4-pentenyl ITC [$]	1084	pungent, peppery, sulphurous, musty	mustard, horseradish-like	nd	5	3	nd	[48]
71	3,5-octadien-2-one [$]	1090	green, pungent	fruity	nd	nd	4 *	nd	-
102	2-isopropyl-3-methoxypyrazine [$]	1091	rotten, potato, vegetative	pea-like, earthy, bean-like	6	nd	6	nd	[60]
15	pentyl ITC [$]	1095	cabbage, green, rotten	green	5 *	4 *	6 *	nd	-
143	<unknown>	1095	cucumber, floral, flowers	-	nd	nd	nd	3	-
54	nonanal	1101	fatty, green	green, fatty	nd	nd	4	5 *	[40]
37	2-phenylethanol	1114	floral	floral	3 *	nd	nd	4 *	[58]
144	<unknown>	1118	spicy, chemical	-	3	nd	nd	nd	-
145	<unknown>	1122	sulphury, gas-like	-	4	nd	2	nd	-
146	<unknown>	1133	sulphury, garlic	-	nd	3	2	nd	-
147	<unknown>	1147	petrol, aromatic	-	nd	nd	nd	6	-
148	<unknown>	1153	cucumber	-	nd	nd	nd	6	-
149	<unknown>	1153	spicy, cinnamon-like, nutty	-	nd	6	nd	nd	-
150	<unknown>	1155	fresh cucumber, rotten, vegetable	-	2	5	nd	nd	-
16	1-isothiocyanato-4-methylpentane [$]	1162	musty †	-	2	4 *	2	nd	-
106	2-*sec*-butyl-3-methoxypyrazine [$]	1173	earthy, rotten potatoes, vegetable-like	musty, green, pea-like, bell pepper-like	3 *	nd	7	nd	-
59	(*Z*)-3-hexenyl butanoate	1181	green, wine-like	green, fruity	4	nd	nd	nd	[61]
18	<unidentified ITC>	1192	radish, green	-	nd	4	nd	nd	-
151	<unknown>	1198	grassy, fruity, chemical, dried fruit	-	nd	nd	nd	6	-

Table 3. *Cont.*

Compound Number	Compound	LRI [a]	Odour Description		Odour Intensity [c]				References
			This Study	Previous Description(s) [b]	Salad Rocket	Wasabi	Horseradish	Watercress	
79	5-(methylsulfanyl) pentanenitrile [$]	1200	radish	broccoli-like, cabbage-like	2	nd	nd	nd	[62]
152	<unknown>	1200	sweet, floral, violets, perfume	-	-	3	4	-	-
60	methyl salicylate	1201	medicinal, camphorous	wintergreen, mint	nd	nd	4 *	nd	[40]
153	<unknown>	1203	peppery, green, earthy	-	nd	5	nd	5	-
107	phenethyl isocyanate	1224	ground pepper, pungent, horseradish †	-	nd	nd	4 *	nd	-
81	benzenepropanenitrile [$]	1242	herbal, green, floral †	-	nd	3	3	4	[21]
154	<unknown>	1249	liquorice, medicinal	-	nd	4	nd	nd	-
155	<unknown>	1266	cooked, cabbage, sulphur	-	nd	nd	3	nd	-
156	<unknown>	1275	green, radish, potato	-	nd	nd	nd	4	-
157	<unknown>	1279	soapy, pungent	-	nd	nd	nd	5	-
158	<unknown>	1285	soapy, grassy, floral	-	nd	2	nd	6	-
19	iberverin [$]	1314	horseradish, radish, vegetative	vegetative, horseradish-like, gooseberry-like	nd	3	5	nd	[63]
159	<unknown>	1318	medicinal, soapy	-	nd	nd	nd	5	-
111	6-methylquinoline [$]	1334	hydrogen sulphide, egg	tobacco, fecal	nd	nd	3 *	nd	-
20	sativin [$]	1349	burnt, rubbery, soily †	rocket-like	3	nd	nd	nd	[12]
21	octyl ITC [$]	1370	green, vegetative †	-	nd	nd	nd	5 *	[64]
22	benzyl ITC [$]	1378	rotten grass, cooked	watercress-like	nd	3 *	4	-	[48]
61	ethyl decanoate	1391	green, waxy	waxy, apple	nd	nd	4 *	nd	[65]
23	erucin [$]	1441	radishy	radish-like, cabbage-like	2	nd	nd	nd	[12]
160	<unknown>	1478	minty, cooling, fresh	-	nd	nd	4	nd	-
24	phenethyl ITC [$]	1480	radish, gooseberry, sweet	horseradish-like, gooseberry-like	nd	nd	6	7	[48]
98	β-ionone	1495	soapy, fusty	floral, woody, fruity	2	nd	nd	6	[41]
113	benzyl tiglate [$]	1500	musty	earthy, mushroom-like	nd	nd	4 *	nd	-

[a] Linear retention index on a HP-5MS column; for identification please check Table 2. [b] Odour description of compound present in the Good Scents online database: http://www.thegoodscentscompany.com/ (accessed on 1 April 2021) and literature sources; † tentative new odour description. [c] Average of intensities recorded by three assessors evaluating each sample in duplicate (scoring scale: weak = 3, medium = 5, strong = 7); nd, not detected; * newly reported for species; [$] compound tentatively identified.

3.1.1. 'Salad' Rocket

A total of 57 volatile compounds were identified or tentatively identified in the headspace of *E. sativa* leaf samples. Nineteen compounds are newly reported for this species, some of which make up relatively large portions of the total volatile compounds' bouquet (Table 2). Compounds with the greatest relative abundances were (*Z*)-3-hexen-1-ol (**31**, 40.9%), (*E*)-2-hexenal (**44**, 14.3%), hexanal (**43**, 8.6%), erucin (**23**, 7.2%), and 1-isothiocyanato-4-methylpentane (**16**, 5%; Figure 2). These observations are broadly in agreement with previous studies of 'salad' rocket [3,12].

Despite its high relative abundance, (*Z*)-3-hexen-1-ol produced only a weak, green, radishy aroma (Table 3) in rocket leaves. (*E*)-2-Hexenal by comparison was 2.9-fold less abundant in relative terms but produced a slightly stronger aroma, described by assessors as green, and apple-like. Hexanal (**43**) produced a pungent, green, grassy aroma of relatively high intensity, which has not been previously described in rocket to our knowledge.

2-Isopropyl-3-methoxypyrazine (**102**) by comparison was of low relative abundance in the rocket headspace (0.3%; Table 2) but was found to have one of the strongest aromas in rocket (rotten, potato-like, vegetative; Table 3). 1-Isothiocyanato-4-methylpentane had a weak aroma and was given a tentative new description of 'musty', as no previous studies have reported an odour for this compound.

The ITC erucin (**23**) is known for its anticarcinogenic properties, but its aroma was only recently described [12]. In agreement with a previous report, this compound produced a radishy aroma of weak intensity. Interestingly, the compound previously associated with characteristic "rocket-like" aroma (sativin, **20**) [12] was relatively weak-smelling. In this study, the compound was found to have a burnt, rubbery, and soil-like aroma. This suggests that sativin is not the main driver of pungency or aroma in 'salad' rocket. Other compounds such as ethyl vinyl ketone (**65**; [41]), hexanal, 3-butenyl ITC (**11**) [48,54], and several unknown compounds (see final paragraph in this section) all had descriptions of pungency at higher intensities than sativin (Table 3). It may also be likely that no single compound is responsible for this attribute of rocket aroma but rather several.

Pentyl ITC (**15**, 0.6%) not previously identified in 'salad' rocket produced a strong odour, which was characterised as cabbage-like, green, and rotten (Table 3, Figure 2). As will be discussed in Section 3.2, we have tentatively identified pentyl GSL (Table 1) as a significant and previously unreported component of the GSL profile of 'salad' rocket, which gives rise to this ITC compound.

Other compounds not previously identified in 'salad' rocket included 2-phenylethanol, 4-heptenal and 2-*sec*-butyl-3-methoxypyrazine. 2-Phenylethanol (**37**, 0.2%) was noted to impart a floral aroma at a medium-weak intensity (Table 3). This compound is derived from phenylalanine and has been found to contribute to aroma and flavour in many foods, such as tomatoes [66]. 4-Heptenal (**45**) occurred in rocket leaves with a relative abundance of 0.1% (Table 2). Despite this low amount in terms of the overall volatile profile, the compound was perceived at a medium intensity by the assessors (Table 3) and described as grassy and green. This compound has been variously described as mushroom-like [50], fatty and fishy [67], and potatoey [68]. The variation in these descriptions may be associated with the isomerisation of the compound, which could not be resolved in this study. 2-*sec*-Butyl-3-methoxypyrazine (**106**, 0.2%) has been reported in several Brassicales species, such as white mustard, rapeseed [60], and horseradish [25], but not in *E. sativa* (Figure 2). It has been variously described as having a pea-like, musty, green, and bell pepper-like aroma. Assessors described the compound as earthy, similar to rotten potatoes, and vegetable-like. Despite its relatively low abundance, it was perceived as a medium-weak smelling compound in the sample headspace.

A total of 13 unidentified odorants were also detected in the headspace of rocket by GC-O (Table 3). These varied in intensity but all were distinguished and reported by assessors. Odour descriptions for these compounds were buttery (**116**), sulphury (**117**, **122**, **124**, **141**, **145**), horseradish-like, rancid (**117**), cabbage-like (**122**), rotten onion (**124**), mustard, pungent (**128**), oniony (**128**, **138**), green, sour apples (**131**), soily, earthy (**139**), gas (**141, 145**), burnt (**141**), roasted (**141, 142**), smoky (**142**), spicy, chemical (**144**), fresh cucumber, rotten, and vegetable-like (**150**).

3.1.2. Wasabi

A total of 43 compounds were identified or tentatively identified in the headspace of wasabi roots (Table 2) with 30 compounds newly described for the species, making this a significant step forward in the understanding of aroma composition in wasabi roots. Two peaks of near-identical spectra were observed and identified as allyl ITC (**5/6**, 8.6%/52.1%) and were the most abundant compounds, which agrees with previous observations [7]. It is unknown why two distinct peaks were formed in this manner, and further investigation may be required to determine the isomeric differences responsible for the separation. 4-Pentenyl ITC (**14**, 12.4%) was also found to be high in terms of overall relative abundance.

Despite having near identical spectra, compounds **5** and **6** presented distinct differences in aroma and intensity. Peak 6 was characterised as being very pungent (Table 3) and having garlic, mustard, and horseradish-like qualities. Allyl ITC is one of the most well characterised ITCs and is well known for these properties [1,48,63] (Figure 2). Peak **5** by contrast had no discernible aroma in wasabi but was apparent in horseradish (see Section 3.1.3). 4-Pentenyl ITC likewise exhibited a pungent aroma and strong odour intensity but also had peppery, sulphurous, and musty notes. This compound is commonly reported in Brassicaceae crops [69]. 3-Butenyl ITC (**11**, 4.2%) was scored as a high odour intensity compound, despite its much lower relative abundance and was described as having a pungent, green, and aromatic odour.

Several other GHP odorants are newly reported in wasabi including cyclopropane ITC, isoamyl ITC, pentyl ITC, 1-isothiocyanato-4-methylpentane, and benzyl ITC. Cyclopropane ITC (**7**, 0.1%) is likely to be a cyclic reaction product of allyl ITC and has been previously reported in brown mustard [49] (Figure 2). To our knowledge, no odour description of this compound has been previously made, but assessors described it as sulphurous, horseradish, garlic, and onion-like (Table 3). The high odour intensity score indicates that it is a significant component of wasabi aroma. Isoamyl ITC (**13**, 0.4%) was described as having a pungent grassy aroma and being of high odour intensity. This compound is not commonly reported in Brassicales species, but it is used as a food additive [70]. Most ITC compounds are noted for their sulphurous and mustard-like potency in Brassicales; however, the contribution of grassy aroma ITCs to volatile compound bouquets has not been previously appreciated or fully understood. Pentyl ITC (**15**, 0.1%) and 1-isothiocyanato-4-methylpentane (**16**, <0.1%) were observed and shared the same odour characteristics as in 'salad' rocket (see Section 3.1.1.). Benzyl ITC (**22**, <0.1%) was reported to have a rotten grass and cooked aroma of medium-weak intensity. Similar to isoamyl ITC, this compound is not regularly reported as a constituent of Brassicales headspace, but these data indicate that even in very low relative abundance, it is odour active.

Another interesting compound was also found in wasabi headspace: cyclohexyl isocyanate (**101**, <0.1%; Figure 2). This has been previously reported in black mustard [53], though it is unclear if it is related to or derived from GHPs. Assessors perceived this odour having a medium-strong intensity and described it as peppery, cooked, and potato-like (Table 3). This is a tentative new odour description for this compound, and our data suggest it to be an important constituent of wasabi aroma.

Two additional compounds not previously identified in wasabi were 1-octen-3-ol and eucalyptol. 1-Octen-3-ol (**34**, <0.1%), despite its very low relative intensity in root tissue headspace (Table 2, Figure 2), exhibited a high odour intensity imparting a mushroom-like odour in agreement with previous descriptions [50]. Eucalyptol (**97**, <0.1%), previously observed in Brassicales crops [18] but not in wasabi (Figure 2), was found to have a medium-strong, characteristic eucalyptus, and mint aroma, and it is likely an important component of the overall volatile bouquet.

Similar to rocket, 17 unidentified odorants were detected in wasabi root samples (Table 3). Reported aromas were sulphury (**114, 117, 123, 146**), buttery (**116**), horseradish-like (**117, 119**), rancid (**117**), mustard (**119, 128**), onion (**121, 128**), garlic (**123, 146**), pungent (**128**), cooked, roasted chicken, chicken soup (**129**), potato (**133**), spicy, cinnamon-like, nutty (**149**), fresh cucumber, rotten, vegetable-like (**150**), radish (**79**), green (**79, 153**), sweet (**152**), floral (**152, 158**), violets, perfume (**152**), peppery, earthy (**153**), liquorice, medicinal (**154**), soapy, and grassy (**158**), confirming our statement that wasabi's volatile profile is far from complete.

3.1.3. Horseradish

A total of 75 compounds were identified or tentatively identified in the headspace of horseradish roots, 38 of which are newly reported (Table 2). As with wasabi, the peaks with the highest relative abundances were dominated by GHPs. Compounds with the highest relative abundances were allyl ITC (**5/6**, 7.4%/39.3%), phenethyl ITC (**24**, 32.5%), thiiraneacetonitrile (**77**, 3.6%), 4-pentenyl ITC (**14**, 2.3%), (*Z*)-3-hexen-1-ol (**31**, 1.9%), and *sec*-butyl ITC (**9**, 1.8%; Figure 2).

As stated in Section 3.1.2, allyl ITC was identified as two distinct peaks (**5** and **6**). As in wasabi, **6** was of the greatest abundance and odour intensity, producing a very strong, pungent, garlic, mustard, and horseradish-like aroma (Table 3), whereas **5** produced a medium intensity, pungent horseradish smell. Thus far, the presence of two peaks has not been addressed or explained satisfactorily within the literature, with only one previous paper reporting the same phenomenon of separate and distinct allyl ITC peaks [71]. *Sec*-butyl ITC (**9**) produced a medium-weak intensity aroma that was vegetative and radish-like (Table 3). The compound was also present in wasabi at a medium intensity. The compound has been previously reported in horseradish as having green, chemical, and mustard like aromas [21], and it is known to activate the human Transient Receptor Potential Ankyrin 1 (TRPA1). This receptor is known to act in response to environmental irritants, and several ITCs identified in this study are known to activate it to varying degrees (isopropyl ITC, **3**; isobutyl ITC, **10**; allyl ITC, **5/6**; 3-butenyl ITC, **11**; 4-pentenyl ITC, **14**; benzyl ITC, **22**; phenylethyl ITC, **24**; [72]). Phenylethyl ITC (**24**) is known to be a key constituent of horseradish aroma, and our data are in agreement with previous reports [73]. Assessors described the compound as radish and gooseberry-like, with a sweet note. It had a high odour intensity and contributed significantly to the odour profile of roots. Likewise, 4-pentenyl ITC (**14**) was observed to have the same odour attributes as previous reports [1] and those found for wasabi in this study, but at a lower intensity. By contrast, pentyl ITC (**15**) was present at much lower relative intensities to other GHPs (0.2%, Table 2) but produced a strong, green, rotten, and cabbage-like aroma.

Thiiraneacetonitrile (**77**) has been previously reported in horseradish [22] and is an epithionitrile hydrolysis product of sinigrin. To our knowledge, no previous studies have described the odour of this compound. We found it to have a sweaty, gas-like aroma of medium intensity (Table 3).

2-Isopropyl-3-methoxypyrazine (**102**, 0.1%) and 2-*sec*-butyl-3-methoxypyrazine (**106**, 0.4%) have been previously described and characterised in horseradish roots [25] as having green and pepper-like aromas. Our data agree with previous reports but found the compounds to be of very high aroma intensity, despite relatively low abundances within the headspace (Table 3). Assessors described the compounds as rotten, earthy, potato-like, and vegetative.

We report several compounds previously unidentified in horseradish including GHPs, isocyanates, alcohols, aldehydes, and a ketone and ester. As in wasabi root, cyclopropane ITC (**7**, 0.1%) produced an intense aroma containing horseradish, garlic, onion, and sulphur notes (Table 3). Therefore, it is likely to be a significant contributor to root odour and the volatile profile, despite its very low abundance, which may be a reason why it has not been previously detected and/or reported.

As discussed in Section 3.1.2, it is unknown if the presence of isocyanates is linked with GSLs and their hydrolysis products. Allyl isocyanate (**99**, <0.1%) and phenethyl isocyanate (**107**, 0.4%) were both observed for the first time in horseradish roots. Given that high abundances of allyl ITCs (**5/6**) and phenethyl ITC (**24**) were observed, it seems likely that isocyanates may be derived from them and/or directly from parent GSLs. Isocyanates are not commonly reported in the literature, and their formation may be because of as-yet-unstudied enzymatic or post-hydrolysis modification processes. Allyl isocyanate was described as having a weak musty and burnt plastic aroma; and phenethyl isocyanate was described as being pungent, with ground pepper and horseradish-like quality at a medium intensity. We are not aware of any previous odour descriptions for these compounds, so these are tentative new characterisations.

1-Octen-3-ol (**34**, <0.1%) was identified, and as in wasabi root, it produced a mushroom-like aroma of medium-weak intensity (Table 3). Benzyl alcohol (**36**, 0.1%) has been previously reported in *Brassica oleracea* [74] and rapeseed [58], but not in horseradish. It was characterised as having a medium intensity aroma, described as fruity, medicinal, and wine-like. Its low abundance but relatively high odour intensity may make it a subtle but key constituent of the root aroma profile. Two aldehyde compounds were also found to contribute to odour within the headspace. 2-Pentenal (**41**, <0.1%) produced a green, apple-like aroma [43], and 2,4-hexadienal (**47**, <0.1%) produced a green, rotten smell, both of medium-weak intensity (Table 3; [51]). A ketone, 3,5-octadien-2-one (**71**, 0.1%) was also tentatively identified (Table 2) and described as having a pungent green aroma of medium intensity (Table 3).

In the esters group, methyl salicylate (**60**, <0.1%), a common compound throughout the plant kingdom, has previously been identified in Brassicales as part of systemic acquired resistance response to herbivory [75], and it was identified for the first time as a volatile constituent of horseradish headspace (Table 2). Its aroma is characteristic of, and present in, plants such as wintergreen. Assessors described its odour as medicinal and camphorous at a medium intensity (Table 3). Ethyl decanoate (**61**, <0.1%), known to be present in *B. oleracea* [59], was described by the assessors as green and waxy (Table 3; [65]). Benzyl tiglate (**113**, <0.1%) was reported at low relative abundance, but a perceptible musty aroma was apparent for this compound.

The presence of *D*-limonene (**96**, 0.1%) is reported for the first time in horseradish root. It is known to be a constituent of *B. oleracea* headspace [76], but its sensory contribution to Brassicales is not well defined. Assessors found this compound to have a medium-weak intensity aroma of lemon and being vegetable-like (Table 3 [57]). This agrees with previous descriptions of the odour properties of the compound.

Finally, 6-methylquinoline (**111**, <0.1%), an aromatic compound, produced an unpleasant hydrogen sulphide and egg-like aroma of medium-weak intensity, and it has not been previously reported.

Similar to the other species tested, 15 unidentified odorants were detected in horseradish root samples (Table 3) and were of weak to medium intensity. Reported aromas were sulphury (**114, 117, 123, 141, 145, 146, 155**), buttery (**116**), horseradish-like (**117, 118**), rancid (**117**), rotten (**120**), cabbage-like (**120, 155**), garlic, (**123, 146**), mustard, pungent, oniony (**128**), cooked (**129, 155**), roasted chicken, chicken soup (**129**), nutty, spicy (**130**), gas (**141, 145**), burnt, roasted (**141**), sweet, floral, violets, perfume (**152**), minty, cooling, and fresh (**160**).

3.1.4. Watercress

A total of 42 compounds were identified or tentatively identified in the headspace of watercress, 26 of which are newly reported (Table 2). The headspace profile was dominated by alcohol and ITC compounds: (*Z*)-3-hexen-1-ol (**31**, 36%), phenethyl ITC (**24**, 30.7%), 1-penten-3-ol (**25**, 8.1%), and (*Z*)-2-penten-1-ol (**28**, 4.8%; Figure 2).

1-Penten-3-ol is a compound present widely in Brassicales species [77]. It exhibited a high intensity in watercress leaves, producing a sulphurous and cabbage-like aroma. These attributes are often attributed to ITCs and other sulphur-containing compounds; however, our data suggest that some of these characteristics in watercress could be attributed to this alcohol. (*Z*)-3-Hexen-1-ol by comparison was much higher in relative abundance but produced a medium intensity aroma that was green and radishy (Table 3).

Phenethyl ITC (aroma attributes described in Section 3.1.2.) had one of the highest intensity aromas in watercress, along with phenylacetaldehyde (**52**, 0.2%). The latter, similar to phenethyl ITC, is derived from phenylalanine, but occurs in much lower abundance (Table 2, Figure 2). Its aroma was described as honey-sweet (Table 3) and is likely a significant contributor to watercress odour that has previously gone unrecognised.

Other compounds contributing high odour intensities despite low relative abundances were 3-pentanone (**66**, 1.9%) and β-ionone (**98**, 0.4%; Figure 2). 3-Pentanone is regularly reported in Brassicales [3] and was described as high intensity, green, grassy, and floral smelling (Table 3, Figure 2). β-Ionone is common to many plant species as a degradation product of carotenoids [78], and it was described as soapy and fusty by assessors, with a high intensity.

Several sulphur-containing compounds, aldehydes, alcohols, ketones, not previously identified in watercress were identified. Methyl thiocyanate (**2**, 0.5%) imparted a sulphury and oniony note, with a medium intensity. This compound is known to be a GSL hydrolysis product of GCP (methyl GSL), but as will be discussed in Section 3.2, this compound was not detected in the UPLC-MS/MS analysis. Therefore, we suggest that it is not directly derived from this GSL and may be a degradation product of other GSL hydrolysis products within the tissues and headspace of the tested Brassicaceae.

Cyclopentyl-1-thiaethane (**8**, 0.5%) was detected, and uniquely present in watercress compared with the other three species tested (Table 2). Little is known about this compound in a biological context. It produced a sweaty, sulphury, medium-weak intensity aroma that is tentatively described in this species for the first time (Table 3).

Five aldehydes, two alcohol and two ketones, are newly reported for watercress, which produced perceptible odours within the headspace bouquet: 2-pentenal (**41**, 0.5%), hexanal (**43**, 1.2%), 4-heptenal (**45**, 0.5%), heptenal (**46**, 0.1%), and nonanal (**54**, 0.3%). Both heptenal and nonanal exhibited fatty, green aromas of medium intensity (Table 3) and are common to other Brassicaceae species [77]. 1-Octen-3-ol (**34**, 0.1%) produced a medium strength mushroom-like aroma (as described in Section 3.1.3.), and 2-phenylethanol (**37**, 2.6%) produced a floral scent. 6-Methyl-5-hepten-2-one (**68**, 0.2%) has been previously observed in 'wild' rocket and described as having a citrus aroma [55,79]. In this study, it was also identified with this characteristic, but also as floral and perfume-like, in both watercress and 'salad' rocket. It produced a medium-strong aroma in watercress. 2,2,6-Trimethylcyclohexanone (**69**, 0.1%) has been variously described as thujonic, menthol-like, and camphorous [80]. Here, it was described by assessors as imparting floral and green odours, with citrus notes.

One ester, (*Z*)-pent-2-en-1-yl acetate (**57**, 0.1%), was only detected in watercress, and it produced a medium-weak aroma. It was described as sulphury and rotten, and we are not aware of any previous odour attributes associated with this compound. As such, this is a tentative first description.

Octyl ITC (**21**, 0.2%) has been previously identified in horseradish [64] but not watercress to our knowledge. Its exact derivation and parent GSL are unclear in the literature, though watercress has been reported to contain glucohirsutin (GHS, (*RS*)-8methylsulfinyl)octyl GSL; [38]). This will be discussed further in Sections 3.2 and 3.3. Assessors found the compound to be of medium aroma strength having a green and vegetative character. Again, we are unaware of previous odour descriptions for this compound.

Finally, there were 29 unidentified odorants detected by GC-O, making this the highest number of the four species analysed (Table 3). Aromas described by assessors were sulphury (**114, 124, 125, 126, 141**), cooked onions (**115, 125**), buttery (**116**), rotten, cabbage-like (**120**), rotten onion (**124**), pungent (**125, 128, 157**), oniony (**126, 128**), green (**127, 131, 137, 153, 156**), parsley (**127**), mustard (**128**), cooked, roasted chicken, chicken soup (**129**), sour apples (**131**), apples (**132**), grass (**132, 134, 151. 158**), potato (**133, 156**), earthy (**135, 153**), musty (**135**), petrol, aromatic (**135, 147**), bread-like (**136**), medicinal (**140, 159**), floral (**140, 143, 158**), gas, burnt, roasted (**141, 142**), smoky (**142**), cucumber (**143, 148**), flowers (**143**), fruity, chemical, dried fruit (**151**), peppery (**153**), radish (**156**), and soapy (**157, 158,**

159). This indicates that the volatile profile of watercress is far from complete, and further research is required to elucidate these compounds.

3.2. Non-Volatile Compounds (Glucosinolates)

GSL composition and concentrations for 'salad' rocket, wasabi, and watercress are presented in Table 4. Due to an unforeseen termination of supply, it was not possible to include horseradish roots in this analysis.

Table 4. Glucosinolate concentrations of 'salad' rocket, wasabi, and watercress determined by UPLC-MS/MS.

Glucosinolate [a]	Abbreviation	Concentration [b]		
		Salad Rocket	Wasabi	Watercress
glucoiberin	GIB	0.007 ± 0.001	0.003 ± <0.001	nd
pentyl GSL [$]	PEN	4.915 ± 1.633	nd	nd
progoitrin	PRO	0.074 ± 0.067	0.001 ± <0.001	nd
sinigrin	SIN	0.002 ± 0.001	11.121 ± 0.247	0.001 ± <0.001
isobutyl GSL [$]	ISO	nd	3.382 ± 0.762	nd
glucoraphanin	GRA	1.761 ± 0.508	0.001 ± <0.001	0.006 ± <0.001
glucorucolamine	GRM	7.571 ± 1.208	nd	nd
glucoalyssin	GAL	0.215 ± 0.044	2.123 ± 0.477	0.470 ± 0.026
glucoputranjivin [$]	GPJ	0.058 ± 0.024	2.289 ± 0.515	nd
gluconapin	GNP	0.001 ± 0.001	0.010 ± 0.001	nd
diglucothiobeinin	DGTB	4.622 ± 1.144	nd	nd
glucoberteroin	GBT	0.412 ± 0.086	nd	nd
4-hydroxyglucobrassicin	4HGB	0.012 ± 0.002	0.062 ± 0.002	0.006 ± <0.001
glucocochlearin	GCL	nd	2.184 ± 0.166	nd
glucosativin	GSV	6.639 ± 2.402	nd	nd
7-(methylsulfinyl)heptyl GSL	7MSH	nd	4.55 ± 0.393	21.472 ± 1.219
glucobrassicanapin	GBN	nd	0.146 ± 0.005	nd
dimeric 4-mercaptobutyl GSL	DMB	78.861 ± 8.384	nd	nd
glucobarbarin	GBB	nd	nd	0.730 ± 0.037
glucotropaeolin	GTP	0.027 ± 0.002	0.001 ± <0.001	0.001 ± <0.001
glucoerucin	GER	0.733 ± 0.088	nd	nd
glucobrassicin	GBC	0.027 ± 0.008	0.001 ± <0.001	0.118 ± 0.006
4-methoxyglucobrassicin	4MGB	0.366 ± 0.113	nd	2.001 ± 0.104
gluconasturtiin	GNT	0.001 ± <0.001	nd	1.514 ± 0.040
neoglucobrassicin	NGB	2.682 ± 0.433	1.525 ± 0.229	1.596 ± 0.037
4-methylpentyl GSL	4MP	1.293 ± 0.836	nd	nd
hexy GSL	HEX	0.225 ± 0.071	nd	0.434 ± 0.031
7-(methylthio)heptyl GSL	7MTH	nd	0.132 ± 0.035	2.663 ± 0.104
butyl GSL [$]	BUT	0.301 ± 0.081	nd	0.606 ± 0.359
Total		111.098 ± 14.633	27.532 ± 2.831	31.645 ± 1.353

[a] GSL: glucosinolate; [$] = tentative identification. [b] Concentration in μmol g^{-1} dry weight; means are from six replicates for salad rocket, five replicates for wasabi and eight replicates for watercress; nd, not detected.

3.2.1. 'Salad' Rocket

'Salad' rocket contained the highest dry weight concentrations of GSLs (111.1 ± 14.6 μmol g^{-1} dw). This was predominantly due to high amounts of DMB (78.9 ± 8.4 μmol g^{-1} dw). Other GSLs of note included GSV (6.6 ± 2.4 μmol g^{-1} dw), GRM (7.6 ± 1.2 μmol g^{-1} dw), and DGTB (4.6 ± 1.1 μmol g^{-1} dw), which are unique to the genera *Eruca* and *Diplotaxis*. Other routinely reported GSLs for this species were GRA (1.8 ± 0.5 μmol g^{-1} dw) and NGB (2.7 ± 0.4 μmol g^{-1} dw).

Interestingly, several other GSLs that have not been, or are rarely reported for the species, were also detected; some in relatively high concentrations: GIB (<0.1 ± <0.1 μmol

g^{-1} dw), PEN (4.9 ± 1.6 μmol g^{-1} dw), GPJ (<0.1 ± <0.1 μmol g^{-1} dw), GBT (0.4 ± <0.1 μmol g^{-1} dw), GTP (<0.1 ± <0.1 μmol g^{-1} dw), 4MP (1.3 ± 0.8 μmol g^{-1} dw), HEX (0.2 ± <0.1 μmol g^{-1} dw), and BUT (0.3 ± <0.1 μmol g^{-1} dw).

Of note is the high abundance of PEN (*m/z* 388). It seems unlikely that a GSL of such relatively high concentration has gone undetected in previous analyses. Therefore, we postulate that previous studies may have attributed the negative ion mass incorrectly to that of PRO, which is also *m/z* 388 (Table 1). The authentic standard of PRO did not match the retention time or MS/MS spectra of PEN, and it was found in only very low concentrations by comparison. While PEN is only a tentative identification (due to the possibility of other isomeric GSLs such as glucojiaputin and 3-methylbutyl GSL), the presence of pentyl ITC (**17**) within the headspace of rocket makes this the most likely identification. See Section 3.3 for further discussion.

3.2.2. Wasabi

Sixteen GSL compounds were identified in wasabi roots, totaling 27.5 ± 2.8 μmol g^{-1} dw (Table 4). The most abundant compound was SIN (11.1 ± 0.2 μmol g^{-1} dw), which agrees with previous studies [81]. Wasabi is known to have a diverse GSL profile, and we observed relatively high abundances for ISO (3.4 ± 0.8 μmol g^{-1} dw), GAL (2.1 ± 0.5 μmol g^{-1} dw), GPJ (2.3 ± 0.5 μmol g^{-1} dw), GCL (2.2 ± 0.2 μmol g^{-1} dw), 7MSH (4.6 ± 0.4 μmol g^{-1} dw), and NGB (1.5 ± 0.2 μmol g^{-1} dw). Other compounds occurring in low abundance that are not frequently reported were GIB and GRA.

3.2.3. Watercress

Fourteen GSLs were found in watercress leaves, amounting to 31.6 ± 1.4 μmol g^{-1} dw (Table 4). In most previous studies of this species, GNT (1.5 ± <0.1 μmol g^{-1} dw) has been found to have the greatest abundance [4]; however, our analysis revealed that 7MSH had the highest total concentration (21.5 ± 1.2 μmol g^{-1} dw), dominating the overall profile in these samples. There were also relatively high concentrations of 7MTH (2.7 ± 0.1 μmol g^{-1} dw), and the indolic GSLs 4MGB (2 ± 0.1 μmol g^{-1} dw), and NGB (1.6 ± <0.1 μmol g^{-1} dw). Minor amounts of SIN, GRA, and GTP were also observed, and they are not frequently reported in this species.

3.3. Discrepancies between Identified Glucosinolate Hydrolysis Products and Glucosinolate Profile Precursors

There is often an 'elephant in the room' regarding volatile GSL hydrolysis products and reported GSL profiles in Brassicales crops: there are often GSLs found with no corresponding hydrolysis products, or more troublingly, hydrolysis products observed but no GSL precursor. Table 5 presents a list of the GSL-derived compounds identified within the headspace of 'salad' rocket, wasabi, and watercress, alongside their expected GSL precursors. It is apparent from our data that the present study is no exception when it comes to discrepancies of this nature, and there is a need to find a robust solution to prevent the inaccurate reporting of both GSLs and their volatile hydrolysis products.

As discussed in previous sections, compounds such as methyl thiocyanate (**2**) may be produced from degradation of other hydrolysis products. Others such as the presence of *sec*-butyl ITC (**9**) in rocket, butyl ITC (**12**) in wasabi, and octyl ITC (**21**) in watercress cannot be so easily explained. There are several explanations with varying levels of likelihood: firstly, the most likely is that the ITCs and other GSL hydrolysis products have been identified incorrectly, and that they belong to other parent GSLs present within the analysed tissues. Despite researchers' best efforts to obtain authentic standards to test and match MS profiles, this is not always possible or affordable, and so there is a heavy reliance upon reference libraries. These are often incomplete and not always accurate. It is also possible that the high temperatures utilised in GC-MS cause thermolytic reactions to occur in GHPs, thus yielding compounds not 'naturally' produced by tissues. The next possibility is that the GSLs responsible for producing hydrolysis products are below the LOD of the MS/MS method. In this case, this is unlikely, as the LOD for each standard

GSL was low, and it is likely much more sensitive and accurate than volatile compound measurements by GC-MS. Thirdly and perhaps least likely is that there is some as-yet-unknown mechanism(s) by which GSL hydrolysis products are modified post-hydrolysis. This is speculation, but there have been recent reports of previously unknown tautomeric rearrangements [82] and enzymatic actions [28] upon hydrolysis products that do not preclude this as an impossibility. Indeed, the dynamics of reactions occurring within the headspace of Brassicales is virtually unstudied, and so some may be produced through the degradation or rearrangement of structurally similar compounds. This is an area that requires much more detailed scrutiny.

Table 5. Identified volatile glucosinolate hydrolysis products in the headspace of 'salad' rocket, wasabi, and watercress, and the presence/absence of their respective glucosinolate precursor.

Precursor Glucosinolate	Glucosinolate Hydrolysis Product (Compound No.)	Glucosinolate Observed?			Hydrolysis Product Observed?		
		Salad Rocket	Wasabi	Watercress	Salad Rocket	Wasabi	Watercress
sinigrin	3-butenenitrile (**74**) $	√	√	√	x	√	x
	allyl thiocyanate (**4**) $	√	√	√	x	√	x
	allyl ITC (**5/6**) $	√	√	√	x	√	x
	cyclopropane ITC (**7**) $	√	√	√	x	√	x
	thiiraneacetonitrile (**77**) $	√	√	√	x	x	x
glucocapparin	methyl thiocyanate (**2**) $	x	x	x	√ *	x	√ *
glucoputranjivin $	isopropyl ITC (**3**) $	√	√	x	√	x	x
glucocochlearin $	*sec*-butyl ITC (**9**) $	x	√	x	√ *	√	x
4-methylpentyl GSL $	5-methylhexanenitrile (**75**) $	√	x	x	√	x	x
	1-isothiocyanato-4-methylpentane (**16**) $	√	x	x	√	√ *	√ *
isobutyl GSL	isobutyl ITC (**10**) $	x	√	x	x	√	x
<unknown>	6-heptenenitrile (**76**) $	-	-	-	x	√ *	x
gluconapin	3-butenyl ITC (**11**) $	√	√	x	√	√	x
butyl GSL	butyl ITC (**12**) $	√	x	√	√	√ *	x
3-methylbutyl GSL	isoamyl ITC (**13**) $	x	x	x	√ *	√ *	√ *
glucobrassicanapin	4-pentenyl ITC (**14**) $	x	√	x	x	√	x
pentyl GSL	pentyl ITC (**15**) $	√	x	x	√	√ *	x
glucotropaeolin	phenylacetonitrile (**78**) $	√	√	√	x	x	x
	benzyl ITC (**22**) $	√	√	√	x	√	x
hexyl GSL	cyclohexyl ITC (**17**) $	√	x	√	x	√ *	x
glucoberteroin	5-(methylsulfanyl)pentanenitrile (**79**) $	√	x	x	√	x	x
glucoerucin	4-(methylthio)butanenitrile (**80**) $	√	x	x	x	x	x
	erucin (**23**) $	√	x	x	√	√ *	x
gluconasturtiin	benzenepropanenitrile (**81**) $	√	x	√	x	x	√
glucoiberverin	iberverin (**19**) $	x	x	x	√ *	√ *	x
glucosativin	sativin (**20**) $	√	x	x	√	x	x
<unknown>	octyl ITC (**21**) $	x	x	x	x	x	√ *
gluconasturtiin	phenylethyl ITC (**24**) $	√	x	√	x	√ *	√

* Hydrolysis product observed but not glucosinolate precursor; $ tentatively identified.

4. Conclusions

This study has highlighted numerous newly identified and tentatively identified volatile compounds present in the headspace of 'salad' rocket, wasabi, horseradish, and watercress. Many of these appear to contribute strongly to the aroma profiles of each respective crop. We have also found 46 aroma traits present in the headspace of the samples that have no association with the identified compounds. This suggests that there are many as-yet undiscovered odour-active compounds present within Brassicales headspace.

Our data also highlight the need for more detailed studies on the volatilome of Brassicales species, and that the sole focus should not be upon GSL hydrolysis products. While accounting for a large proportion of the respective volatile profiles and odour active components of species, we have identified numerous instances where non-GSL derived compounds have odour intensities greater than those that are GSL-derived, such as 1-penten-3-ol (**25**), phenylacetaldehyde (**52**), 2-*sec*-butyl-3-methoxypyrazine (**106**), and β-ionone (**98**). Several non-GSL derived compounds also share similar pungent odour characteristics with GHPs, indicating that the latter may not be the only source of these sensations in Brassicales crops. There is also a clear need for the improvement of mass spectral libraries and the availability of GSL and GHP standards in order to overcome discrepancies between GSL profiles and the reported volatiles derived therefrom.

Supplementary Materials: The following are available online at https://www.mdpi.com/article/10.3390/foods10051055/s1, Sup-plementary Data S1. Compound fragmentation spectra with corresponding library matches (where available) and GC-MS chromatograms.

Author Contributions: Conceptualization, L.B. and S.L.; methodology, L.B. and S.L.; software, L.B., E.K., O.O.O. and S.L.; validation, L.B., O.O.O. and S.L; formal analysis, L.B., E.K. and S.L.; investigation, L.B., E.K. and S.L.; resources, L.B., E.K., O.O.O. and S.L.; data curation, L.B., E.K., O.O.O. and S.L.; writing—original draft preparation, L.B.; writing—review and editing, L.B., E.K., O.O.O. and S.L.; visualization, L.B., O.O.O. and S.L.; supervision, L.B. and S.L.; project administration, L.B. and S.L. All authors have read and agreed to the published version of the manuscript.

Funding: This research received no external funding.

Data Availability Statement: The data presented in this study is available on request from the corresponding author.

Acknowledgments: The authors would like to thank Richard Tudor of Elsoms Seeds Ltd. (Spalding, UK) for providing *Eruca sativa* seeds.

Conflicts of Interest: The authors declare no conflict of interest.

References

1. Bell, L.; Oloyede, O.O.; Lignou, S.; Wagstaff, C.; Methven, L. Taste and flavor perceptions of glucosinolates, isothiocyanates, and related compounds. *Mol. Nutr. Food Res.* **2018**, *62*, 1700990. [CrossRef] [PubMed]
2. Ravindran, P.N.; Pillai, G.S.; Divakaran, M. Other herbs and spices: Mango ginger to wasabi. In *Handbook of Herbs and Spices*, 2nd ed.; Peter, K.V., Ed.; Woodhead Publishing: Cambridge, UK, 2012; pp. 557–582. [CrossRef]
3. Bell, L.; Spadafora, N.D.; Müller, C.T.; Wagstaff, C.; Rogers, H.J. Use of TD-GC–TOF-MS to assess volatile composition during post-harvest storage in seven accessions of rocket salad (*Eruca sativa*). *Food Chem.* **2016**, *194*, 626–636. [CrossRef]
4. Giallourou, N.; Oruna-Concha, M.J.; Harbourne, N. Effects of domestic processing methods on the phytochemical content of watercress (*Nasturtium officinale*). *Food Chem.* **2016**, *212*, 411–419. [CrossRef]
5. Sultana, T.; Savage, G.P.; Mcneil, D.L.; Porter, G.P.; Clark, B. Comparison of flavour compounds in wasabi and horseradish. *J. Food Agric. Environ.* **2003**, *1*, 117–121.
6. Kroener, E.-M.; Buettner, A. Unravelling important odorants in horseradish (*Armoracia rusticana*). *Food Chem.* **2017**, *232*, 455–465. [CrossRef]
7. Depree, J.A.; Howard, T.M.; Savage, G.P. Flavour and pharmaceutical properties of the volatile sulphur compounds of Wasabi (*Wasabia japonica*). *Food Res. Int.* **1999**, *31*, 329–337. [CrossRef]
8. Bell, L.; Wagstaff, C. Glucosinolates, myrosinase hydrolysis products, and flavonols found in rocket (*Eruca sativa* and *Diplotaxis tenuifolia*). *J. Agric. Food Chem.* **2014**, *62*, 4481–4492. [CrossRef]
9. Palaniswamy, U.R. Watercress: A salad crop with chemopreventive potential. *Horttechnology* **2001**, *11*, 622–626. [CrossRef]
10. Siegmund, B. Biogenesis of aroma compounds: Flavour formation in fruits and vegetables. In *Flavour Development, Analysis and Perception in Food and Beverages*; Parker, J.K., Elmore, J.S., Methven, L., Eds.; Woodhead Publishing: Cambridge, UK, 2015; pp. 127–149. [CrossRef]
11. Kim, J.I.; Dolan, W.L.; Anderson, N.A.; Chapple, C. Indole glucosinolate biosynthesis limits phenylpropanoid accumulation in *Arabidopsis thaliana*. *Plant Cell* **2015**, *27*, 1529–1546. [CrossRef] [PubMed]
12. Raffo, A.; Masci, M.; Moneta, E.; Nicoli, S.; Sánchez del Pulgar, J.; Paoletti, F.; Pulgar, D. Characterization of volatiles and identification of odor-active compounds of rocket leaves. *Food Chem.* **2018**, *240*, 1161–1170. [CrossRef]
13. Blazevic, I.; Mastelic, J. Free and bound volatiles of rocket (*Eruca sativa* Mill.). *Flavour Fragrance J.* **2008**, *23*, 278–285. [CrossRef]
14. Jirovetz, L.; Smith, D.; Buchbauer, G. Aroma compound analysis of *Eruca sativa* (Brassicaceae) SPME headspace leaf samples using GC, GC−MS, and olfactometry. *J. Agric. Food Chem.* **2002**, *50*, 4643–4646. [CrossRef] [PubMed]
15. Miyazawa, M.; Maehara, T.; Kurose, K. Composition of the essential oil from the leaves of Eruca sativa. *Flavour Fragrance J.* **2002**, *17*, 187–190. [CrossRef]
16. Nakanishi, A.; Miyazawa, N.; Haraguchi, K.; Watanabe, H.; Kurobayashi, Y.; Nammoku, T. Determination of the absolute configuration of a novel odour-active lactone, cis-3-methyl-4-decanolide, in wasabi (*Wasabia japonica* Matsum.). *Flavour Fragrance J.* **2014**, *29*, 220–227. [CrossRef]

17. Kameoka, H.; Hashimoto, S. Volatile flavor components from wild *Wasabia japonica* Matsum. (Wasabi) and *Nasturtium officinale* R. Br. (Orandagarashi). *J. Agric. Chem. Soc. Japan* **1982**, *56*, 441–443.
18. Liu, Y.; Zhang, H.; Umashankar, S.; Liang, X.; Lee, H.; Swarup, S.; Ong, C. Characterization of plant volatiles reveals distinct metabolic profiles and pathways among 12 Brassicaceae vegetables. *Metabolites* **2018**, *8*, 94. [CrossRef]
19. Silva, M.F.; Campos, V.P.; Barros, A.F.; Terra, W.C.; Pedroso, M.P.; Gomes, V.A.; Ribeiro, C.R.; Silva, F.J. Volatile emissions of watercress (*Nasturtium officinale*) leaves and passion fruit (*Passiflora edulis*) seeds against Meloidogyne incognita. *Pest Manag. Sci.* **2020**, *76*, 1413–1421. [CrossRef] [PubMed]
20. Spence, R.-M.M.; Tucknott, O.G. An apparatus for the comparative collection of headspace volatiles of watercress. *J. Sci. Food Agric.* **1983**, *34*, 412–416. [CrossRef]
21. Agneta, R.; Möllers, C.; Rivelli, A.R. Horseradish (*Armoracia rusticana*), a neglected medical and condiment species with a relevant glucosinolate profile: A review. *Genet. Resour. Crop Evol.* **2013**, *60*, 1923–1943. [CrossRef]
22. Dekić, M.S.; Radulović, N.S.; Stojanović, N.M.; Randjelović, P.J.; Stojanović-Radić, Z.Z.; Najman, S.; Stojanović, S. Spasmolytic, antimicrobial and cytotoxic activities of 5-phenylpentyl isothiocyanate, a new glucosinolate autolysis product from horseradish (*Armoracia rusticana* P. Gaertn., B. Mey. & Scherb., Brassicaceae). *Food Chem.* **2017**, *232*, 329–339. [CrossRef]
23. Gilbert, J.; Nursten, H.E. Volatile constituents of horseradish roots. *J. Sci. Food Agric.* **1972**, *23*, 527–539. [CrossRef]
24. Pilipczuk, T.; Kusznierewicz, B.; Chmiel, T.; Przychodzeń, W.; Bartoszek, A. Simultaneous determination of individual isothiocyanates in plant samples by HPLC-DAD-MS following SPE and derivatization with N -acetyl- l -cysteine. *Food Chem.* **2017**, *214*, 587–596. [CrossRef]
25. Kroener, E.-M.; Buettner, A. Sensory-analytical comparison of the aroma of different horseradish varieties (*Armoracia rusticana*). *Front. Chem.* **2018**, *6*. [CrossRef] [PubMed]
26. Pasini, F.; Verardo, V.; Caboni, M.F.; D'Antuono, L.F. Determination of glucosinolates and phenolic compounds in rocket salad by HPLC-DAD–MS: Evaluation of *Eruca sativa* Mill. and *Diplotaxis tenuifolia* L. genetic resources. *Food Chem.* **2012**, *133*, 1025–1033. [CrossRef]
27. Bell, L.; Wagstaff, C. Rocket science: A review of phytochemical & health-related research in Eruca & Diplotaxis species. *Food Chem. X* **2019**, *1*, 100002. [CrossRef] [PubMed]
28. Agerbirk, N.; Matthes, A.; Erthmann, P.Ø.; Ugolini, L.; Cinti, S.; Lazaridi, E.; Nuzillard, J.-M.; Müller, C.; Bak, S.; Rollin, P.; et al. Glucosinolate turnover in Brassicales species to an oxazolidin-2-one, formed via the 2-thione and without formation of thioamide. *Phytochemistry* **2018**, *153*, 79–93. [CrossRef]
29. Bell, L.; Oruna-Concha, M.J.; Wagstaff, C. Identification and quantification of glucosinolate and flavonol compounds in rocket salad (*Eruca sativa*, *Eruca vesicaria* and *Diplotaxis tenuifolia*) by LC–MS: Highlighting the potential for improving nutritional value of rocket crops. *Food Chem.* **2015**, *172*, 852–861. [CrossRef]
30. Rochfort, S.J.; Trenerry, V.C.; Imsic, M.; Panozzo, J.; Jones, R. Class targeted metabolomics: ESI ion trap screening methods for glucosinolates based on MSn fragmentation. *Phytochemistry* **2008**, *69*, 1671–1679. [CrossRef] [PubMed]
31. Andini, S.; Dekker, P.; Gruppen, H.; Araya-Cloutier, C.; Vincken, J.-P. Modulation of glucosinolate composition in Brassicaceae seeds by germination and fungal elicitation. *J. Agric. Food Chem.* **2019**, *67*, 12770–12779. [CrossRef]
32. Cataldi, T.R.I.; Rubino, A.; Lelario, F.; Bufo, S.A. Naturally occurring glucosinolates in plant extracts of rocket salad (*Eruca sativa* L.) identified by liquid chromatography coupled with negative ion electrospray ionization and quadrupole ion-trap mass spectrometry. *Rapid Commun. Mass Spectrom.* **2007**, *21*, 2374–2388. [CrossRef] [PubMed]
33. Agneta, R.; Rivelli, A.R.; Ventrella, E.; Lelario, F.; Sarli, G.; Bufo, S.A. Investigation of glucosinolate profile and qualitative aspects in sprouts and roots of horseradish (*Armoracia rusticana*) using LC-ESI–hybrid linear ion trap with fourier transform ion cyclotron resonance mass spectrometry and infrared multiphoton dissociation. *J. Agric. Food Chem.* **2012**, *60*, 7474–7482. [CrossRef] [PubMed]
34. Maldini, M.; Foddai, M.; Natella, F.; Petretto, G.L.; Rourke, J.P.; Chessa, M.; Pintore, G. Identification and quantification of glucosinolates in different tissues of *Raphanus raphanistrum* by liquid chromatography tandem-mass spectrometry. *J. Food Compos. Anal.* **2017**, *61*, 20–27. [CrossRef]
35. Shi, H.; Zhao, Y.; Sun, J.; Yu, L.; Chen, P. Chemical profiling of glucosinolates in cruciferous vegetables-based dietary supplements using ultra-high performance liquid chromatography coupled to tandem high resolution mass spectrometry. *J. Food Compos. Anal.* **2017**, *61*, 67–72. [CrossRef]
36. Mellon, F.A.; Bennett, R.N.; Holst, B.; Williamson, G. Intact glucosinolate analysis in plant extracts by programmed cone voltage electrospray LC/MS: Performance and comparison with LC/MS/MS methods. *Anal. Biochem.* **2002**, *306*, 83–91. [CrossRef] [PubMed]
37. Bianco, G.; Agerbirk, N.; Losito, I.; Cataldi, T.R.I. Acylated glucosinolates with diverse acyl groups investigated by high resolution mass spectrometry and infrared multiphoton dissociation. *Phytochemistry* **2014**, *100*, 92–102. [CrossRef] [PubMed]
38. Blažević, I.; Montaut, S.; Burčul, F.; Olsen, C.; Burow, M.; Rollin, P.; Agerbirk, N. Glucosinolate structural diversity, identification, chemical synthesis and metabolism in plants. *Phytochemistry* **2019**, *169*, 112100. [CrossRef] [PubMed]
39. Al-Gendy, A.A.; Lockwood, G.B. GC-MS analysis of volatile hydrolysis products from glucosinolates in Farsetia aegyptia var. ovalis. *Flavour Fragrance J.* **2003**, *18*, 148–152. [CrossRef]
40. Wang, C.; Xing, J.; Chin, C.-K.; Ho, C.-T.; Martin, C.E. Modification of fatty acids changes the flavor volatiles in tomato leaves. *Phytochemistry* **2001**, *58*, 227–232. [CrossRef]

41. Buttery, R.G.; Teranishi, R.; Ling, L.C.; Turnbaugh, J.G. Quantitative and sensory studies on tomato paste volatiles. *J. Agric. Food Chem.* **1990**, *38*, 336–340. [CrossRef]
42. Deasy, W.; Shepherd, T.; Alexander, C.J.; Birch, A.N.E.; Evans, K.A. Development and validation of a SPME-GC-MS method for in situ passive sampling of root volatiles from glasshouse-grown broccoli plants undergoing blow-ground herbivory by larvae of cabbage root fly, *Delia radicum* L. *Phytochem. Anal.* **2016**, *27*, 375–393. [CrossRef]
43. Moshonas, M.G.; Shaw, P.E. Some newly found orange essence components including trans-2-pentenal. *J. Food Sci.* **1973**, *38*, 360–361. [CrossRef]
44. Berger, R.G.; Drawert, F.; Kollmannsberger, H. The flavour of cape gooseberry (*Physalis peruviana* L.). *Z. Für Lebensm. -Unters. und Forsch.* **1989**, *188*, 122–126. [CrossRef]
45. Hatanaka, A.; Harada, T. Formation of cis-3-hexenal, trans-2-hexenal and cis-3-hexenol in macerated *Thea sinensis* leaves. *Phytochemistry* **1973**, *12*, 2341–2346. [CrossRef]
46. D'Auria, J.C.; Pichersky, E.; Schaub, A.; Hansel, A.; Gershenzon, J. Characterization of a BAHD acyltransferase responsible for producing the green leaf volatile (Z)-3-hexen-1-yl acetate in Arabidopsis thaliana. *Plant J.* **2007**, *49*, 194–207. [CrossRef]
47. Angerosa, F. Virgin olive oil odour notes: Their relationships with volatile compounds from the lipoxygenase pathway and secoiridoid compounds. *Food Chem.* **2000**, *68*, 283–287. [CrossRef]
48. Kato, M.; Imayoshi, Y.; Iwabuchi, H.; Shimomura, K. Kinetic changes in glucosinolate-derived volatiles by heat-treatment and myrosinase activity in nakajimana (*Brassica rapa* L. cv. *nakajimana*). *J. Agric. Food Chem.* **2011**, *59*, 11034–11039. [CrossRef]
49. Sharma, A.; Rai, P.K.; Prasad, S. GC–MS detection and determination of major volatile compounds in *Brassica juncea* L. leaves and seeds. *Microchem. J.* **2018**, *138*, 488–493. [CrossRef]
50. Song, H.; Liu, J. GC-O-MS technique and its applications in food flavor analysis. *Food Res. Int.* **2018**, *114*, 187–198. [CrossRef]
51. Aparicio, R.; Morales, M.T.; Alonso, V. Authentication of European virgin olive oils by their chemical compounds, sensory attributes, and consumers' attitudes. *J. Agric. Food Chem.* **1997**, *45*, 1076–1083. [CrossRef]
52. Hossain, S.; Bergkvist, G.; Berglund, K.; Glinwood, R.; Kabouw, P.; Mårtensson, A.; Persson, P. Concentration- and time-dependent effects of isothiocyanates produced from Brassicaceae shoot tissues on the pea root rot pathogen *Aphanomyces euteiches*. *J. Agric. Food Chem.* **2014**, *62*, 4584–4591. [CrossRef]
53. Kask, K.; Kännaste, A.; Talts, E.; Copolovici, L.; Niinemets, Ü. How specialized volatiles respond to chronic and short-term physiological and shock heat stress in *Brassica nigra*. *Plant Cell Environ.* **2016**, *39*, 2027–2042. [CrossRef] [PubMed]
54. Bruce, T.J.A. Glucosinolates in oilseed rape: Secondary metabolites that influence interactions with herbivores and their natural enemies. *Ann. Appl. Biol.* **2014**, *164*, 348–353. [CrossRef]
55. Hansen, M.; Laustsen, A.M.; Olsen, C.E.; Poll, L.; SØRensen, H. Chemical and sensory quality of Broccoli (*Brassica oleracea* L. Var *italica*). *J. Food Qual.* **1997**, *20*, 441–459. [CrossRef]
56. Hanschen, F.S.; Kaufmann, M.; Kupke, F.; Hackl, T.; Kroh, L.W.; Rohn, S.; Schreiner, M. Brassica vegetables as sources of epithionitriles: Novel secondary products formed during cooking. *Food Chem.* **2018**, *245*, 564–569. [CrossRef] [PubMed]
57. Fernandes, F.; Pereira, D.M.; Guedes de Pinho, P.; Valentão, P.; Pereira, J.A.; Bento, A.; Andrade, P.B. Headspace solid-phase microextraction and gas chromatography/ion trap-mass spectrometry applied to a living system: Pieris brassicae fed with kale. *Food Chem.* **2010**, *119*, 1681–1693. [CrossRef]
58. Bartlet, E.; Blight, M.M.; Lane, P.; Williams, I.H. The responses of the cabbage seed weevil *Ceutorhynchus assimilis* to volatile compounds from oilseed rape in a linear track olfactometer. *Entomol. Exp. Appl.* **1997**, *85*, 257–262. [CrossRef]
59. Fernandes, F.; Guedes de Pinho, P.; Valentão, P.; Pereira, J.A.; Andrade, P.B. Volatile constituents throughout *Brassica oleracea* L. Var. *acephala* germination. *J. Agric. Food Chem.* **2009**, *57*, 6795–6802. [CrossRef]
60. Ortner, E.; Granvogl, M.; Schieberle, P. Elucidation of thermally induced changes in key odorants of white mustard seeds (*Sinapis alba* L.) and rapeseeds (*Brassica napus* L.) using molecular sensory science. *J. Agric. Food Chem.* **2016**, *64*, 8179–8190. [CrossRef] [PubMed]
61. Smid, H.M.; Van Loon, J.J.A.; Posthumus, M.A.; Vet, L.E.M. GC-EAG-analysis of volatiles from brussels sprouts plants damaged by two species of Pieris caterpillars: Olfactory receptive range of a specialist and a generalist parasitoid wasp species. *Chemoecology* **2002**, *12*, 169–176. [CrossRef]
62. Klopsch, R.; Witzel, K.; Börner, A.; Schreiner, M.; Hanschen, F.S. Metabolic profiling of glucosinolates and their hydrolysis products in a germplasm collection of *Brassica rapa* turnips. *Food Res. Int.* **2017**, *100*, 392–403. [CrossRef]
63. Angelino, D.; Jeffery, E. Glucosinolate hydrolysis and bioavailability of resulting isothiocyanates: Focus on glucoraphanin. *J. Funct. Foods* **2014**, *7*, 67–76. [CrossRef]
64. Petrović, S.; Drobac, M.; Ušjak, L.; Filipović, V.; Milenković, M.; Niketić, M. Volatiles of roots of wild-growing and cultivated *Armoracia macrocarpa* and their antimicrobial activity, in comparison to horseradish. *A. rusticana*. *Ind. Crops Prod.* **2017**, *109*, 398–403. [CrossRef]
65. Tohar, N.; Mohd, M.A.; Jantan, I.; Awang, K. A comparative study of the essential oils of the genus Plumeria Linn. from Malaysia. *Flavour Fragr. J.* **2006**, *21*, 859–863. [CrossRef]
66. Tieman, D.; Taylor, M.; Schauer, N.; Fernie, A.R.; Hanson, A.D.; Klee, H.J. Tomato aromatic amino acid decarboxylases participate in synthesis of the flavor volatiles 2-phenylethanol and 2-phenylacetaldehyde. *Proc. Natl. Acad. Sci. USA* **2006**, *103*, 8287–8292. [CrossRef]

67. Triqui, R.; Reineccius, G. Changes in flavor profiles with ripening of anchovy (*Engraulis encrasicholus*). *J. Agric. Food Chem.* **1995**, *43*, 1883–1889. [CrossRef]
68. Parker, J.K. Introduction to aroma compounds in foods. In *Flavour Development, Analysis and Perception in Food and Beverages*; Parker, J.K., Elmore, J.S., Methven, L., Eds.; Woodhead Publishing: Cambridge, UK, 2015; pp. 3–30. [CrossRef]
69. Schreiner, M.; Krumbein, A.; Ruppel, S. Interaction between plants and bacteria: Glucosinolates and phyllospheric colonization of cruciferous vegetables by Enterobacter radicincitans DSM 16656. *J. Mol. Microbiol. Biotechnol.* **2009**, *17*, 124–135. [CrossRef]
70. Deng, Q.; Zinoviadou, K.G.; Galanakis, C.M.; Orlien, V.; Grimi, N.; Vorobiev, E.; Lebovka, N.; Barba, F.J. The effects of conventional and non-conventional processing on glucosinolates and its derived forms, isothiocyanates: Extraction, degradation, and applications. *Food Eng. Rev.* **2015**, *7*, 357–381. [CrossRef]
71. Park, H.-W.; Choi, K.-D.; Shin, I.-S. Antimicrobial activity of isothiocyanates (ITCs) extracted from horseradish (*Armoracia rusticana*) root against oral microorganisms. *Biocontrol Sci.* **2013**, *18*, 163–168. [CrossRef] [PubMed]
72. Terada, Y.; Masuda, H.; Watanabe, T. Structure–activity relationship study on isothiocyanates: Comparison of TRPA1-activating ability between allyl isothiocyanate and specific flavor components of wasabi, horseradish, and white mustard. *J. Nat. Prod.* **2015**, *78*, 1937–1941. [CrossRef]
73. Masuda, H.; Harada, Y.; Tanaka, K.; Nakajima, M.; Tabeta, H. Characteristic odorants of wasabi (*Wasabia japonica matum*), Japanese horseradish, in comparison with those of horseradish (*Armoracia rusticana*). In *Biotechnology for Improved Foods and Flavors*; American Chemical Society: Washington, DC, USA, 1996; Volume 637, pp. 67–78.
74. Lv, J.; Wu, J.; Zuo, J.; Fan, L.; Shi, J.; Gao, L.; Li, M.; Wang, Q. Effect of Se treatment on the volatile compounds in broccoli. *Food Chem.* **2017**, *216*, 225–233. [CrossRef]
75. Blight, M.M.; Pickett, J.A.; Wadhams, L.J.; Woodcock, C.M. Antennal perception of oilseed rape, *Brassica napus* (Brassicaceae), volatiles by the cabbage seed weevil *Ceutorhynchus assimilis* (Coleoptera, Curculionidae). *J. Chem. Ecol.* **1995**, *21*, 1649–1664. [CrossRef]
76. Chen, H.-Z.; Zhang, M.; Guo, Z. Discrimination of fresh-cut broccoli freshness by volatiles using electronic nose and gas chromatography-mass spectrometry. *Postharvest Biol. Technol.* **2019**, *148*, 168–175. [CrossRef]
77. Müller-Maatsch, J.; Gurtner, K.; Carle, R.; Björn Steingass, C. Investigation into the removal of glucosinolates and volatiles from anthocyanin-rich extracts of red cabbage. *Food Chem.* **2019**, *278*, 406–414. [CrossRef] [PubMed]
78. Davies, K. *Annual Plant Reviews, Plant Pigments and Their Manipulation*; Blackwell Publishing: Oxford, UK, 2004; Volume 14.
79. Mastrandrea, L.; Amodio, M.L.; Pati, S.; Colelli, G. Effect of modified atmosphere packaging and temperature abuse on flavor related volatile compounds of rocket leaves (*Diplotaxis tenuifolia* L.). *J. Food Sci. Technol.* **2017**, *54*, 2433–2442. [CrossRef] [PubMed]
80. Silva Souza, M.A.; Peres, L.E.; Freschi, J.R.; Purgatto, E.; Lajolo, F.M.; Hassimotto, N.M. Changes in flavonoid and carotenoid profiles alter volatile organic compounds in purple and orange cherry tomatoes obtained by allele introgression. *J. Sci. Food Agric.* **2020**, *100*, 1662–1670. [CrossRef]
81. Li, L.; Lee, W.; Lee, W.J.; Auh, J.H.; Kim, S.S.; Yoon, J. Extraction of allyl isothiocyanate from wasabi (*Wasabia japonica Matsum*) using supercritical carbon dioxide. *Food Sci. Biotechnol.* **2010**, *19*, 405–410. [CrossRef]
82. Fechner, J.; Kaufmann, M.; Herz, C.; Eisenschmidt, D.; Lamy, E.; Kroh, L.W.; Hanschen, F.S. The major glucosinolate hydrolysis product in rocket (*Eruca sativa* L.), sativin, is 1,3-thiazepane-2-thione: Elucidation of structure, bioactivity, and stability compared to other rocket isothiocyanates. *Food Chem.* **2018**, *261*, 57–65. [CrossRef]

foods

MDPI

Article

High Glucosinolate Content in Rocket Leaves (*Diplotaxis tenuifolia* and *Eruca sativa*) after Multiple Harvests Is Associated with Increased Bitterness, Pungency, and Reduced Consumer Liking

Luke Bell [1,*], Stella Lignou [2] and Carol Wagstaff [2]

1 School of Agriculture, Policy & Development, University of Reading, Whiteknights, P.O. Box 237, Reading, Berkshire RG6 6AR, UK

2 School of Chemistry Food & Pharmacy, University of Reading, Whiteknights, P.O. Box 226, Reading, Berkshire RG6 6AP, UK; s.lignou@reading.ac.uk (S.L.); c.wagstaff@reading.ac.uk (C.W.)

* Correspondence: luke.bell@reading.ac.uk

Received: 28 October 2020; Accepted: 1 December 2020; Published: 3 December 2020

Abstract: Rocket (*Diplotaxis tenuifolia* and *Eruca sativa*) leaves delivered to the UK market are variable in appearance, taste, and flavour over the growing season. This study presents sensory and consumer analyses of rocket produce delivered to the UK over the course of one year, and evaluated the contribution of environmental and cultivation factors upon quality traits and phytochemicals called glucosinolates (GSLs). GSL abundance was positively correlated with higher average growth temperatures during the crop cycle, and perceptions of pepperiness, bitterness, and hotness. This in turn was associated with reduced liking, and corresponded to low consumer acceptance. Conversely, leaves with greater sugar content were perceived as more sweet, and had a higher correlation with consumer acceptance of the test panel. First cut leaves of rocket were favoured more by consumers, with multiple leaf cuts associated with low acceptance and higher glucosinolate concentrations. Our data suggest that the practice of harvesting rocket crops multiple times reduces consumer acceptability due to increases in GSLs, and the associated bitter, hot, and peppery perceptions some of their hydrolysis products produce. This may have significant implications for cultivation practices during seasonal transitions, where leaves typically receive multiple harvests and longer growth cycles.

Keywords: glucosinolates; rucola; arugula; *Diplotaxis*; *Eruca*; bitter taste; flavour; postharvest

1. Introduction

Rocket (also known as arugula and rucola) salad species such as *Diplotaxis tenuifolia* and *Eruca sativa* are leafy vegetables of the order Brassicales, and are popular throughout the world [1]. They are commonly sold in bags of loose leaves, or as part of a leafy salad mixture with other crops, such as lettuce, spinach, and watercress [2]. Previous studies have evaluated sensory properties of rocket leaves [3–8] in conjunction with phytochemical compositions, and in one instance, consumer preferences according to human taste receptor genotype [9]. One factor not accounted for in any of these studies is the temporal variability of rocket produce over the course of a growing season, and the inherent environmental variability associated with this.

Rocket salad is in demand year-round in the UK; however, no British region is suitable for its continuous cultivation. As such, produce is typically sourced from several different countries throughout a year, according to the season [10]. In the UK, the vast majority of rocket is imported from Italy, with only seasonal summer rocket production possible in the south of England, as winters are too cold, wet, and humid for viable winter growth. In Italy, rocket is grown in the north and east during summer months; typically, in regions such as the Veneto, Lazio, and Emilia-Romagna.

In winter, there is a shift in cultivation towards the south where temperatures typically remain higher, and humidity lower; for example to Campania and Apulia. In times of high demand, rocket is sourced for the UK market from other European countries (e.g., Spain), Northern Africa (e.g., Morocco), or even as far as the United States and India.

It has been well documented in the literature that crop quality and respiration rates are influenced by seasonality [11]. This is intrinsically linked to growing temperature, as metabolic rates tend to be higher under warmer conditions. Growth temperature therefore plays a distinct role in determining shelf life longevity and visual acceptability of leaves [12]. There is also evidence suggesting that pungency is increased during spring and summer months [13], though this has not been shown quantitatively in rocket. Conversely, sugars have been shown to be reduced in some Brassicales species under "high" growth temperatures (>20 °C) [14], which has implications for sensory traits. Anecdotal evidence suggests that source and season significantly affects the quality and consistency of rocket leaves. This presents a problem for producers and supermarkets as leaves are typically marketed as generic products, but the quality of the produce is not consistent.

A previous study by Bell et al. [9] highlighted that pungency (or hotness) of rocket leaves is the main driver of rocket liking. Excessive pungency is typically rejected by consumers, and most people in that study preferred milder and sweeter leaves. There was a significant plant genotypic component determining the pungency of leaves, but it remains unstudied how pungency and consumer preference may be affected by changes in climate and seasonal growth. The pungency of rocket leaves is due to the presence of glucosinolates (GSLs) within tissues, which are hydrolysed by myrosinase enzymes to produce isothiocyanates (ITCs) and numerous other products. One compound present in both *E. sativa* and *D. tenuifolia* is glucosativin (4-mercaptobutyl GSL; GSV) and its dimer (dimeric 4-mercaptobutyl GSL; DMB). These produce the ITC, 4-mercaptobutyl ITC, which undergoes spontaneous tautomeric rearrangement post-hydrolysis to form 1,3-thiazepane-2-thione (sativin; SAT) [15]. It is unknown if it is the ITC or tautomer that is responsible for pungency and flavour, but olfactometry has shown a distinctive rocket-like aroma associated with the hydrolysis product of GSV [16]. The inherent variability of GSL biosynthesis in response to growth environment affects the presence and abundance of GSV and health related GSLs (such as glucoraphanin, GRA; and glucoerucin, GER) [17], and therefore may impact upon sensory properties of leaves and consumer acceptance.

An often-neglected component of taste and flavour perception in vegetables is sugar content and composition. Sugar content of Brassicales leaves is known to be variable according to growth conditions [18]. In combination with effects on GSL composition, sugars may therefore have a strong influence upon sensory traits and consumer preference throughout a growing season.

We present phytochemical, sensory analysis, and consumer preference and perception data from a year-long study of rocket produce sourced from commercial farms delivered to UK-based processors and supermarkets. We hypothesised that GSL and sugar content would significantly affect sensory properties of rocket leaves at different times of the year. This in turn would affect consumer perceptions of leaves and their preference of rocket would change seasonally. We also present the effects of climatic factors (such as temperature) and cultivation practices imposed on rocket crops (such as multiple harvests/cuts of the same crop).

2. Materials and Methods

2.1. Plant Material

Rocket leaf material was sourced and delivered to the University of Reading Sensory Science Centre monthly (with the exceptions of May and December) for one year (2014) by Bakkavor Ltd. (Spalding, UK). This material corresponded to leaf material delivered to a processing facility in England, and the material that would be used in products destined for UK supermarkets and consumers. Each sample batch was harvested, washed, and processed in accordance with industry practice.

For reasons of commercial sensitivity, the exact locations of growers that supplied material for this study will not be named. Only the country of origin will be detailed, along with the growth environment (polytunnel, glasshouse, or open field) and cultivar used. It should be noted that as with many other crops, cultivars are selected by growers according to season and performance under specific conditions and climates. As such, the cultivars tested each month were not always the same. This was intentional to properly evaluate the level of consistency of produce as it exists within the UK supply chain, not to evaluate the changes of any single cultivar over the year or between different countries. Data were provided by growers relating to the length of each crop cycle (i.e., how long each of the samples had been grown for), the number of cuts each sample had undergone, and the percentage of dry matter of each batch at intake. See Table 1 for a summary of samples tested in each respective month.

2.2. Temperature Data

Temperature data for each of the rocket growing sites were supplied by Bakkavor Ltd. Data were logged throughout the cropping cycle of plants via on-farm weather stations. Four measurements were provided: the average daily temperature for the entire cropping cycle (referred to as "avg. temp"), the maximum temperature of the week preceding harvest (referred to as "max. temp. week"), the minimum temperature of the week preceding harvest (referred to as "min. temp. week"), and the average temperature in the week preceding harvest (referred to as "avg. temp. week").

2.3. Sensory Analysis

A panel of 12 previously trained and experienced assessors (ten female, two male) evaluated the samples at the Sensory Science Centre (University of Reading, Reading, UK). All training and monitoring of the sensory panel was in accordance with ISO 8586:2012 and ISO 11132:2012 standards. The panellists used the consensus vocabulary developed by Bell et al. [4] to describe the samples; the developed terms included appearance, odour, taste, flavour, mouthfeel, and after-effects attributes. Panellists were familiarized with the evaluated sensory attributes prior to each monthly scoring session, with reference standards when required.

Each month the panellists rated the samples individually in isolated well ventilated, temperature-controlled booths (22 °C), under artificial daylight. Attribute intensity was scored on 15 cm unstructured line scales (data scaled 0–100). Samples were presented monadically in a balanced presentation order, coded with three-digit random codes. Cold water and natural unflavoured yoghurt were provided for palate cleansing, and warm water for washing fingers between samples. Data were collected in duplicate using Compusense (version 5.5, Guelph, ON, Canada).

2.4. Consumer Analysis

Consumer recruitment and assessments were conducted as per the protocols of Bell et al. [9]. Evaluations were held on a bimonthly basis. Briefly, consumers were recruited from in and around the University of Reading and asked to attend as many of the evaluation sessions as possible over the course of the year-long study. A total of 55 consumers (out of 101) attended every session of the study.

Volunteers were presented with three leaves and asked to score their liking of leaf appearance, taste, and "overall" liking of each sample on a line scale (0–10), and based on their own individual experience of rocket sensory attributes. Volunteers were also asked to score their perceptions of bitterness, hotness (pungency), sweetness, and pepperiness. Scores were entered into a general labelled magnitude scale (gLMS) ranging from "not detectable", "weak", "moderate", "strong", "very strong", to "strongest imaginable". Data were subsequently converted to antilog values and normalized for statistical analysis [19]. Samples were presented monadically in a randomized, balanced presentation order, with three-digit random codes in duplicate. Data were collected using Compusense (version 5.5, Guelph, ON, Canada).

Table 1. Cultivation and origin details of rocket samples used for sensory and consumer analyses.

Month	Cultivar	Symbol Code	Species	Country of Origin	Growth Environment	Cut Number	Crop Cycle Length (Days)	Dry Matter (%)
January	Yeti	◆✦	*D. tenuifolia*	Italy	Polytunnel	2nd	60	8.1
	Yeti	◆★	*D. tenuifolia*	Italy	Polytunnel	2nd	71	8.3
	Tricia	◆⅄	*D. tenuifolia*	Italy	Polytunnel	2nd	24	9.5
	Wildfire	■✦	*D. tenuifolia*	Italy	Polytunnel	1st	51	8.7
	Fast Grow	❖★	*D. tenuifolia*	Italy	-	2nd+	70	8.1
							55.2	**8.5**
February	Venere	❖⅄	*D. tenuifolia*	Italy	Polytunnel	2nd+	27	8.9
	Wildfire	❖⅄	*D. tenuifolia*	Italy	Polytunnel	2nd+	23	13.4
	Selezione Enza	❖✶	*D. tenuifolia*	Italy	Polytunnel	2nd+	124	8.8
	Tricia	❖⅄	*D. tenuifolia*	Italy	Polytunnel	2nd+	24	7.9
	Fast Grow	◆★	*D. tenuifolia*	Italy	-	2nd	85	9.4
							56.6	**9.7**
March	Yeti	❖✶	*D. tenuifolia*	Italy	Polytunnel	2nd+	137	-
	Wildfire	■✦	*D. tenuifolia*	USA	Open field	1st	31	7.9
	Fast Grow	■-	*D. tenuifolia*	Italy	Polytunnel	1st	-	-
	Voyager	❖★	*D. tenuifolia*	India	Open field	2nd+	85	-
	Fast Grow	❖✶	*D. tenuifolia*	Italy	-	2nd+	132	-
							96.3	-
April	Wildfire	❖✶	*D. tenuifolia*	Italy	Polytunnel	2nd+	180	9.3
	Venere	◆✦	*D. tenuifolia*	Italy	Polytunnel	2nd	35	7.8
	Standard	❖✦	*D. tenuifolia*	Spain	Polytunnel	2nd+	40	8.1
	Giove	❖-	*D. tenuifolia*	Italy	Polytunnel	2nd+	-	9.3
	Quaggio	◆✦	*D. tenuifolia*	Italy	Polytunnel	2nd	43	7.7
							74.5	**8.4**
June	Giove	◆⅄	*D. tenuifolia*	Italy	Polytunnel	2nd	30	6.7
	Giove	■⅄	*D. tenuifolia*	Italy	Polytunnel	1st	23	5.9
	Torino	■✦	*D. tenuifolia*	UK	Open field	1st	40	12.3
	Giove	■⅄	*D. tenuifolia*	Italy	Polytunnel	1st	23	7.5
	Extrema	◆✦	*D. tenuifolia*	Italy	Polytunnel	2nd	39	6.4
							31.0	**7.8**

Table 1. *Cont.*

Month	Cultivar	Symbol Code	Species	Country of Origin	Growth Environment	Cut Number	Crop Cycle Length (Days)	Dry Matter (%)
July	Giove	■⅄	*D. tenuifolia*	Italy	Polytunnel	1st	20	8.0
	Giove	■⅄	*D. tenuifolia*	Italy	Polytunnel	1st	20	6.6
	Napoli	■⅄	*D. tenuifolia*	UK	Open field	1st	24	6.1
	Voyager	■✦	*D. tenuifolia*	UK	Open field	1st	44	14.8
	Giove	◆⅄	*D. tenuifolia*	Italy	Polytunnel	2nd	28	7.3
							27.2	**8.6**
August	Giove	■⅄	*D. tenuifolia*	Italy	Polytunnel	1st	24	9.7
	Extrema	◆✦	*D. tenuifolia*	Italy	Polytunnel	2nd	33	8.5
	Voyager	■⅄	*D. tenuifolia*	UK	Glasshouse	1st	24	6.6
							27.0	**8.3**
September	Voyager	■✦	*D. tenuifolia*	UK	Glasshouse	1st	35	8.7
	Shamrock	■✦	*D. tenuifolia*	UK	Open field	1st	38	9.6
	Extrema	◆✦	*D. tenuifolia*	Italy	Polytunnel	2nd	31	7.0
	Extrema	◆✦	*D. tenuifolia*	Italy	Polytunnel	2nd	31	9.7
	Giove	■⅄	*D. tenuifolia*	Italy	Polytunnel	1st	22	5.7
							31.4	**8.1**
October	Selezione Enza	■⅄	*D. tenuifolia*	Italy	Polytunnel	1st	25	5.8
	Selezione Enza	◆✦	*D. tenuifolia*	Italy	Polytunnel	2nd	33	6.5
	Venere	■⅄	*D. tenuifolia*	Italy	Polytunnel	1st	23	6.3
	Multi	◆⅄	*D. tenuifolia*	Italy	Polytunnel	2nd	30	6.2
	Napoli	◆-	*D. tenuifolia*	UK	Glasshouse	2nd	-	8.1
							27.8	**6.6**
November	Yeti	■✦	*D. tenuifolia*	Italy	Polytunnel	1st	31	6.3
	Yeti	■✦	*D. tenuifolia*	Italy	Polytunnel	1st	32	6.5
	Selezione Enza	◆✦	*D. tenuifolia*	Italy	Polytunnel	2nd	60	6.4
	Tokita	■⅄	*E. sativa*	Italy	Polytunnel	1st	24	10.4
	Tokita	■⅄	*E. sativa*	Italy	Polytunnel	1st	24	7.2
							34.2	**7.4**

■ = 1st cut; ◆ = second cut; ❖ = 2nd+ cut; ⅄ = <30 day crop cycle; ✦ = 31–60 day crop cycle; ★ = 61–90 day crop cycle; ✷ = >91 day crop cycle. Symbol codes denoting cut number and crop cycle length are also utilised in subsequent text tables and figures. Numbers in bold are monthly averages. N.B. Hyphens (-) indicate data was not supplied from the grower.

The demographics and characteristics of each respective panel are presented in Supplementary Table S1. The average number of recruits for bimonthly evaluations was 87, with an average age of 35 years old, ranging from 18 to 70. Volunteers were predominantly female (70.7%, on average), which is partly due to the gender balance present within the School of Chemistry, Food and Pharmacy at the University of Reading. We acknowledge that the sample population of consumers may not be representative of the "typical" UK consumer, however it does incorporate a broad range of culturally and ethnically diverse individuals that encompass a wide diversity of potential sensory genotypes.

On average, 50.9% were employed, and 46.9% were students. Of these, 27.2% were food and nutrition students from within the school. The multicultural nature of the staff and student body produced a diverse cohort, with 48.2% identifying as "white", 27.7% as "other" (i.e., non-white and/or European), and 6.7% of Chinese nationality. The remainder were composed of those regarding themselves as African (4.8%), Caribbean (2.3%), Indian (2.7%), or of mixed race (1%). Of the volunteers 3.3% declined to provide a response.

2.5. Phytochemical Analyses

2.5.1. Preparation of Samples

Upon receipt of samples at the University of Reading, a subset of leaves (50 g) was taken for analysis, frozen at −80 °C, and lyophilized prior to extraction. Tissues were then milled into a fine powder using a Wiley Mini Mill (Thomas Scientific, Swedesboro, NJ, USA).

2.5.2. Glucosinolate Analysis

DMB, GER, GRA, and GSV concentrations were determined by Liquid Chromatography Mass Spectrometry (LC–MS) as per the methodology presented by Bell et al. [20]. Three separate biological replicate extractions were performed on each sample, with three technical replicates analysed by LC–MS ($n = 9$). The data for each sample were then averaged to give a representative concentration of each assessment month. Individual sample averages were retained and used for subsequent PCA. Individual cultivar results for each respective month of the study can be found in Supplementary Data File S1.

2.5.3. Sugar Analysis

Concentrations of fructose, galactose, and glucose were determined by extraction and analysis by capillary electrophoresis (CE) according to the methodology presented by Bell et al. [4]. The same level of replication as for the analysis of GSLs was used for each sample ($n = 9$) and averaged to produce a representative monthly concentration for presentation. As above, individual sample averages were retained and used for subsequent PCA. Individual cultivar results for each respective month of the study can be found in Supplementary Data File S1.

2.6. Statistical Analysis

2.6.1. Panellist Performance

Data were collated and panellist performance evaluated using SenPAQ (v5.01; Qi Statistics, Reading, UK). For each monthly assessment, scores were averaged and used for further statistical analysis. Discrimination, repeatability, and consistency were checked for all assessors.

2.6.2. Analysis of Variance

Shapiro–Wilk normality tests were conducted for all sensory and consumer variables. All of which were concluded to fit with a normal distribution and allow for statistical comparison using a parametric test. Analysis of variance (ANOVA) was performed on each data set (sensory, consumer, and phytochemical) and supplied temperature data. Each test was performed using XLSTAT

(Addinsoft, Paris, France) with a protected post-hoc Tukey's honest significant difference (HSD) test (p-values ≤ 0.05). Only attributes with statistically significant differences were selected for presentation.

2.6.3. Agglomerative Hierarchical Clustering

Agglomerative hierarchical clustering (AHC) was conducted on the consumer liking data using XLSTAT. This approach was used to cluster consumers who had similar liking patterns (for taste and overall liking) for each of the bimonthly panels. Dissimilarity of responses was determined by Euclidean distance, and agglomeration using Ward's method (set to automatic truncation).

2.6.4. Principal Component and Correlation Analysis

Consumer liking and perception response data were used to extract principal components (PCs; with Varimax rotation) and we performed correlation analyses (Pearson, $n - 1$). Phytochemical, temperature, and agronomic data for each sample were regressed as supplementary variables within the PCA model. Variables such as month, cultivar (variety), cut, and country of origin were regressed as qualitative variables to generate categorical centroids within the model. Seven PCs were extracted with the first four components containing a cumulative 98.3% of variability. PCs 1 and 4 had eigenvalues of 4.1 and 0.3, respectively) and were selected for presentation after Varimax rotation. Correlation matrices of all attributes used in the analysis were produced at the 5%, 1%. and 0.1% significance levels, and are summarized in Supplementary Data File S1.

3. Results and Discussion

3.1. Monthly Differences in Rocket Agronomic Practices

The majority of rocket supplied to the UK market is *D. tenuifolia*, with *E. sativa* making up a small amount. The latter is usually supplied in winter months due to its faster establishment, early vigour, and cold tolerance [10]. In this study *E. sativa* was only supplied in November (Table 1).

Cultivation practices varied distinctly between countries, and indeed between individual growers, based on local cultural practices and individual experience. In Italy, produce destined for the UK market is typically cultivated under polytunnel or glass, year-round; whereas UK grown material is either grown in open field or under glass (Table 1). One dominant reason for this difference is that the wetter and more humid climate of the UK can cause severe fungal pathogen outbreaks. The reduced airflow within polytunnels typically exacerbates this problem, and so open field is preferred to minimize losses.

The length of crop cycles depends on the season, though there are large differences between individual growers and countries (Table 1). Cycle length is longer in the winter and spring months, with much faster growth and regrowth in summer and autumn. The shortest average crop cycle in this study was 27 days (August), and the longest 96 (March). The extremes of the overall range (Table 1) can vary from 23 (June) to 180 days (April).

As establishment of rocket crops is more difficult in winter months, Italian growers favour repeated harvests until warmer weather arrives. It is not unusual for >5 cuts to be taken from a single sowing. During the experiment, sourced material came not only from Italy, but the USA and India (Table 1) in order to meet shortfalls in demand. During the summer season, UK rocket enters the market and typically has short growth cycles and receives only one cut. The humid climate does not favour regrowth, as damaged leaves become infected with fungal pathogens and are unsaleable.

The length of crop cycle and cut number have important implications for rocket taste, flavour, and acceptability. It is widely acknowledged that the more harvests a rocket crop undergoes, the more pungent and aromatic it becomes, due to the initiation of wound response and increases in secondary metabolites, such as GSLs [21]. However, no quantitative research has been conducted to evaluate consumer preferences for first, second, or multiple cut leaf material of rocket. As will be discussed in the following sections, cut number is a key determinant of taste and flavour perception, and liking of leaves at different times of the year.

3.2. *Monthly Variation in Rocket Growth Temperature*

Due to the seasonal distribution of rocket production geographically throughout a growing season, crops may be exposed to a range of temperature maxima and minima. Supplementary Figure S1 presents an average of the temperatures recorded at each farm location, giving a representative value of all growing sites for each month.

The highest average temperature across the growing season was 21 °C in August, with the highest average temperature in the week preceding harvest being 21.6 °C. The highest average maximum and minimum temperatures in the week preceding harvest were also in August; 27 °C and 15.6 °C, respectively. Lowest average temperatures were observed in January (avg. temp. 10 °C, max. temp. week 14.8 °C, min. temp. week 4.6 °C, and avg. temp week 9.8 °C). The significant differences observed in monthly temperatures correspond to distinct changes in phytochemical content, sensory perceptions, and consumer acceptance.

3.3. *Phytochemical Composition and Monthly Variability*

3.3.1. Glucosinolates

The monthly average GSL concentrations of rocket leaves are presented in Figure 1a. For individual cultivar concentrations, see Supplementary Data File S1. The data show a very large amount of variability over the course of the year. This lack of consistency likely plays a significant role in the perceived quality changes in rocket produce by processors, supermarkets and consumers.

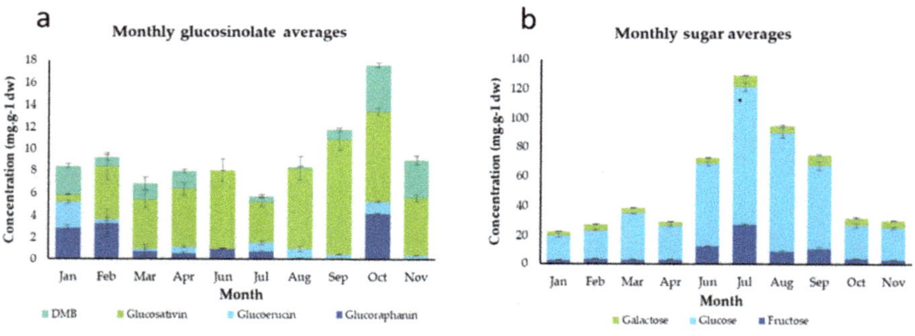

c

Month	Glucoraphanin	Glucoerucin	Glucosativin	DMB	Total GSLs	Fructose	Glucose	Galactose	Total sugars	Sugar:GSL ratio
Jan	b	e	a	e	ab	a	a	a	a	a
Feb	bc	abc	b	bcd	bc	ab	a	ab	a	a
Mar	a	ab	b	d	ab	a	a	ab	a	a
Apr/May	a	bc	bc	d	ab	a	a	a	a	a
Jun	a	a	cd	a	ab	c	b	b	b	a
Jul	a	cd	b	abc	a	d	c	c	c	b
Aug	a	cd	cd	ab	ab	abc	bc	b	bc	a
Sep	a	bc	e	cd	c	bc	b	c	b	a
Oct	c	d	d	g	d	ab	a	b	a	a
Nov	a	abc	bc	f	bc	ab	a	b	a	a
p-value	< 0.0001	< 0.0001	< 0.0001	< 0.0001	< 0.0001	< 0.0001	< 0.0001	< 0.0001	< 0.0001	< 0.0001

Figure 1. Average glucosinolate concentrations (**a**), sugar concentrations (**b**), and analysis of variance (ANOVA) pairwise comparisons (post-hoc Tukey's honest significant difference; (**c**) of rocket leaves observed on a monthly basis. Significant differences of glucosinolate concentrations between each sampling month are indicated by differing lower case letters within each column. Concentrations are expressed as mg·g^{-1} of dry weight. Error bars represent standard error of the mean of each compound. See insets for compound colour coding (**a**,**b**). For individual cultivar composition data of samples received each month see Supplementary Data File S1.

Figure 1c details the significant differences between each growing month for GSL composition. Total concentrations of the four major GSLs of rocket were highest in October (17.6 ± 0.6 mg g^{-1} dw) and lowest in July (5.8 ± 0.7 mg g^{-1} dw). GRA concentrations were significantly higher in October (4.2 ± 0.1 mg g^{-1} dw) and February (3.3 ± 1.3 mg g^{-1} dw) compared with the months from March to September. GER concentrations were significantly higher in January (2.3 ± 0.1 mg g^{-1} dw) than at any other time of the study year. Previous studies in broccoli sprouts [22] have shown that cooler temperatures (<16 °C) increase the concentrations of methylthioalkyl GSLs, such as GRA and GER. This may be due to a abiotic stress response and upregulation of secondary metabolite biosynthesis, causing greater concentrations of these health related GSLs. GER, GRA, and their respective hydrolysis products are not known to have any significant odour or flavour, but the elevations observed in winter months suggests that cultivation in lower-temperature climates may improve rocket nutritional potential. Both sulforaphane (SF) and erucin (ERU; isothiocyanate hydrolysis products of GRA and GER, respectively) are known to be effective against some forms of cancer [23,24].

Concentrations of DMB were also significantly higher in January (2.6 ± 0.2 mg g^{-1} dw), October (4.2 ± 0.2 mg g^{-1} dw), and November (3.4 ± 0.4 mg g^{-1} dw), whereas amounts of the monomer GSV were highest in September (10.4 ± 1.0 mg g^{-1} dw). The relationship between GSV, DMB, and sensory properties is not understood, but previous studies have noted associations between GSV content and pungency (likely due to hydrolysis producing SAT), but not DMB [4]. The significant variations in monomer and dimer forms across the year suggest that there is some as-yet-unknown mechanism by which the two are interconverted [15]; possibly on a genetic and enzymatic level. This process may dictate the levels of pungency found in leaves.

3.3.2. Sugars

The pattern of sugar accumulation in rocket leaves was much more distinct than for GSLs. Total concentrations were significantly higher from June to September (Figure 1b,c) indicating a strong relationship with seasonal climate. This could conceivably be linked to temperature and light intensity duration and quality during summer months. A study on broccoli [25] previously observed that glucose and fructose concentrations were significantly elevated under higher temperature conditions, for example.

Glucose was the dominant monosaccharide in rocket leaves, and concentrations were significantly higher from June to September (peaking in July, 93.9 ± 3.0 mg g^{-1} dw). Fructose accumulations also followed this pattern, with 27.5 ± 0.9 mg g^{-1} dw in July, compared to only 2.5 ± 0.2 mg g^{-1} dw in January. These data are strong evidence for the role of season and climate in the generation of sugars in rocket leaves; and as will be discussed, this has implications for preference and quality of leaves.

3.4. *Sensory Profiling Monthly Variability*

3.4.1. Appearance Traits

Leaf size and uniformity of size were the only two appearance attributes tested that varied significantly between monthly assessments of rocket produce (p = 0.005 and <0.0001, respectively; Supplementary Figure S2 and Table 2). Leaf size was significantly smaller in January compared with April, June, August, and October. Similarly uniformity of size was significantly lower in January than any other month (with the exception of February). Combined with the low average temperatures (Supplementary Figure S1) at this time of year, it is likely that the colder temperatures and reduced light levels (short days) in Italy at this time of year result in slower, and more uneven growth rates [26] compared to other times of the year.

Table 2. Analysis of variance (ANOVA), Tukey's honest significant difference (HSD), and pairwise comparison data for sensory attributes of rocket salad leaves supplied to the United Kingdom over the course of one year.

Attribute	Month										
	January	February	March	April	June	July	August	September	October	November	*p*-Value
Appearance											
Leaf size	a	ab	ab	b	b	ab	b	ab	b	ab	0.005
Uniformity of size	a	ab	b	b	b	b	b	b	b	b	<0.0001
Odour											
Green	b	ab	a	ab	a	a	ab	ab	ab	a	0.013
Stalky	bc	abc	bc	ab	a	ab	ab	c	abc	ab	<0.0001
Earthy	c	ab	ab	abc	a	abc	ab	ab	bc	abc	<0.0001
Peppery	b	a	a	a	a	a	ab	a	a	a	<0.0001
Sweet	a	a	ab	ab	ab	b	b	ab	ab	b	<0.0001
Mustard	a	a	a	a	a	a	a	a	a	a	0.029
Mouthfeel											
Crisp	a	abc	bc	c	c	c	c	c	c	ab	<0.0001
Drying	ab	ab	a	a	a	ab	ab	b	b	a	<0.0001
Taste											
Sour	c	abc	a	bc	ab	abc	c	abc	abc	a	<0.0001
Savoury	d	ab	bcd	abcd	a	abc	cd	bcd	abcd	d	<0.0001
Flavour											
Stalky	ab	a	ab	ab	a	ab	a	b	ab	a	0.001
Peppery	b	ab	a	a	a	a	ab	ab	a	a	0.001
Earthy	c	ab	abc	abc	a	bc	abc	abc	abc	abc	0.005
Aftereffects											
Sweet	ab	a	ab	ab	ab	b	ab	ab	ab	ab	0.024
Sour	b	ab	a	ab	ab	ab	ab	ab	ab	a	0.006
Savoury	cd	ab	cd	ab	a	abc	bcd	bcd	bcd	d	<0.0001
Peppery	b	ab	ab	ab	ab	ab	ab	ab	ab	a	0.014
Green	ab	a	ab	a	a	ab	ab	ab	b	ab	0.008
Earthy	a	a	a	a	a	a	a	a	a	a	0.031

Different letters within each row indicate significant differences between attributes in each sampling month (a = the lowest level of significant difference). Refer to Supplementary Figure S2 for data and standard errors.

3.4.2. Odour Traits

The odour attributes of rocket leaves defined as green, stalky, earthy, peppery, sweet, and mustard were all found to vary significantly between assessment months (Supplementary Figure S2 and Table 2). Green, peppery, and earthy odours were observed to be elevated, on average, in January, whereas stalky and sweet odours were scored higher in July, August, and September. Volatile profiles are known to be influenced by seasonal variations, and storage conditions [27], and so differences between the UK and Italian climates likely play a role in determining the intensity of these odours.

3.4.3. Taste and Flavour Traits

Sour taste, savoury taste, stalky flavour, peppery flavour, and earthy flavour of rocket leaves were found to vary significantly between months (Supplementary Figure S2 and Table 2). Sour and savoury taste scores were highly variable between months, with no distinct pattern emerging according to seasonality. As with aroma attributes, stalky, peppery, and earthy flavours were each scored highest in September and January.

3.4.4. Mouthfeel Traits

Significant variation was observed between monthly assessments of rocket for crisp and drying mouthfeels. Leaves tested in January were significantly less crispy than those received from March to October (Supplementary Figure S2 and Table 2). Soluble sugars are known to help maintain turgidity of leaves [28], and the low concentrations accumulated at this time of year may therefore be related to mouthfeel quality.

Drying sensation was perceived as significantly more in September and October than the months from March to June, and November. Little is known about the cause of drying sensation caused by rocket leaves, but one possible explanation is the presence of polyphenols [29], which have been observed to increase significantly under heat stress conditions [30].

3.4.5. Aftereffect Traits

Aftereffect attributes with significant monthly variation are presented in Supplementary Figure S2 and Table 2. Of note are sweet and peppery aftereffects, which have previously been associated with improved consumer acceptance [9]. Sweet aftereffects were significantly higher in July, corresponding to the peak of glucose and fructose concentrations within leaves.

Peppery aftereffects were significantly higher in January, in agreement with the aroma and flavour scores for this attribute. Some GSL hydrolysis products are known to have different aromas at different concentrations [31], and the low abundances of GSV in January (Figure 1a) would suggest that SAT production may also be reduced, and correspond to reduced pungency and increased pepperiness.

3.5. Correlation Analysis of Sensory Attributes

3.5.1. Growing Temperature

Correlation analyses and significances are presented in Supplementary Data File S1. Average crop cycle temperature, the minimum, and average temperatures in the week preceding harvest were significantly correlated with sweet odour of leaves (all r = >0.462; p = <0.0001). Abiotic stress is known to promote formation of secondary metabolites in many plant species [32] and so higher growth temperatures may promote the synthesis of aldehydes that impart sweet odour, as have been identified in other Brassicales species [33].

All temperature data were also significantly correlated with crisp mouthfeel (r = 0.527; p = <0.0001). Previous research and modelling of rapeseed plants has shown that growth temperature significantly impacts leaf morphology; particularly leaf length and thickness [34]. This may partly explain why rocket leaves are perceived as crispier in summer and autumn months compared with winter (Supplementary Figure S2).

3.5.2. Cultivation Practice

One of the largest differences observed between months was the length of the crop cycle (Table 1). Correlation analysis (Supplementary Data File S1) found that the length of the crop cycle was significantly associated with key sensory traits potentially linked with consumer acceptance. These were: bitter taste ($r = 0.3$; $p = 0.043$) and bitter aftereffects ($r = 0.325$; $p = 0.027$). Sweet aftereffects were also significantly and negatively correlated with the length of crop cycle ($r = -0.316$; $p = 0.032$). The low sugar:GSL ratio in samples with longer crop cycles might explain some of these correlations. With lower sugar concentrations in the winter/early spring months (Figure 1b), GSLs and their hydrolysis products may be perceived more strongly with the masking effect of sugars reduced.

3.5.3. Glucosinolates

Individual GSL concentrations are known to be associated with sensory attributes of rocket species [3]. GRA is a compound not known to impart any taste or flavour [31], but correlation analysis revealed significant negative associations with sweet odour ($r = -0.543$; $p = <0.0001$), taste ($r = -0.402$; $p = 0.005$), and aftereffects ($r = -0.304$; $p = 0.035$). The abundance of GRA was negatively correlated with the average growth temperature ($r = -0.344$; $p = 0.017$) and max. temperature in the week preceding harvest ($r = -0.306$; $p = 0.035$). These two points indicate that GRA biosynthesis is lower in samples grown in months with higher temperatures, which also corresponds to increased sugar concentrations (Figure 1).

GER is similar to GRA in the respect that it is not known to impart taste [31], however its hydrolysis product erucin (ERU) has been described as having a "radish-like" aroma [16]. In this study, several previously unobserved associations were found. GER itself is negatively correlated with pungent odour ($r = -0.299$; $p = 0.039$), but positively with green and peppery odours ($r = 0.459$; $p = 0.001$, and $r = 0.364$; $p = 0.011$, respectively) and flavours ($r = 0.367$; $p = 0.01$, and $r = 0.337$; $p = 0.019$, respectively). While these data are not conclusive of a causative relationship with these attributes, it does suggest that occurrence of GER in high concentrations may elicit, or be associated with, perceptions of pepperiness and green attributes, and is worth studying in greater detail in future studies.

GSV exists in a monomer and dimer form (DMB), and typically makes up the largest proportion of the GSL profile of rocket [20]. A previous study found that its hydrolysis product SAT has a "rocket-like" aroma [16]. While this may be considered a somewhat subjective description, it is speculated that SAT is responsible for the perceived pungency of rocket leaves. The data in this study agreed with this hypothesis, as GSV concentrations were significantly correlated with pungent aroma ($r = 0.393$; $p = 0.006$). It was however also negatively correlated with peppery odour ($r = -0.295$; $p = 0.042$), suggesting that the two attributes are separate, with only GSV being indirectly responsible for pungency.

Correlations of DMB with sensory attributes were distinct and separate from the monomer, suggesting that concentrations of the two forms are influenced by the environment and as-yet-unknown genetic regulation, possibly in response to abiotic stress. It is unknown if DMB itself imparts taste or flavour, but its abundance was positively correlated with savoury taste ($r = 0.323$; $p = 0.025$) and aftereffects ($r = 0.391$; $p = 0.006$). This is in agreement with previous sensory and consumer studies of rocket [4,9]. It was also observed that GSV was significantly correlated with each of the four temperature measurements used in the analysis (Supplementary Data File S1; Supplementary Figure S1) whereas DMB was negatively correlated with the max. temperature in the week preceding harvest ($r = -0.311$; $p = 0.031$). This suggests that the relative abundances of the monomer and dimer forms of GSV had an environmental component, with greater concentrations of GSV present in hotter months.

3.5.4. Sugars

Total sugars, fructose, and glucose concentrations were significantly correlated with dry matter percentage (r = 0.494; p = 0.001, r = 0.622; p = <0.0001, and r = 0.439; p = 0.003, respectively). This suggests that this physical property of leaves may be indicative of a dry matter concentration effect. This is reflected in several negative correlations with moistness mouthfeel (r = −0.385; p = 0.007, r = −0.515; p = 0.000, and r = −0.332; p = 0.021, respectively). Only galactose concentrations were significantly correlated with sweet taste (r = 0.39; p = 0.006), and fructose and galactose with sweet aftereffects (r = 0.303; p = 0.036, and r = 0.308; p = 0.033, respectively). Despite the significantly higher sugar concentrations in summer months (Figure 1b,c) there were no significant correlations with growth temperature.

The sugar:GSL ratio was also similarly correlated with the aforementioned mouthfeel effects (Supplementary Data File S1) and dry matter content (r = 0.571; p = <0.0001); but only sweet aftereffects (r = 0.393; p = 0.006) and not sweet taste. This association is not as strong as found in previous studies of rocket [4].

3.6. Consumer Acceptability and Perception

3.6.1. Liking of Taste

Consumer liking of taste and the results of AHC are presented in Supplementary Table S2. Liking of taste is defined as liking associated with taste and flavour attributes alone (bitterness, sweetness, pepperiness, and hotness), irrespective of appearance traits. Three clusters were identified in each respective month, except for March, where four clusters were observed. The largest clusters in each month consistently scored cultivars higher for taste liking than the overall cohort and monthly averages. This indicates that for most consumers, the taste of rocket is acceptable year-round, with average scores consistently >6.0.

March and April/May had significantly lower taste liking scores than any of the other months. Highest average taste liking was in January, which is contrary to our hypothesis that rocket liking would be greater during summer months. Average scores for July, September, and November were also relatively high (6.0, 6.2, and 6.2, respectively), indicating that in terms of consumer taste liking, spring months show a distinct reduction in acceptability.

Crop cycles of rocket in spring are also typically very long (96.3 days, average) with successive cuts (>2), potentially producing very pungent and bitter leaves. Figure 2 presents consumer perception data of bitterness, hotness, and pepperiness. All these attributes were scored highest in March, with bitterness being a dominant attribute until July. Sweetness perception by comparison remained relatively unchanged, peaking in July. The reason for increased taste liking in January may therefore be explained by the significantly lower perception of hotness of leaves relative to spring and summer months.

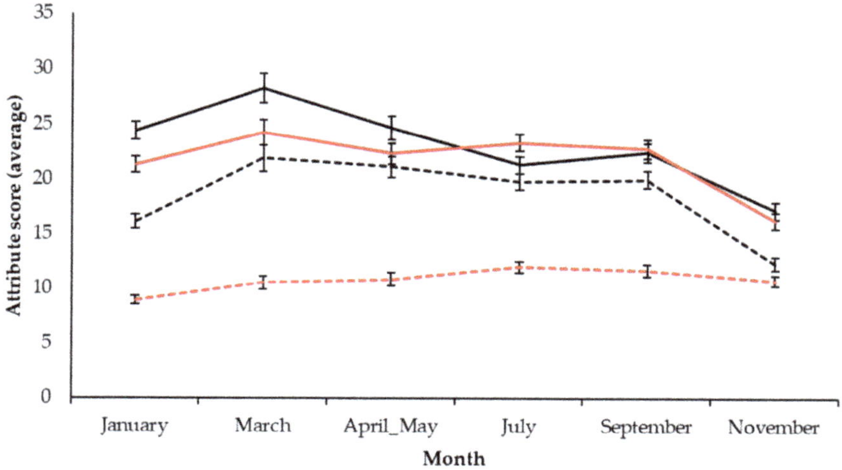

		January	March	April_May	July	September	November	Pr > F(Model)
Bitter perception	——	bc	c	bc	b	b	a	< 0.0001
Hotness perception	----	b	c	c	c	c	a	< 0.0001
Sweetness perception	----	a	ab	ab	b	b	ab	0.000
Pepper perception	——	b	b	b	b	b	a	< 0.0001

Figure 2. Consumer perceptions of bitter, hotness, sweetness, and pepper attributes of rocket leaves on a bimonthly basis over a growing season. Inset table presents the results of analysis of variance (ANOVA) pairwise comparisons (post-hoc Tukey's honest significant difference). Significant differences for each perception attribute are indicated by differing lower case letters within rows. See inset for colour coding of attributes. Values are presented as normalized averages of each respective consumer assessment. See Supplementary Table S1 for the numbers of participants in each respective consumer panel.

3.6.2. Overall Liking

Table 3 presents AHC data and average monthly scores for overall rocket leaf liking. Overall liking encompasses liking of both taste and appearance attributes. Analysis identified three groups for each respective month, except for January, where five clusters were observed. Appearance of leafy salads is known to be a significant factor in consumer purchase intent and liking [2] and the sensory panels determined significantly smaller leaf size and uniformity of shape in January (Supplementary Figure S2). This disparity between cultivars seems to be compensated for by higher taste liking (Supplementary Table S2), suggesting that appearance liking may be secondary to taste liking for some consumers; for example in cluster 4 January (n = 43) where scores were all consistently higher than the total cohort average.

The consistency of cultivars was extremely variable during the March and April/May panels. "Fast Grow" (■-; 6.8) was scored significantly higher on average than all the other samples in March and preferred by all three cluster groups. A similar pattern of inconsistency was observed in the autumn months of September and November. This further suggests that seasonal transitions result in more variable rocket produce.

Table 3. Overall liking of rocket cultivars for the clusters of consumers obtained from agglomerative hierarchical clustering.

Month	Cluster	Cultivar					Average
		Tricia ◆⅄	**Yeti ◆✦**	**Wildfire ■✦**	**Fast Grow ❖★**	**Yeti ◆★**	
January	1 (n = 15)	3.9	5.8	4.1	4.0	5.3	4.6
	2 (n = 6)	7.5	6.8	2.8	7.8	7.2	6.4
	3 (n = 18)	6.2	3.6	6.9	6.6	6.5	6.0
	4 (n = 43)	7.6	7.0	7.3	6.7	7.0	7.1
	5 (n = 19)	4.8	7.2	6.2	6.5	6.9	6.3
	All	6.3 ns	6.7 ns	6.3 ns	6.3 ns	6.3 ns	6.4 D
March		Yeti ❖✶	Wildfire ■✦	Fast Grow ■-	Fast Grow ❖✶	Voyager ❖★	Average
	1 (n = 29)	6.8	6.3	7.0	6.3	5.3	6.3
	2 (n = 20)	2.8	4.6	7.3	5.7	5.1	5.1
	3 (n = 6)	4.0	2.2	4.2	2.7	3.8	3.4
	All	5.0 a	5.2 a	6.8 b	5.7 a	5.1 a	5.6 A
April/May		Venere ◆✦	Wildfire ❖✶	Standard ❖✦	Giove ❖-	Quaggio ◆✦	Average
	1 (n = 10)	5.3	5.1	6.0	4.6	2.0	4.6
	2 (n = 39)	6.7	4.2	6.1	3.6	6.8	5.5
	3 (n = 41)	5.9	6.1	5.7	7.0	7.1	6.4
	All	6.2 b	5.2 a	5.9 ab	5.3 a	6.4 b	5.8 AB
July		Giove ■⅄	Giove ■⅄	Giove ■⅄	Napoli ■⅄	Voyager ■✦	Average
	1 (n = 43)	6.2	6.2	3.9	5.4	5.2	5.4
	2 (n = 12)	3.5	3.8	6.3	6.0	5.3	5.0
	3 (n = 45)	6.2	6.6	7.2	6.8	7.4	6.8
	All	5.9 ns	6.1 ns	5.7 ns	6.1 ns	6.2 ns	6.0 BC
September		Voyager ■✦	Shamrock ■✦	Extrema ◆✦	Extrema ◆✦	Giove ■⅄	Average
	1 (n = 65)	6.8	7.0	6.9	7.2	6.1	6.8
	2 (n = 9)	4.6	4.8	4.0	6.6	2.6	4.5
	3 (n = 15)	5.3	5.9	4.6	3.3	6.6	5.1
	All	6.3 ab	6.6 b	6.2 ab	6.4 ab	5.8 a	6.3 CD
November		Yeti ■✦	Yeti ■✦	Selezione Enza ◆✦	Tokita ■⅄	Tokita ■⅄	Average
	1 (n = 40)	7.2	6.8	7.3	6.6	7.4	7.1
	2 (n = 19)	5.2	5.1	6.3	4.3	4.1	5.0
	3 (n = 27)	7.0	6.4	3.7	6.0	6.0	5.8
	All	6.7 b	6.3 ab	5.9 a	5.9 a	6.2 ab	6.2 CD

Letters indicate ANOVA pairwise comparison significances (Tukey's HSD). Lower case letters within the "All" rows of each month refer to individual cultivar scores across each consumer panel month. Upper case letters in the "Average" column refer to significant differences in monthly average scores. Where letters are different a significant difference at the p = ≤0.0001 level was observed. ■ = 1st cut; ◆ = second cut; ❖ = 2nd + cut; ⅄ = <30 day crop cycle; ✦ = 31–60 day crop cycle; ★ = 61–90 day crop cycle; ✶ = >91 day crop cycle; ns = no significant difference. N.B. Hyphens (-) indicate data was not supplied from the grower.

3.7. Principal Component Analysis

3.7.1. Relationships between Consumer Liking and Perceptions

Despite sensory panels not detecting significant differences in sweet or bitter tastes, consumers were able to do so, and this significantly affected their liking for rocket throughout the growing season. This is likely due to the increased diversity of taste receptor profiles present within the population compared with the sensory panel [35].

PCA of the data sets from each consumer panel month revealed a distinct separation between sweetness perception and hotness, bitterness, and pepperiness perceptions along PC1 (Figure 3a). Taste and overall liking are in turn more positively associated with sweetness perception in the upper left quadrant along the PC4 axis, which is in agreement with previous observations in rocket [9]. These data are therefore strong evidence that most consumers are likely to reject rocket if it is too pungent and bitter, as is found in samples received in March and April/May.

3.7.2. Influence of Growing Temperatures on Consumer Preference and Perceptions

Despite the higher sugar concentrations in July, this does not colocalize with overall/taste liking within the PCA (Figure 3b). The months of March and April/May are in fact most negatively associated with consumer liking, and January and November positively associated with these. Therefore, rocket produced at cooler temperatures is more likely to be preferred by consumers, as bitterness and hotness perceptions are likely to be lower in these months (Figure 3a).

3.7.3. Influence of Cultivation Practice on Consumer Liking and Perceptions

Figure 3b shows that cut number also explains some separation for taste and overall liking. Rocket leaves that were of first cut are generally more common in the upper left quadrant of the PCA plot, and more closely associated with taste liking and sweetness perception (Figure 3). Leaves with more than two cuts separate in the opposite direction towards the lower right quadrant, in the direction of bitter/hotness/pepperiness perception. Anecdotal evidence of traditional cultivation practices by growers has suggested that second cuts (and above) are preferred, because leaves are more uniform, more greatly serrated in shape, and have a more intense flavour. These assertions are in agreement with this study; however none are associated with positive consumer liking or taste liking of rocket leaves. While a subset of consumers may prefer cultivars with increased hotness (as seen in AHC analysis; Supplementary Table S2 and Table 3) consumers generally do not like this attribute, and prefer milder, sweeter leaves. Thus, conventional agronomic practices of harvesting multiple cuts of rocket may be detrimental to consumer acceptance; particularly in spring months when crop cycles are longer, and the growing season is transitioning.

There is also a significant gap in research more generally about the response of rocket species and cultivars to differences in growth environments and cultivation practices. In this study cultivars were supplied from various growers, and this variable was not controlled so as to assess "real-world" differences in rocket consistency as supplied to consumers. Future studies should aim to assess the variability of multiple cultivars across growing regions, and sample multiple cuts. Such studies are logistically difficult to organise, however it would provide valuable information on how environment influences GSL and hydrolysis product formation on a genotypic basis and how performance of cultivars varies according to the environment. Controlled environment studies have begun to explore these effects [17], but none have to date accounted for variances in soil composition or climatic conditions in the field.

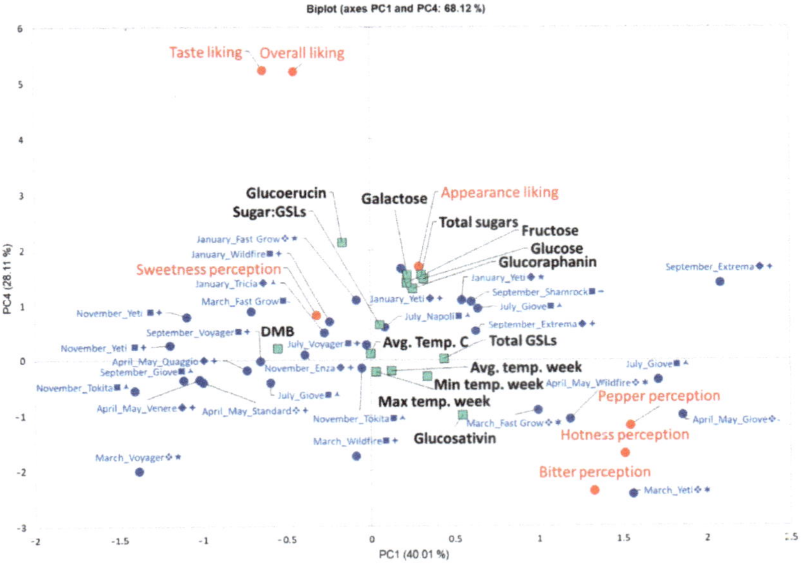

(**a**)

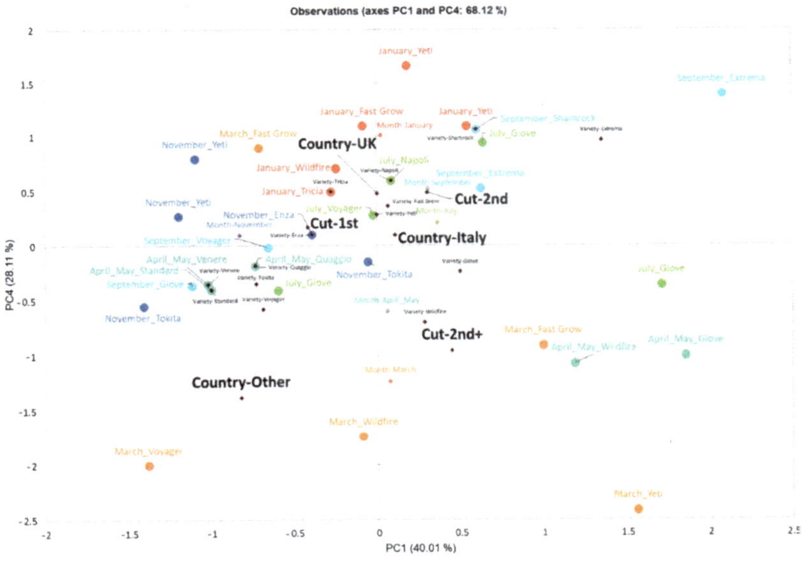

(**b**)

Figure 3. Principal component analysis of consumer liking and perception data of rocket cultivars over the course of one growing season. Biplots display principal components (PCs) 1 and 4, which represent 68.1% of variation within the data; (**a**) factor loadings plot and (**b**) factor scores plot. In (**a**) red circles represent consumer liking and perception attributes, blue circles average monthly samples, and teal squares supplementary phytochemical, and cultivation temperature. Variable label symbols (refer to Table 1): ■ = 1st cut; ◆ = second cut; ❖ = 2nd + cut; ⅄ = <30 day crop cycle; ✦ = 31–60 day crop cycle; ★ = 61–90 day crop cycle; ✱ = >91 day crop cycle. In (**b**), see inset for monthly colour coding of samples. Black and red diamonds indicate supplementary variable centroids.

3.7.4. Influence of Glucosinolate Contents on Consumer Liking and Perceptions

Higher concentrations of GER are associated with higher taste and overall liking of rocket leaves (Figure 3a). This trend is in opposition to the abundance of GSV, which is in the bottom right quadrant with bitter/hotness/pepper perceptions. DMB is also separated from GSV and negatively associated with hotness, bitterness, and pepperiness perceptions. This is new evidence that suggests the ratio between monomer and dimer forms of GSV may play a significant role in determining consumer acceptability of rocket leaves. Nothing is known of the genetic mechanisms responsible for the biosynthesis of GSV and DMB, or the mechanisms responsible for determining their relative abundances; but it is generally accepted that SAT (derived from GSV and responsible for pungency) is produced from the monomer form [15]. This may therefore explain the association between GSV and perception traits and indicates that DMB has no objectionable taste of its own.

4. Conclusions

This study has found evidence for significant sensorial variability in rocket leaves produced over the course of a growing season as a result of varied cultivation practices and growing locations. This in turn results in variations in consumer liking, which may influence purchase intents and repurchase of rocket leaf products. Seasonal practices, such as growth temperature and the number of cuts crops received, underlie changes in phytochemical composition and may result in the production of overly pungent leaves that consumers are likely to reject. To produce more consistent and acceptable rocket leaves, the practice of multiple harvests should be reserved for developing products targeted at those consumers who like high pungency leaves. First cuts tend to be milder and could therefore be marketed to a wider set of consumers that prefer sweeter leaves and low levels of pungency and bitterness.

Supplementary Materials: The following are available online at http://www.mdpi.com/2304-8158/9/12/1799/s1. Supplementary data file S1: Pearson $n - 1$ correlation tests at three levels of significance ($p = \leq 0.05$, ≤ 0.01, and ≤ 0.001) and month-by-month individual cultivar glucosinolate and sugar concentrations of rocket samples used in sensory and consumer analyses. Figure S1: Monthly average temperature data of rocket crop cultivation areas over the course of one year, and analysis of variance (ANOVA) pairwise comparisons (post-hoc Tukey's honest significant difference). Averages represent temperatures at multiple growing locations in a given month. Significant differences are indicated by differing lower case letters. Max. temp. week, min. temp. week, and avg. temp. week refer to average temperature values in the week preceding harvest of each rocket crop. Figure S2: Sensory analysis results for appearance (a), odour (b), taste (c), flavour (d), mouthfeel (e), and aftereffect (f) traits of rocket leaves analysed monthly over the course of a growing season. See insets for individual line chart colour coding of attributes. Values are presented as averages of each respective monthly sensory assessment (n = 12 panellists). Table S1: Consumer demographics and characteristics for each of the rocket study panel months. Table S2: Taste liking of rocket cultivars for the clusters of consumers obtained from agglomerative hierarchical clustering.

Author Contributions: Conceptualization, C.W.; methodology, L.B. and S.L.; formal analysis, L.B. & S.L.; investigation, S.L.; resources, C.W.; data curation, L.B.; writing—original draft preparation, L.B.; writing—review and editing, S.L. and C.W.; visualization, L.B.; supervision, C.W.; project administration, C.W.; funding acquisition, C.W. All authors have read and agreed to the published version of the manuscript.

Funding: This research was funded by Bakkavor Group Ltd. (Spalding, UK).

Acknowledgments: The authors would like to thank Lorraine Shaw and Chris Jeffes of Bakkavor Ltd. (Spalding, UK) for arranging supply and delivery of rocket leaves. Special thanks to Suzanne Bird for performing glucosinolate and sugar analyses, and help running sensory and consumer panels.

Conflicts of Interest: This work was conducted with funds provided by Bakkavor Ltd. (Spalding, UK). The authors declare that the research was conducted in the absence of any financial relationships that could be construed as a potential conflict of interest. The funders had no role in the design of the study; in the collection, analyses, or interpretation of data; or in the writing of the manuscript.

Abbreviations

GSL	glucosinolate
ITC	isothiocyanate
GSV	glucosativin
DMB	dimeric 4-mercaptobutyl glucosinolate
SAT	sativin
GRA	glucoraphanin
GER	glucoerucin
gLMS	general Labelled Magnitude Scale
LC–MS	liquid chromatography mass spectrometry
CE	capillary electrophoresis
PCA	Principal Component Analysis
ANOVA	analysis of variance
AHC	agglomerative hierarchical clustering
SF	sulforaphane
ERU	erucin

References

1. Hall, M.K.D.; Jobling, J.J.; Rogers, G.S. Factors affecting growth of perennial wall rocket and annual garden rocket. *Int. J. Veg. Sci.* **2012**, *18*, 393–411. [CrossRef]
2. Spence, C. Gastrophysics: Nudging consumers toward eating more leafy (salad) greens. *Food Qual. Prefer.* **2020**, *80*, 103800. [CrossRef]
3. D'Antuono, L.F.; Elementi, S.; Neri, R. Exploring new potential health-promoting vegetables: Glucosinolates and sensory attributes of rocket salads and related *Diplotaxis* and *Eruca* species. *J. Sci. Food Agric.* **2009**, *89*, 713–722. [CrossRef]
4. Bell, L.; Methven, L.; Signore, A.; Oruna-Concha, M.J.; Wagstaff, C. Analysis of seven salad rocket (*Eruca sativa*) Accessions: The relationships between sensory attributes and volatile and non-volatile compounds. *Food Chem.* **2017**, *218*, 181–191. [CrossRef] [PubMed]
5. Løkke, M.M.; Seefeldt, H.F.; Skov, T.; Edelenbos, M. Color and textural quality of packaged wild rocket measured by multispectral imaging. *Postharvest Biol. Technol.* **2013**, *75*, 86–95. [CrossRef]
6. Lokke, M.M.; Seefeldt, H.F.; Edelenbos, M. Freshness and sensory quality of packaged wild rocket. *Postharvest Biol. Technol.* **2012**, *73*, 99–106. [CrossRef]
7. Martínez-Sánchez, A.; Allende, A.; Bennett, R.N.; Ferreres, F.; Gil, M.I. Microbial, nutritional and sensory quality of rocket leaves as affected by different sanitizers. *Postharvest Biol. Technol.* **2006**, *42*, 86–97. [CrossRef]
8. Pasini, F.; Verardo, V.; Cerretani, L.; Caboni, M.F.; D'Antuono, L.F. Rocket salad (*Diplotaxis* and *Eruca* spp.) sensory analysis and relation with glucosinolate and phenolic content. *J. Sci. Food Agric.* **2011**, *91*, 2858–2864. [CrossRef]
9. Bell, L.; Methven, L.; Wagstaff, C. The influence of phytochemical composition and resulting sensory attributes on preference for salad rocket (*Eruca sativa*) accessions by consumers of varying TAS2R38 diplotype. *Food Chem.* **2017**, *222*, 6–17. [CrossRef]
10. Hall, M.; Jobling, J.J.; Rogers, G.S. Fundamental differences between perennial wall rocket and annual garden rocket influence the commercial year-round supply of these crops. *J. Agric. Sci.* **2015**, *7*, 1–7. [CrossRef]
11. Gil, M.I. Preharvest factors and fresh-cut quality of leafy vegetables. *Acta Hortic.* **2016**, 57–64. [CrossRef]
12. Hall, M.K.D.; Jobling, J.J.; Rogers, G.S. Influence of storage temperature on the seasonal shelf life of perennial wall rocket and annual garden rocket. *Int. J. Veg. Sci.* **2013**, *19*, 83–95. [CrossRef]
13. Coogan, R.C.; Wills, R.B.H.; Nguyen, V.Q. Pungency levels of white radish (*Raphanus sativus* L.) grown in different seasons in Australia. *Food Chem.* **2001**, *72*, 1–3. [CrossRef]
14. Johansen, T.J.; Hagen, S.F.; Bengtsson, G.B.; Mølmann, J.A.B. Growth temperature affects sensory quality and contents of glucosinolates, vitamin C and sugars in swede roots (*Brassica napus* L. ssp. *rapifera* Metzg.). *Food Chem.* **2016**, *196*, 228–235. [CrossRef]

15. Fechner, J.; Kaufmann, M.; Herz, C.; Eisenschmidt, D.; Lamy, E.; Kroh, L.W.; Hanschen, F.S. The major glucosinolate hydrolysis product in rocket (*Eruca sativa* L.), sativin, is 1,3-thiazepane-2-thione: Elucidation of structure, bioactivity, and stability compared to other rocket isothiocyanates. *Food Chem.* **2018**, *261*, 57–65. [CrossRef]
16. Raffo, A.; Masci, M.; Moneta, E.; Nicoli, S.; del Pulgar, J.S.; Paoletti, F.; Pulgar, D. Characterization of volatiles and identification of odor-active compounds of rocket leaves. *Food Chem.* **2018**, *240*, 1161–1170. [CrossRef]
17. Jasper, J.; Wagstaff, C.; Bell, L. Growth temperature influences postharvest glucosinolate concentrations and hydrolysis product formation in first and second cuts of rocket salad. *Postharvest Biol. Technol.* **2020**, *163*, 111157. [CrossRef]
18. Metallo, R.M.; Kopsell, D.A.; Sams, C.E.; Bumgarner, N.R. Influence of blue/red vs. white LED light treatments on biomass, shoot morphology, and quality parameters of hydroponically grown kale. *Sci. Hortic. (Amsterdam)* **2018**, *235*, 189–197. [CrossRef]
19. Bartoshuk, L.M.; Duffy, V.B.; Fast, K.; Green, B.G.; Prutkin, J.; Snyder, D.J. Labeled scales (e.g., category, Likert, VAS) and invalid across-group comparisons: What we have learned from genetic variation in taste. *Food Qual. Prefer.* **2003**, *14*, 125–138. [CrossRef]
20. Bell, L.; Oruna-Concha, M.J.; Wagstaff, C. Identification and quantification of glucosinolate and flavonol compounds in rocket salad (*Eruca sativa, Eruca vesicaria* and *Diplotaxis tenuifolia*) by LC-MS: Highlighting the potential for improving nutritional value of rocket crops. *Food Chem.* **2015**, *172*, 852–861. [CrossRef]
21. Torres-Contreras, A.M.; Senés-Guerrero, C.; Pacheco, A.; González-Agüero, M.; Ramos-Parra, P.A.; Cisneros-Zevallos, L.; Jacobo-Velázquez, D.A. Genes differentially expressed in broccoli as an early and late response to wounding stress. *Postharvest Biol. Technol.* **2018**, *145*, 172–182. [CrossRef]
22. Pereira, F.M.V.; Rosa, E.; Fahey, J.W.; Stephenson, K.K.; Carvalho, R.; Aires, A. Influence of temperature and ontogeny on the levels of glucosinolates in broccoli (*Brassica oleracea* var. *italica*) sprouts and their effect on the induction of mammalian phase 2 enzymes. *J. Agric. Food Chem.* **2002**, *50*, 6239–6244. [CrossRef] [PubMed]
23. Houghton, C.A.; Fassett, R.G.; Coombes, J.S. Sulforaphane: Translational research from laboratory bench to clinic. *Nutr. Rev.* **2013**, *71*, 709–726. [CrossRef] [PubMed]
24. Cartea, M.E.; Velasco, P. Glucosinolates in Brassica foods: Bioavailability in food and significance for human health. *Phytochem. Rev.* **2008**, *7*, 213–229. [CrossRef]
25. Steindal, A.L.H.; Mølmann, J.; Bengtsson, G.B.; Johansen, T.J. Influence of day length and temperature on the content of health-related compounds in Broccoli (*Brassica oleracea* L. var. *italica*). *J. Agric. Food Chem.* **2013**, *61*, 10779–10786. [CrossRef] [PubMed]
26. Balasubramanian, S.; Schwartz, C.; Singh, A.; Warthmann, N.; Kim, M.C.; Maloof, J.N.; Loudet, O.; Trainer, G.T.; Dabi, T.; Borevitz, J.O.; et al. QTL mapping in new *Arabidopsis thaliana* Advanced Intercross-recombinant inbred lines. *PLoS ONE* **2009**, *4*, e4318. [CrossRef]
27. Lytou, A.E.; Panagou, E.Z.; Nychas, G.-J.E. Volatilomics for food quality and authentication. *Curr. Opin. Food Sci.* **2019**, *28*, 88–95. [CrossRef]
28. Sami, F.; Yusuf, M.; Faizan, M.; Faraz, A.; Hayat, S. Role of sugars under abiotic stress. *Plant Physiol. Biochem.* **2016**, *109*, 54–61. [CrossRef]
29. Drewnowski, A.; Gomez-Carneros, C. Bitter taste, phytonutrients, and the consumer: A review. *Am. J. Clin. Nutr.* **2000**, *72*, 1424–1435. [CrossRef]
30. Ancillotti, C.; Bogani, P.; Biricolti, S.; Calistri, E.; Checchini, L.; Ciofi, L.; Gonnelli, C.; Bubba, M. Changes in polyphenol and sugar concentrations in wild type and genetically modified *Nicotiana langsdorffii* Weinmann in response to water and heat stress. *Plant Physiol. Biochem.* **2015**, *97*, 52–61. [CrossRef]
31. Bell, L.; Oloyede, O.O.; Lignou, S.; Wagstaff, C.; Methven, L. Taste and flavour perceptions of glucosinolates, isothiocyanates, and related compounds. *Mol. Nutr. Food Res.* **2018**, *62*, 1700990. [CrossRef] [PubMed]
32. Toscano, S.; Trivellini, A.; Cocetta, G.; Bulgari, R.; Francini, A.; Romano, D.; Ferrante, A. Effect of pre-harvest abiotic stresses on the accumulation of bioactive compounds in horticultural produce. *Front. Plant Sci.* **2019**, *10*, 1212. [CrossRef] [PubMed]
33. Kroener, E.-M.; Buettner, A. Unravelling important odorants in horseradish (*Armoracia rusticana*). *Food Chem.* **2017**, *232*, 455–465. [CrossRef] [PubMed]
34. Tian, T.; Wu, L.; Henke, M.; Ali, B.; Zhou, W.; Buck-Sorlin, G. Modeling allometric relationships in leaves of young rapeseed (*Brassica napus* L.) grown at different temperature treatments. *Front. Plant Sci.* **2017**, *8*, 313. [CrossRef]

35. Reed, D.R.; Tanaka, T.; McDaniel, A.H. Diverse tastes: Genetics of sweet and bitter perception. *Physiol. Behav.* **2006**, *88*, 215–226. [CrossRef]

Publisher's Note: MDPI stays neutral with regard to jurisdictional claims in published maps and institutional affiliations.

foods

MDPI

Article

The Relationship between Glucosinolates and the Sensory Characteristics of Steamed-Pureed Turnip (*Brassica Rapa* subsp. *Rapa* L.)

Nurfarhana Diana Mohd Nor [1,2], Stella Lignou [2], Luke Bell [3], Carmel Houston-Price [4], Kate Harvey [4] and Lisa Methven [2,*]

1 Department of Early Childhood Education, Faculty of Human Development, Sultan Idris Education University, Tanjong Malim 35900, Perak, Malaysia; farhanadiana@fpm.upsi.edu.my
2 Sensory Science Centre, Department of Food and Nutritional Sciences, University of Reading, Whiteknights, Reading RG6 6DZ, UK; s.lignou@reading.ac.uk
3 School of Agriculture, Policy & Development, University of Reading, Whiteknights, Reading RG6 6EU, UK; luke.bell@reading.ac.uk
4 School of Psychology and Clinical Language Sciences, University of Reading, Early Gate, Whiteknights, Reading RG6 6AL, UK; c.houston-price@reading.ac.uk (C.H.-P.); k.n.harvey@reading.ac.uk (K.H.)
* Correspondence: l.methven@reading.ac.uk; Tel.: +44-0118-378-8714

Received: 14 October 2020; Accepted: 19 November 2020; Published: 23 November 2020

Abstract: Glucosinolates (GSLs) are phytochemical compounds that can be found in *Brassica* vegetables. Seven separate batches of steamed-pureed turnip were assessed for GSL content using liquid chromatography mass spectrometry (LC-MS) and for sensory attributes by sensory profiling (carried out by a trained sensory panel). Twelve individual GSLs, which included 7 aliphatic, 4 indole and 1 arylaliphatic GSL, were identified across all batches. There were significant differences in individual GSL content between batches, with gluconasturtiin as the most abundant GSL. The total GSL content ranged from 16.07 to 44.74 μmol g^{-1} dry weight (DW). Sensory profiling concluded there were positive correlations between GSLs and bitter taste and negative correlations between GSLs (except glucobrassicanapin) and sweet taste. The batches, which had been purchased across different seasons, all led to cooked turnip that contained substantial levels of GSLs which were subsequently all rated as bitter.

Keywords: glucosinolates; turnip; *Brassica*; bitter taste; *Brassicaceae*; vegetable

1. Introduction

Brassica vegetables such as turnip, cabbage, broccoli and cauliflower are rich with sulphur-containing glucosinolate compounds (GSLs) [1]. These compounds are water-soluble and have a role in plant defence against pests and diseases [2]. GSLs can be structurally classified into aliphatic, arylaliphatic and indole types [1]. Kim and Park [3] discussed that the degradation products of GSLs possess anticarcinogenic properties, reducing risks of certain cancers in humans. Glucoraphanin, glucobrassicin and gluconasturtiin are among the GSLs that have hydrolysis products shown to have anti-cancer properties, and these are all found in turnip [4].

GSLs are, amongst other compounds, partly responsible for the taste characteristics of *Brassica* vegetables. Individual GSLs such as sinigrin, gluconapin, progoitrin and neoglucobrassicin have been associated with bitter taste [5,6]. Furthermore, Bell et al. [7] reported that GSLs were also correlated with earthy, pepper, mustard flavour and pungency in rocket varieties (*Eruca sativa* Mill.).

GSL contents in *Brassica* vegetables are influenced by many factors, such as environmental conditions and genetic variability between cultivars. The abundance of GSLs in plants is varied, depending on the type of plant species, developmental stage and plant part (root, shoot, seeds and leaves) [8–10]. Concerning cultivars, Kabouw et al. [10] showed that there was a significant difference in GSL content between white cabbage cultivars (*Brassica oleracea* var. *capitata* L.), and Zhu et al. [11] reported significant differences in GSL content between pak choi cultivars. In addition, nutrient supply contributes to the concentration of GSLs in plants. GSL content increases with an adequate supply of sulphur [12], however nitrogen in the absence of sulphur and also selenium supply have been shown to result in a decrease of GSL content [13,14], whereas nitrogen with a sufficient sulphur supply may either increase GSL content or have no effect [14]. Such variations in GSLs can lead to distinctive sensory characteristics of *Brassica* vegetables [15], which are thought to influence their consumption frequency [16].

GSL content in *Brassica* vegetables is also affected when they are handled and prepared before consumption. GSLs undergo hydrolysis to produce breakdown products when the plant cells are wounded [17]. Preparation processes, including cooking and cutting, trigger myrosinase enzymes in plant cells to hydrolyse GSLs and produce isothiocyanates (ITCs) plus other breakdown products; including nitriles, thiocyanates, epithionitriles, oxazolidine-2-thiones and epithioalkanes [15]. A review by Nugrahedi [18] concluded that boiling and blanching significantly reduced GSL content in *Brassica* vegetables due to leaching of compounds. On the other hand, steaming, microwaving and stir-frying may limit GSL loss compared to boiling.

Turnips (*Brassica rapa* subsp. *Rapa* L.) are a traditional vegetable grown in the UK that are no longer frequently consumed by UK consumers in comparison to other *Brassica* vegetables, such as broccoli, cauliflower and cabbage. In 1992, turnip (together with swede) accounted for 5400 hectares of production whereas this had dropped to less than 2700 hectares by 2017. Although the field area for cauliflower fell over the same period, it remained higher than turnip at over 9200 hectares in 2017 [19]. However, turnip could provide a beneficial source of glucosinolates if incorporated more regularly into the diet. As a vegetable that is predominantly consumed cooked, it is the GSL and sensory profile of cooked turnips that are of relevance to the consumer.

Realising that GSL content in commercial turnip may vary between cultivars, growth conditions, seasons and cooking batches, the aim of this study was to evaluate the variability in GSL content and resulting differences in sensory perception, as purchased commercially and as the vegetable would be consumed by consumers. Numerous research papers concerning *Brassica* vegetables focus on the raw vegetable rather than the material as consumed, and where studies focus on cooking, they recommend minimal cooking to preserve GSL content. Minimal processing is not suitable for a hard root vegetable such turnip, and therefore, it is important to establish whether more rigorous cooking and preparation does successfully deliver beneficial GSL to consumers. To achieve our aim, seven batches of steamed-pureed turnip were prepared and subsequently analysed for GSLs (identification and quantification using liquid chromatography mass spectrometry; LC-MS) and sensory profile (trained sensory panel). The hypothesis was that each batch of steamed-pureed turnip would contain substantial amounts of GSLs and have a perceivable bitter taste, regardless of any differences between batches.

2. Materials and Methods

2.1. Turnip Sample and Preparation

Seven batches of steamed-pureed turnip were used in this study. Steaming was chosen rather than boiling to reduce loss of GSLs leaching into cooking water. Pureeing was chosen to produce a homogenous sample, and also steam and puree are among many methods used to prepare turnips at home. Turnips (grown in the UK and the Netherlands) were bought from two local stores in Reading (UK), from December 2015 to June 2016, and each batch was cooked on a different day (Table 1).

Table 1. Purchase date and the origin of turnips for each batch.

Batch	Purchase Date	Origin of Turnip
B1	December 2015	The UK
B2	December 2015	The Netherlands
B3	February 2016	The UK (75%) and The Netherlands (25%) [1]
B4	April 2016	The UK (24%) and The Netherlands (76%) [1]
B5	April 2016	The UK
B6	June 2016	The Netherlands
B7	June 2016	The UK

[1] For B3, 75% turnips came from the UK and 25% came from the Netherlands; for B4 24% turnips came from the UK and 76% came from the Netherlands.

The root was used in the preparation of the samples; prior to cooking, turnips were peeled, and stems and tails removed, then washed and sliced to a thickness of approximately 0.5 cm. Between 8.2 and 13.6 kg raw turnips were used to make each batch of cooked turnip. For each cooking cycle, approximately 2.4 kg of sliced turnips were placed into an electric 3-tier steamer (Tefal; 800 g in each tier), with 1 L of water added to the base of the steamer and steamed for 25 min. Sliced turnips from tier 1 were transferred to tier 3 and vice versa (to ensure equal heat circulation), water was added again up to 1 L and steamed for another 25 min to ensure the root was soft enough to be blended. The internal temperature of the steamer was ~64 °C. Turnips were then blended using a hand blender (Russell Hobbs) for approximately 5 min until the texture was smooth. All cooked turnips were then placed into plastic containers, labelled, and stored in a freezer at −18 °C.

Prior to GSL extraction, samples were frozen (−80 °C) and then freeze-dried for 5 days (Stokes freeze dryer, F.J Stokes Corporation, Philadelphia, USA). The dried samples were ground (pestle and mortar) and then sieved (20 mesh) to ensure a fine powder.

2.2. Reagents and Chemicals

All chemicals used were of LC-MS grade and purchased from Sigma-Aldrich (Poole, UK), unless otherwise stated.

2.3. Glucosinolates Extraction

The extraction method was adapted from [20]. Three replicates of each batch were prepared as follows: 40 mg of ground steamed-pureed turnip powder was heated in a dry-block at 75 °C for 2 min to ensure inactivation of any remaining active myrosinase enzyme. Preheated 70% (*v*/*v*) methanol (1.2 mL; 70 °C) was added and the sample placed in a water bath for 20 min at 70 °C. Samples were then centrifuged for 10 min (10,000 rpm, 18 °C) to collect loose material into a pellet. The supernatant was then filtered through 0.22 μm Acrodisc syringe filters with Supor membrane (hydrophilic polyethersulfone; VWR, Lutterworth, UK) and frozen (−80 °C) in Eppendorf tubes until analysis by LC-MS.

2.4. LC-MS Analysis

LC-MS analysis method was adapted from [21]. Sinigrin hydrate was used as an external reference standard for quantification of GSL compounds. Preparation was as presented by Jin et al. [22]. LC-MS analysis was performed in the negative ion mode on an Agilent 1260 Infinity Series LC system (Stockport, UK) equipped with a binary pump, degasser, autosampler, column heater, diode array detector, coupled to an Agilent 6120 Series single quadrupole mass spectrometer. Separation of compounds was achieved on a Gemini 3 μm C_{18} 110 Å (150 × 4.6 mm) column (with Security Guard column, C_{18}; (4 mm × 3 mm); Phenomenex, Macclesfield, UK). GSLs were separated during a 40 min chromatographic run, with 5 min post-run sequence. Mobile phases consisted of ammonium formate (0.1%; A) and acetonitrile (B) with the following gradient timetable: (i) 0 min (A-B, 95:5, *v*/*v*); (ii) 0–13 min

(A-B, 95:5, *v*/*v*); (iii) 13–18 min (A-B, 40:60, *v*/*v*); (iv) 18–26 min (A-B, 40:60, *v*/*v*), 26–30 min (A-B, 95:5, *v*/*v*); (v) 30–40 min (A-B, 95:5, *v*/*v*). The diode array detector recorded spectra at 229 nm. The flow rate was optimised for the system at 0.4 $mLmin^{-1}$, with a column temperature of 30 °C, with 25 μL of sample injected into the system. Quantification was conducted at a wavelength of 229 nm.

MS analysis settings were as follows: API-ES was carried out at atmospheric pressure in negative ion mode (scan range m/z 100–1500 Da). Nebulizer pressure was set at 50 psi, gas-drying temperature at 350 °C, and capillary voltage at 2000 V.

Compounds were identified using MS through both spectra available in the literature [23,24] or from GSL standards in our own laboratory and by comparing relative retention times with those published in the literature [25]. Semi-quantification was carried out using UV absorbance (diode array detector; DAD) peak area data and relating that to the external sinigrin standard (regression: y = 26.7X + 52.6; r^2 = 0.998). Relative response factors (RRFs) were used in the calculation of GSL concentrations where available [23]; however, they were assumed to be 1.00 if such data was not available in the literature [25] or from our laboratory standards. All data were analysed using Agilent OpenLAB CDS ChemStation Edition for LC-MS (Agilent, version A.02.10).

2.5. Sensory Analysis

Sensory analysis was carried out by nine sensory trained panellists, each with a minimum of six months experience, using sensory profiling. The panel developed a consensus vocabulary for the seven batches of steamed-pureed turnip concerning aroma, taste and flavour (Table 2). Spinach, mashed potato, sucrose (granulated sugar) and quinine sulphate solutions were used as references to help the panel to standardise the vocabulary. During duplicate sample evaluations, samples were presented in a balanced sequential order, and each characteristic was scored on a line scales (0–100), using Compusense Cloud Software (Ontario, Canada). Line scales were unstructured except for the sweet and bitter attributes where a structured scale was used. Table 2 shows the levels of reference standards used for these two attributes. The panel tasted and scored the reference standards; their mean values for these standards were used as anchors on the scale. For sweet, the anchor positions for the four standards were 13.8, 29.1, 57.6 and 80.6, respectively. For bitter taste, the anchor positions were 8.1, 23.0, 38.9, 63.2 and 82.6, respectively. Evaluation sessions were conducted in a sensory room within the Sensory Science Centre at the Department of Food and Nutritional Sciences, Reading, UK. Each panellist sat in an individual booth equipped with artificial daylight and with room temperature controlled (approximately 22 °C).

Table 2. Definition of sensory characteristics associated with 7 batches of steamed-pureed turnip and references used during vocabulary development.

Sensory Characteristic	Definition
	Aroma
Apple	Aroma associated with apple
Cooked swede	Aroma associated with cooked swede
Green vegetable	Aroma associated with green vegetable (spinach)
Sweetcorn	Aroma associated with sweetcorn
Savoury	Aroma associated with savoury food
Sweet	Aroma associated with sweet food
Earthy	Aroma associated with earth or soil
Starchy	Aroma associated with starchy food (mashed potato)
Tannin	Aroma associated with tea
Wet	Aroma associated with musty
	Taste
Salty	Taste associated with sodium chloride
Umami	Taste associated with monosodium glutamate
Sweet	Taste associated with sucrose solution (0.5%, 1.0%, 2.0% and 2.6%)
Bitter	Taste associated with quinine sulphate solution (0.00005%, 0.0001%, 0.0002%, 0.0004% and 0.0006%)
	Flavour
Earthy	Flavour associated with earth or soil
Tannin	Flavour associated with tea
Apple	Flavour associated with apple
Starchy	Flavour associated with starchy food (mashed potato)

2.6. Statistical Analysis

The analytical results presented are the mean of three replicates (n = 3) for each sample. One-way ANOVA was used for comparison of GSL content between batches of steamed-pureed turnip. A principal component analysis (PCA) was carried out to relate GSLs with sensory characteristics. GSL data were projected onto the PCA with the mean sensory data as supplementary variables; Pearson's correlation was used. These tests were performed by using XLStat (Addinsoft, Paris, France).

Sensory profile data were tested using two-way ANOVA in SENPAQ (Qi Statistics Ltd.. Reading, UK) where the main effects (sample and assessor) were tested against the sample by assessor interaction, with sample as fixed effect and assessor as random effect. All significant differences between samples were assessed by using Tukey's HSD post hoc test at a significance level of 5%.

3. Results

3.1. Identification and Quantification of Glucosinolates

Twelve individual GSLs were detected across all batches of steamed-pureed turnip (Figure 1), and the concentration of each of GSL varied significantly between batches (Table 3). There were 7 aliphatic GSLs (progoitrin, glucoalyssin, gluconapin, glucobrassicanapin, gluconapoleiferin, glucoerucin, and glucoberteroin), 4 indole GSLs (4-hydroxyglucobrassicin, glucobrassicin, 4-methoxyglucobrassicin, and neoglucobrassicin) and 1 arylaliphatic GSL (gluconasturtiin). Glucoalyssin was only detected in batches B1 and B2, while no glucoerucin was detected in B5. Gluconasturtiin was the most abundant GSL across all batches. Total GSL concentration ranged from 16.07 to 44.74 μmol g^{-1} DW.

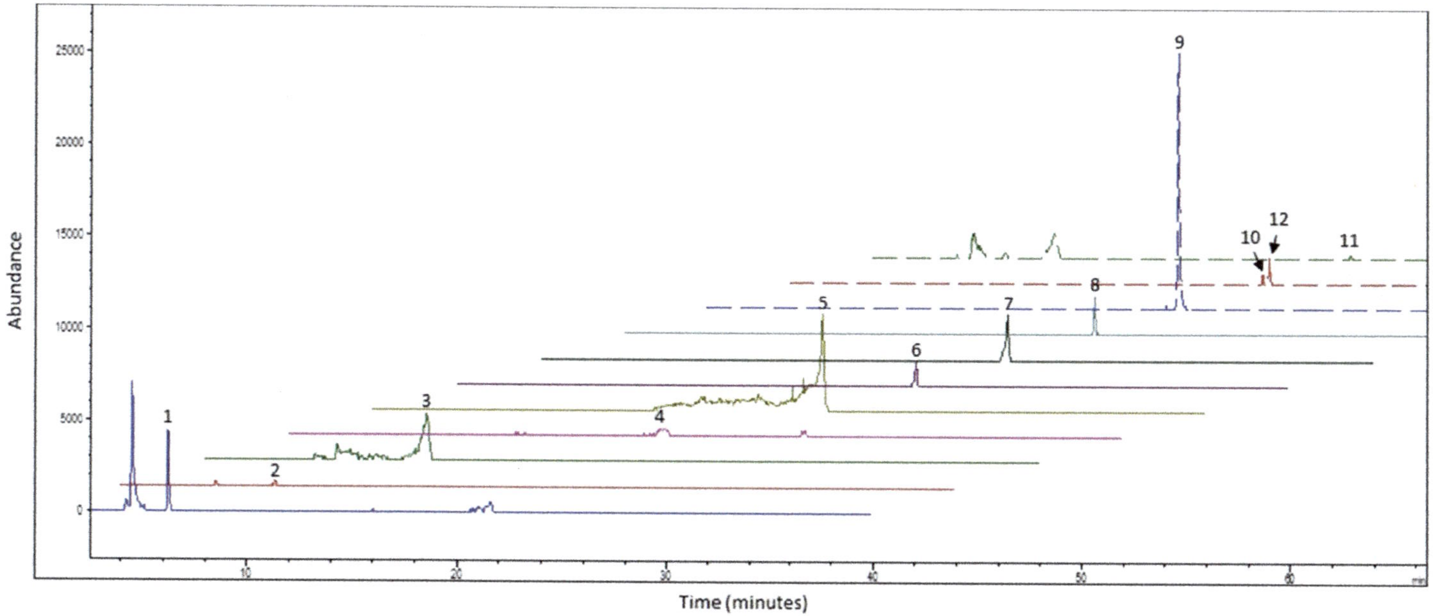

Figure 1. Example LC-MS extracted ion chromatogram of glucosinolates from steamed-pureed turnips. Peak identity: (1) progoitrin; (2) glucoalyssin; (3) gluconapin; (4) 4-hydroxy-glucobrassicin; (5) glucobrassicanapin; (6) glucoerucin; (7) glucobrassicin; (8) glucoberteroin; (9) gluconasturtiin; (10) 4-methoxyglucobrassicin; (11) Gluconapoleiferin and (12) Neoglucobrassicin, corresponding with glucosinolates listed in Table 3 and their extracted mass ion.

Table 3. Mean concentration of glucosinolates in seven batches of steamed-pureed turnip (B1 to B7). Results are expressed as µmolg-1 DW ± standard deviation. Different superscript letters indicate significant differences in mean concentration between batches. Abbreviation: ND, not detected.

Peak No.	Glucosinolate	Group	Side Chain	Mass Ion	Batch							p Value
					B1	B2	B3	B4	B5	B6	B7	
1	Progoitrin	Aliphatic	(2R)-2-hydroxy-3-butenyl	388	1.68 ± 0.03 ^{ab}	1.94 ± 0.2 ^{a}	1.73 ± 0.2 ^{ab}	1.76 ± 0.04 ^{ab}	1.34 ± 0.05 ^{b}	1.37 ± 0.14 ^{b}	1.69 ± 0.26 ^{ab}	0.004
2	Glucoalyssin	Aliphatic	5-methylsulfinylpentyl	450	0.10 ± 0.09 ^{ab}	0.14 ± 0.1 ^{a}	ND ^{b}	ND ^{b}	ND ^{b}	ND ^{b}	ND ^{b}	0.01
3	Gluconapin	Aliphatic	3-butenyl	372	1.15 ± 0.34 ^{b}	2.03 ± 0.95 ^{b}	1.22 ± 0.12 ^{b}	0.80 ± 0.25 ^{b}	0.43 ± 0.25 ^{b}	9.49 ± 0.68 ^{a}	11.21 ± 1.4 ^{a}	<0.0001
4	4-hydroxy-glucobrassicin	Indole	4-hydroxy-3-indolylmethyl	463	0.32 ± 0.01 ^{bc}	0.30 ± 0.02 ^{c}	0.18 ± 0.03 ^{d}	0.27 ± 0.01 ^{c}	0.14 ± 0.01 ^{d}	0.37 ± 0.03 ^{ab}	0.39 ± 0.03 ^{a}	<0.0001
5	Glucobrassicanapin	Aliphatic	4-pentenyl	386	3.77 ± 0.26 ^{a}	5.06 ± 0.97 ^{a}	4.76 ± 0.95 ^{a}	3.70 ± 0.04 ^{a}	1.92 ± 0.13 ^{b}	1.29 ± 0.1 ^{b}	1.33 ± 0.1 ^{b}	<0.0001
6	Glucoerucin	Aliphatic	4-methylthiobutyl	420	0.48 ± 0.07 ^{de}	0.84 ± 0.24 ^{cde}	1.46 ± 0.14 ^{c}	1.15 ± 0.55 ^{cd}	ND ^{e}	7.15 ± 0.32 ^{a}	6.27 ± 0.39 ^{b}	<0.0001
7	Glucobrassicin	Indole	3-indolylmethyl	447	0.87 ± 0.02 ^{c}	1.08 ± 0.06 ^{ab}	0.65 ± 0.06 ^{d}	0.70 ± 0.08 ^{cd}	0.90 ± 0.15 ^{bc}	1.13 ± 0.04 ^{a}	1.19 ± 0.07 ^{a}	<0.0001
8	Glucoberteroin	Aliphatic	5-methylthiopentyl	434	1.37 ± 0.12 ^{a}	1.56 ± 0.03 ^{a}	0.95 ± 0.1 ^{c}	1.08 ± 0.07 ^{bc}	0.21 ± 0.06 ^{d}	1.30 ± 0.09 ^{ab}	1.38 ± 0.16 ^{a}	<0.0001
9	Gluconasturtiin	Arylaliphatic	2-phenethyl	422	9.72 ± 0.27 ^{bc}	10.94 ± 0.59 ^{b}	8.96 ± 0.2 ^{c}	9.20 ± 0.57 ^{bc}	9.43 ± 0.1 ^{bc}	19.81 ± 1.5 ^{a}	19.32 ± 0.6 ^{a}	<0.0001
10	4-methoxy-glucobrassicin	Indole	4-methoxy-3-indolylmethyl	477	0.05 ± 0.01 ^{b}	0.07 ± <0.01 ^{a}	0.03 ± <0.01 ^{b}	0.04 ± <0.01 ^{b}	0.05 ± <0.01 ^{b}	0.07 ± 0.02 ^{a}	0.05 ± < 0.01 ^{ab}	<0.001
11	Glucona-poleiferin	Aliphatic	2-hydroxy-4-pentenyl	402	0.72 ± 0.01 ^{e}	1.10 ± 0.02 ^{cd}	1.00 ± 0.21 ^{cd}	0.97 ± 0.04 ^{d}	1.23 ± 0.01 ^{bc}	1.38 ± 0.07 ^{ab}	1.58 ± 0.06 ^{a}	<0.0001
12	Neogluco-brassicin	Indole	N-methoxy-3-indolylmethyl	477	0.26 ± 0.03 ^{b}	0.41 ± 0.03 ^{a}	0.31 ± 0.06 ^{b}	0.30 ± 0.01 ^{b}	0.41 ± 0.03 ^{a}	0.28 ± 0.02 ^{b}	0.34 ± 0.02 ^{ab}	<0.001
	Total glucosinolates				20.48 ± 0.67 ^{bc}	25.46 ± 2.47 ^{b}	21.25 ± 1.97 ^{bc}	19.97 ± 1.47 ^{bc}	16.07 ± 0.46 ^{c}	43.64 ± 2.66 ^{a}	44.74 ± 3.0 ^{a}	<0.0001

3.2. Sensory Characteristics

Table 4 summarises the mean sensory characteristic scores for the seven batches of steamed-pureed turnip. There was a significant difference in wet aroma where batch B2 had a higher score than B7. No other aroma characteristics were significantly different between batches.

For taste characteristics, there was a significant difference in bitter taste between batches, where batch B2 had the highest intensity for bitter taste, whereas B1 and B4 were significantly less intense. All batches were perceived as bitter with mean ratings varying between 30 and 53 in bitter taste intensity of the 0.0002% and below the 0.0004% quinine standard used. Sweetness did not vary significantly between batches; the range of mean scores were between 26 and 35 on the 100-point scale, being in the region of sweetness of the 1% sucrose standard used.

Significant differences between batches can be found for tannin and apple flavour. B2 was significantly higher than B1, B3, B4 and B5 for tannin flavour. B5 was significantly higher than B2, B6 and B7 in terms of apple flavour. There were no significant differences between batches for other characteristics.

Table 4. Mean scores for sensory characteristics for seven batches of steamed-pureed turnip. Different superscript letters indicate significant differences between batches.

Sensory Characteristic	Batch							Significance Between Samples (p-Value)
	B1	B2	B3	B4	B5	B6	B7	
				Aroma				
Apple	2.8	4.3	8.0	4.0	9.1	3.8	2.8	0.34
Cooked Swede	15.7	17.6	13.4	20.8	15.4	21.9	22.3	0.21
Green vegetable	12.8	17.9	12.7	12.5	14.3	14.6	18.2	0.66
Sweetcorn	3.5	5.3	1.4	3.7	1.8	3.2	2.1	0.32
Savoury	18.0	24.0	19.8	21.6	22.8	24.9	26.3	0.06
Sweet	15.1	13.7	16.6	14.7	17.2	15.5	15.2	0.83
Earthy	11.0	12.5	9.7	11.2	9.5	16.6	20.1	0.06
Starchy	18.4	16.7	15.5	14.2	12.9	13.2	12.3	0.35
Tannin	2.1	1.8	2.0	2.4	2.4	1.3	2.9	0.75
Wet	12.2 ^{ab}	14.7 ^{a}	9.7 ^{ab}	8.9 ^{ab}	9.4 ^{ab}	10.4 ^{ab}	8.0 ^{b}	0.04
				Taste				
Salty	6.4	7.1	5.5	10.2	13.6	6.4	7.8	0.05
Umami	14.3	19.6	17.2	19.5	23.3	23.0	15.3	0.08
Sweet	33.1	30.3	31.2	35.3	30.5	34.5	26.3	0.24
Bitter	30.8 ^{c}	53.2 ^{a}	33.1 ^{bc}	30.2 ^{c}	34.5 ^{bc}	40.3 ^{bc}	43.3 ^{ab}	<0.0001
				Flavour				
Earthy	11.8	18.9	11.0	16.2	15.1	18.9	19.9	0.11
Tannin	9.8 ^{b}	20.1 ^{a}	8.3 ^{b}	7.9 ^{b}	9.4 ^{b}	12.3 ^{ab}	15.9 ^{ab}	0.0003
Apple	4.3 ^{abc}	3.3 ^{bc}	8.7 ^{abc}	12.0 ^{ab}	12.6 ^{a}	2.6 ^{c}	1.9 ^{c}	0.0008
Starchy	13.5	14.5	12.3	15.3	14.3	12.6	12.0	0.78

3.3. Principal Component Analysis (PCA)

Principal component analysis (PCA) of the GSL data was carried out to demonstrate the batch separation according to GSLs, and onto this map the sensory data was fitted as supplementary data in order to investigate any correlation of the GSLs with the sensory characteristics (Figure 2). Dimensions 1 and 2 recovered over 78% of the variance in the data. Total GSL and many of the individual GSLs were predominantly located on the right side of PC1, located alongside turnip batches B6 and B7. PC2 was highly correlated with gluoberteroin (r = 0.88) and glucoalyssin (r = 0.84).

The position for the total GSL content strongly correlated with PC1 (r = 0.98) and also to many of the individual GSLs: gluconapin (r = 0.99, p < 0.001), gluconasturtiin (r = 0.98, p < 0.001), glucoerucin (r = 0.97, p < 0.001), 4-hydroxyglucobrassicin (r = 0.82, p = 0.03), glucobrassicin (r = 0.78, p = 0.04) and gluconapoleiferin (r = 0.77, p = 0.04). However, 4 other GSLs strongly correlated with each other and with dimension PC2: glucoberteroin (r = 0.88), glucoalyssin (r = 0.84), progoitrin (r = 0.80) and glucobrassicanapin (r = 0.59).

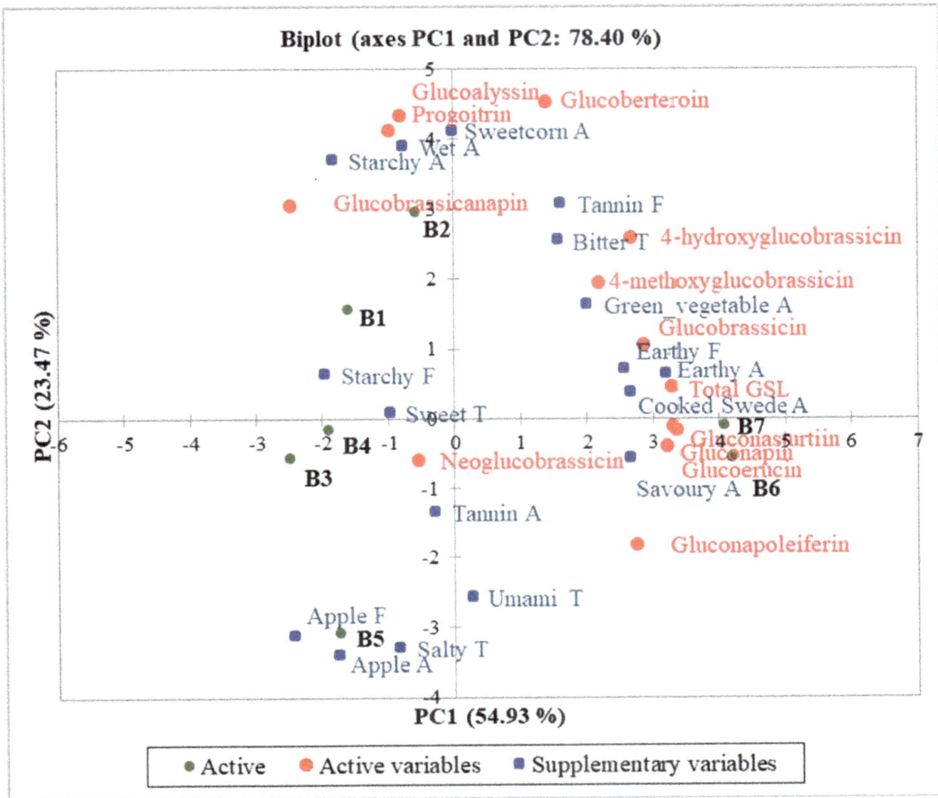

Figure 2. PCA biplot of glucosinolate compounds in 7 batches of steamed-pureed turnip (B1 to B7), with mean ratings of sensory attributes fitted onto the plot as supplementary variables. Abbreviations: A, aroma; T, taste; F, flavour.

There was a clear separation of groups of sensory characteristics on the PC biplot. Earthy (aroma and flavour), cooked swede aroma and savoury aroma were positioned to the right of PC1 and negatively correlated with sweet taste. Bitter taste and tannin flavour were positioned in the top right quadrant of the plot and negatively correlated with apple (aroma and flavour).

As expected, many of the GSLs correlated with bitter taste. The total GSL content was positively, but not significantly, correlated with bitter taste ($r = 0.47$, $p = 0.29$). Of the 12 GSLs quantified, one, glucobrassicanapin, had clearly no association with bitter taste ($r = 0.033$, $p = 0.94$) whereas the correlation coefficient between the other GSLs and bitter taste varied between 0.30 and 0.75. The only significant correlation was 4-methoxyglucobrassicin ($r = 0.82$, $p = 0.02$), while glucobrassicin also had a strong positive correlation ($r = 0.75$, $p = 0.052$), despite the levels of these two GSLs not being particularly high in the turnip batches, indeed very low for 4-methoxyglucobrassicin (Table 3). Such correlations cannot prove which of these GSLs have the greatest contribution to bitter taste, but they do support the hypothesis that the GSLs in turnip contribute to bitter taste. Bitter taste will suppress sweet taste, so it was as expected that all GSLs (except glucobrassicanapin) were negatively correlated with sweet taste ($r = -0.55$ to $r = -0.01$).

B1 and B2 were negatively correlated with B6 and B7; B1 and B2 were separated from B3, B4 and B5 along PC2. Moreover, B6 and B7 were separated from the other batches along PC1, and this was driven by the higher level of total GSL and particularly 4-hydroxyglucobrassicin, 4-methoxyglucobrassicin, glucobrassicin, gluconasturtiin, gluconapin and glucoerucin. These 2 batches were indeed the most bitter tasting, along with B2, which although not as high in total GSL, was highest in glucobrassicanapin. PC2 particularly separated B5 from B2, where B5 was particularly low in all GSLs and higher in apple (aroma and flavour).

4. Discussion

Twelve individual GSLs were detected across all batches. The total GSL content ranged from 16.07 to 44.74 μmol g^{-1} DW with mean value of 27.37 μmol g^{-1} DW. The total content is comparable to findings reported by Zhang et al. [26], (16.4 to 31.4 μmol g^{-1} DW), but lower than those reported by Lee et al. [4], (117.05 μmol g^{-1} DW). Zhang et al. [26] freeze dried the raw turnip roots rather than cooked turnip as in the present study; however, both studies later incubated at 75 °C before extraction with methanol. The results remain comparable as the steaming of turnip in the current study would denature the myrosinase enzyme and limit transformation to hydrolysis products. Other reasons might be because of the similarity in environmental factors that both studies have, as the turnips were sown across different seasons, which then would yield similar GSL content. Contradictory to the Lee et al. [4] study, the turnips were sown and grown in a controlled environment (i.e., temperature-controlled) to minimise seasonal differences, hence the large difference in the GSL content.

In the present study, aliphatic GSLs were the most abundant, representing 48.6% of total GSL content, followed by 45.6% of arylaliphatic GSL and 5.8% of indole GSL. These results are in agreement with other studies which confirm that these compounds are common GSLs in turnip varieties [4,26–28]. Gluconasturtiin was the dominant GSL (45.6%), ranging from 8.96 to 19.81 μmol g^{-1} DW, with a mean value of 12.48 μmol g^{-1} DW. This GSL compound has previously been shown to be the most abundant in turnip greens [28] and turnip roots [26].

There were significant differences in each individual GSL between batches and this is expected as the turnips were bought on different days, across different seasons, and from a variety of suppliers. Although they were all "purple top" turnips, they were potentially of different cultivars. Type of cultivar will affect GSL content; indeed, Kim et al. [29] reported that the GSL content of turnip seeds varied significantly between 12 cultivars. There are many other factors that could also contribute to variability. Kim et al. [30] reported that GSL content in turnip is dependent on harvest times. Subsequent research papers have noted that, in addition to harvest time, growth season could also result in the GSL variation [26,31]. Environmental conditions of different growing sites, such as soil pH, can influence GSL content too [32]. Our PCA plot showed that batches B1 and B2 were similar, as were B4 and B5, and B6 and B7. These similarities can be explained by the month the turnips were purchased. Turnips for batches B1 and B2 were purchased in autumn/winter season, and they were negatively correlated in terms of GSL content and sensory characteristics, with B6 and B7, which were bought in spring/summer season. Although turnips for batch B3 were purchased in a different season from B4 and B5, these three batches were correlated with each other, in terms of GSL content and sensory characteristics. It could be speculated that these three batches may be from the same cultivar of turnip, and the cultivar effect is greater than season effect; however, this cannot be concluded as the turnip cultivar was not controlled for in this study.

In summary, the significant differences in GSL content between cooked turnip batches in this study might be caused by differences in cultivars, seasons or growth conditions. Turnips sold in the UK come from many different countries with different growth conditions. Therefore, variation in GSL content at the point of consumption is expected from turnips purchased in the UK supermarkets at different times of year.

GSLs are among the compounds that are responsible for the sensory characteristics of *Brassica* vegetables. As noted in (Figure 2), most of the GSLs were positively (although not significantly) correlated with bitter taste, the strongest correlations being with 4-methoxyglucobrassicin and glucobrassicin. Although Helland et al. [33] also found 4-methoxyglucobrassicin to be related to the bitterness of swede and turnip, the levels of this compound in the current study were very low. Glucobrassicin was present at higher levels (0.65 to 1.19 μmolg^{-1} dry weight) and was clearly correlated with most bitter turnip batches (B2, B6 and B7). Glucobrassicin has previously been reported to cause bitter taste, alongside 4-hydroxyglucobrassicin, progoitrin, gluconapin and neoglucobrassicin, in turnip, swede, rocket, broccoli and cauliflower, [33–35] which is consistent with the current study where all were positively correlated with the bitter taste in turnip ($r = 0.33$ to $r = 0.75$). The GSL in highest abundance in the cooked turnip was gluconasturtiin; this has a positive but relatively weak correlation to bitter taste ($r = 0.43$). Although this finding does not confirm a relationship between gluconasturtiin and bitterness, it was present at high levels in all batches (compared to other GSLs), and all batches were perceived to be bitter. Bladh et al. [36] previously concluded that it was a hydrolysis product of gluconasturtiin, phenethyl isothiocyanate, that had a strong bitter taste. Although GSLs are accepted to impart bitterness, the correlation between GSLs and perception of bitterness does not confirm causality. Bell et al. [37] reviewed the relationship between GSLs and bitterness and noted that there very few studies where GSLs have been isolated and rated by sensory panels; one specific exception being sinigrin where bitter taste thresholds have been reported. Relating bitter taste to specific GSLs in *Brassica* samples is limited by the high correlation between the quantities of individual GSLs. This Bell et al. [37] review also noted that inconsistencies in relating GSLs to bitter taste can also arise from differences in preparation and cooking methods between studies. In addition, hydrolysis product of GSLs are often not quantified, and therefore, their contribution to bitterness is often not accounted for. A further review by Wieczorek et al. [38] concluded that inconsistencies between studies can also result from differences in consumers' sensitivity to GLS-derived bitter taste. Interactions between taste modalities must also be considered; our results showed that all individual GSLs (except glucobrassicanapin) were negatively correlated with sweet taste. This was similarly reported by Francisco et al. [6], suggesting that bitter taste suppressed sweet taste in the perception of turnip.

Bitter taste was positively correlated with tannin flavour, and two individual GSLs were highly correlated with this attribute: 4-methoxyglucobrassicin and glucobrassicin. In our sensory profile data, batches B2, B6 and B7 were rated the highest in tannin flavour and bitter taste. The tannin flavour is likely to originate from phenolic compounds rather than from the GSLs. Such phenolic compounds (flavonoids, quinic acid derivatives, sinapic acids derivatives and tannins) have been found in turnip [39,40] and are also associated with bitter taste [36]. However, phenolic compounds were not measured in the current study, hence the relationship between bitter taste and phenolic compounds could not be determined.

It the present study, it was also observed that gluconapoleiferin, gluconapin, 4-hydroxyglucobrassicin, glucoerucin, glucobrassicin and gluconasturtiin and total GSL were highly correlated with earthy aroma, and gluconapoleiferin, glucobrassicin and 4-methoxyglucobrassicin were highly correlated with earthy flavour. In comparison, Helland et al. [33] observed that gluconapin, glucoerucin and glucobrassicanapin were positively correlated with earthy aroma. However, there are possible compounds other than GSLs that contribute to aroma and flavour of vegetables, such as the breakdown products of GSLs, which were not measured in this study.

5. Conclusions

The results obtained in this study showed that individual and total GSL varied between different batches of steamed-pureed turnip. The GSL compounds were correlated with aroma, taste and flavour characteristic of turnip. Almost all GSLs positively correlated with bitter taste; however, many GSLs correlated with other GSLs in concentration, which limits interpretation of which have the greatest influence on bitter taste. The strongest correlations with bitter taste were for 4-methoxyglucobrassicin and glucobrassicin; however, these two GSLs were highly correlated and the 4-methoxyglucobrassicin was at particularly low levels, so their individual contribution to bitter taste cannot be confirmed.

Overall, all batches of steamed-pureed turnip demonstrated both bitter and sweet taste, and these two taste characteristics were negatively correlated. It was evident that the bitter taste suppressed the sweet taste of the turnip as the batches containing the least GSL were the sweetest. The impact of this finding is in the conclusion that turnips bought commercially in the UK do provide a substantial amount of GSLs even after rigorous cooking and preparation, and as such cooked turnip could provide the well documented health benefits of GSLs if they were regularly consumed in the diet. However, the cooked product has a consistently bitter taste which may be a barrier to some consumers.

Author Contributions: N.D.M.N., C.H.-P., K.H. and L.M.; methodology, S.L., L.B. and L.M.; data analysis, N.D., C.H.-P., K.H. and L.M.; writing—original draft preparation, N.D.M.N. and L.B.; writing—review and editing, S.L., L.B., C.H.-P., K.H. and L.M.; supervision, C.H.-P., K.H. and L.M. All authors have read and agreed to the published version of the manuscript.

Funding: This research was funded by The Ministry of Higher Education of Malaysia.

Acknowledgments: The authors would like to thank Xirui Zhou and Omobolanle Oloyede of the University of Reading for their guidance and advice.

Conflicts of Interest: There are no conflicts of interest to report.

References

1. Mithen, R.F.; Dekker, M.; Verkerk, R.; Rabot, S.; Johnson, I.T. The nutritional significance, biosynthesis and bioavailability of glucosinolates in human foods. *J. Sci. Food Agric.* **2000**, *80*, 967–984. [CrossRef]
2. Al-Gendy, A.A.; El-gindi, O.D.; Hafez, A.S.; Ateya, A.M. Glucosinolates, volatile constituents and biological activities of Erysimum corinthium Boiss. (Brassicaceae). *Food Chem.* **2010**, *118*, 519–524. [CrossRef]
3. Kim, M.K.; Park, J.H.Y. Cruciferous vegetable intake and the risk of human cancer: Epidemiological evidence. *Proc. Nutr. Soc.* **2009**, *68*, 103–110. [CrossRef] [PubMed]
4. Lee, J.G.; Bonnema, G.; Zhang, N.; Kwak, J.H.; de Vos, R.C.H.; Beekwilder, J. Evaluation of glucosinolate variation in a collection of turnip (Brassica rapa) germplasm by the analysis of intact and desulfo glucosinolates. *J. Agric. Food Chem.* **2013**, *61*, 3984–3993. [CrossRef] [PubMed]
5. Engel, E.; Baty, C.; Le Corre, D.; Souchon, I.; Martin, N. Flavor-active compounds potentially implicated in cooked cauliflower acceptance. *J. Agric. Food Chem.* **2002**, *50*, 6459–6467. [CrossRef]
6. Francisco, M.; Velasco, P.; Romero, Á.; Vázquez, L.; Cartea, M.E. Sensory quality of turnip greens and turnip tops grown in northwestern Spain. *Eur. Food Res. Technol.* **2009**, *230*, 281–290.
7. Bell, L.; Methven, L.; Signore, A.; Oruna-Concha, M.J.; Wagstaff, C. Analysis of seven salad rocket (Eruca sativa) accessions: The relationships between sensory attributes and volatile and non-volatile compounds. *Food Chem.* **2017**, *218*, 181–191. [CrossRef]
8. Bhandari, S.R.; Jo, J.S.; Lee, J.G. Comparison of glucosinolate profiles in different tissues of nine Brassica crops. *Molecules* **2015**, *20*, 15827–15841. [CrossRef]
9. Hanschen, F.S.; Schreiner, M. Isothiocyanates, nitriles, and epithionitriles from glucosinolates are affected by genotype and developmental stage in Brassica oleracea varieties. *Front. Plant Sci.* **2017**, *8*, 1095. [CrossRef]
10. Kabouw, P.; Biere, A.; Van Der Putten, W.H.; Van Dam, N.M. Intra-specific differences in root and shoot glucosinolate profiles among white cabbage (Brassica oleracea var. capitata) cultivars. *J. Agric. Food Chem.* **2010**, *58*, 411–417. [CrossRef]
11. Zhu, B.; Yang, J.; Zhu, Z. Variation in glucosinolates in pak choi cultivars and various organs at different stages of vegetative growth during the harvest period. *J. Zhejiang Univ. Sci. B* **2013**, *14*, 309–317. [CrossRef] [PubMed]
12. Tong, Y.; Gabriel-Neumann, E.; Ngwene, B.; Krumbein, A.; George, E.; Platz, S.; Rohn, S.; Schreiner, M. Topsoil drying combined with increased sulfur supply leads to enhanced aliphatic glucosinolates in Brassica juncea leaves and roots. *Food Chem.* **2014**, *152*, 190–196. [CrossRef] [PubMed]
13. Barickman, T.C.; Kopsell, D.A.; Sams, C.E. Selenium influences glucosinolates and isothiocyanates and increases sulfur uptake in Arabidopsis thaliana and rapid-cycling Brassica oleracea. *J. Agric. Food Chem.* **2013**, *61*, 202–209. [CrossRef] [PubMed]
14. Li, S.; Schonhof, I.; Krumbein, A.; Li, L.; Stützel, H.; Schreiner, M. Glucosinolate concentration in turnip (*Brassica rapa* ssp. *rapifera* L.) roots as affected by nitrogen and sulfur supply. *J. Agric. Food Chem.* **2007**, *55*, 8452–8457. [CrossRef] [PubMed]
15. Traka, M.; Mithen, R. Glucosinolates, isothiocyanates and human health. *Phytochem Rev.* **2009**, *8*, 269–282. [CrossRef]

16. Cox, D.N.; Melo, L.; Zabaras, D.; Delahunty, C.M. Acceptance of health-promoting Brassica vegetables: The influence of taste perception, information and attitudes. *Public Health Nutr.* **2012**, *15*, 1474–1482. [CrossRef]
17. Jia, C.-G.; Xu, C.-J.; Wei, J.; Yuan, J.; Yuan, G.-F.; Wang, B.-L.; Wang, Q.-M. Effect of modified atmosphere packaging on visual quality and glucosinolates of broccoli florets. *Food Chem.* **2009**, *114*, 28–37. [CrossRef]
18. Nugrahedi, P.Y.; Verkerk, R.; Widianarko, B.; Dekker, M. A mechanistic perspective on process-induced changes in glucosinolate content in Brassica vegetables: A review. *Crit. Rev. Food Sci. Nutr.* **2015**, *55*, 823–838. [CrossRef]
19. CEIC. United Kingdom Agricultural Sown Area: By Commodities. Available online: https://www.ceicdata.com/en/united-kingdom/agricultural-sown-area-by-commodities (accessed on 14 February 2020).
20. Bell, L.; Oruna-Concha, M.J.; Wagstaff, C. Identification and quantification of glucosinolate and flavonol compounds in rocket salad (Eruca sativa, Eruca vesicaria and Diplotaxis tenuifolia) by LC-MS: Highlighting the potential for improving nutritional value of rocket crops. *Food Chem.* **2015**, *172*, 852–861. [CrossRef]
21. Jasper, J.; Wagstaff, C.; Bell, L. Growth temperature influences postharvest glucosinolate concentrations and hydrolysis product formation in first and second cuts of rocket salad. *Postharvest Biol. Technol.* **2020**, *163*, 111157.
22. Jin, J.; Koroleva, O.A.; Gibson, T.; Swanston, J.; Magan, J.; Zhang, Y.; Rowland, I.R.; Wagstaff, C. Analysis of phytochemical composition and chemoprotective capacity of rocket (Eruca sativa and Diplotaxis tenuifolia) leafy salad following cultivation in different environments. *J. Agric. Food Chem.* **2009**, *57*, 5227–5234. [CrossRef] [PubMed]
23. Clarke, D.B. Glucosinolates, structures and analysis in food. *Anal. Methods* **2010**, *2*, 310–325. [CrossRef]
24. Cataldi, T.R.I.; Rubino, A.; Lelario, F.; Bufo, S.A. Naturally occurring glucosinolates in plant extracts of rocket salad (*Eruca sativa* L.) identified by liquid chromatography coupled with negative ion electrospray ionization and quadrupole ion-trap mass spectrometry. *Rapid Commun. Mass Spectrom.* **2007**, *21*, 2374–2388. [CrossRef] [PubMed]
25. Lewis, J.; Fenwick, G.R. Glucosinolate content of Brassica vegetables: Analysis of twenty-four cultivars of Calabrese (green sprouting broccoli, *Brassica oleracea* L. var. botrytis subvar. cymosa Lam.). *Food Chem.* **1987**, *25*, 259–268. [CrossRef]
26. Zhang, H.; Schonhof, I.; Krumbein, A.; Gutezeit, B.; Li, L.; Stützel, H.; Schreiner, M. Water supply and growing season influence glucosinolate concentration and composition in turnip root (*Brassica rapa* ssp. *rapifera* L.). *J. Plant Nutr. Soil Sci.* **2008**, *171*, 255–265. [CrossRef]
27. Justen, V.L.; Fritz, V.A.; Cohen, J.D. Seasonal variation in glucosinolate accumulation in turnips grown under photoselective nettings. *Hortic. Environ. Biotechnol.* **2012**, *53*, 108–115. [CrossRef]
28. Padilla, G.; Cartea, M.E.; Velasco, P.; de Haro, A.; Ordás, A. Variation of glucosinolates in vegetable crops of Brassica rapa. *Phytochemistry* **2007**, *68*, 536–545. [CrossRef]
29. Kim, S.-J.; Kawaguchi, S.; Watanabe, Y. Glucosinolates in vegetative tissues and seeds of twelve cultivars of vegetable turnip rape (*Brassica rapa* L.). *Soil Sci. Plant Nutr.* **2003**, *49*, 337–346. [CrossRef]
30. Kim, S.-J.; Ishida, M.; Matsuo, T.; Watanabe, M.; Watanabe, Y. Separation and identification of glucosinolates of vegetable turnip rape by LC / APCI-MS and comparison of their contents in ten cultivars of vegetable turnip rape (*Brassica rapa* L.). *Soil Sci. Plant Nutr.* **2001**, *47*, 167–177. [CrossRef]
31. del Carmen Martínez-Ballesta, M.; Moreno, D.A.; Carvajal, M. The physiological importance of glucosinolates on plant response to abiotic stress in Brassica. *Int. J. Mol. Sci.* **2013**, *14*, 11607–11625.
32. Francisco, M.; Cartea, M.E.; Soengas, P.; Velasco, P. Effect of genotype and environmental conditions on health-promoting compounds in Brassica rapa. *J. Agric. Food Chem.* **2011**, *59*, 2421–2431. [CrossRef] [PubMed]
33. Helland, H.S.; Leufvén, A.; Bengtsson, G.B.; Pettersen, M.K.; Lea, P.; Wold, A.-B. Storage of fresh-cut swede and turnip: Effect of temperature, including sub-zero temperature, and packaging material on sensory attributes, sugars and glucosinolates. *Postharvest Biol. Technol.* **2016**, *111*, 370–379. [CrossRef]
34. Pasini, F.; Verardo, V.; Cerretani, L.; Caboni, M.F.; D'Antuono, L.F. Rocket salad (Diplotaxis and Eruca spp.) sensory analysis and relation with glucosinolate and phenolic content. *J. Sci. Food Agric.* **2011**, *91*, 2858–2864. [CrossRef] [PubMed]
35. Schonhof, I.; Krumbein, A.; Brückner, B. Genotypic effects on glucosinolates and sensory properties of broccoli and cauliflower. *Nahrung/Food* **2004**, *48*, 25–33. [CrossRef]

36. Bladh, K.T.; Olsson, K.M.; Yndgaard, F. Evaluation of glucosinolates in Nordic horseradish (Armoracia Rusticana). *Bot. Lith.* **2013**, *19*, 48–56.
37. Bell, L.; Oloyede, O.O.; Lignou, S.; Wagstaff, C.; Methven, L. Taste and flavor perceptions of glucosinolates, isothiocyanates, and related Compounds. *Mol. Nutr. Food Res.* **2018**, *62*, 1700990. [CrossRef]
38. Wieczorek, M.N.; Walczak, M.; Skrzypczak-Zielińska, M.; Jeleń, H.H. Bitter taste of Brassica vegetables: The role of genetic factors, receptors, isothiocyanates, glucosinolates, and flavor context. *Crit. Rev. Food Sci. Nutr.* **2018**, *58*, 3130–3140. [CrossRef]
39. Francisco, M.; Velasco, P.; Moreno, D.A.; García-Viguera, C.; Cartea, M.E. Cooking methods of Brassica rapa affect the preservation of glucosinolates, phenolics and vitamin C. *Food Res. Int.* **2010**, *43*, 1455–1463. [CrossRef]
40. HassanpourFard, M.; Naseh, G.; Lotfi, N.; Hosseini, S.M.; Hosseini, M. Effects of aqueous extract of turnip leaf (Brassica rapa) in alloxan-induced diabetic rats. *Avicenna J. Phytomed.* **2015**, *5*, 148–156.

Publisher's Note: MDPI stays neutral with regard to jurisdictional claims in published maps and institutional affiliations.

foods

MDPI

Article

The Impact of Domestic Cooking Methods on Myrosinase Stability, Glucosinolates and Their Hydrolysis Products in Different Cabbage (*Brassica oleracea*) Accessions

Omobolanle O. Oloyede *, Carol Wagstaff and Lisa Methven

Department of Food and Nutritional Sciences, Harry Nursten Building, University of Reading, Whiteknights, Reading RG6 6DZ, UK; c.wagstaff@reading.ac.uk (C.W.); l.methven@reading.ac.uk (L.M.)
* Correspondence: bola.oloyede@reading.ac.uk; Tel.: +44-(0)113-378-3606

Abstract: Glucosinolate hydrolysis products are responsible for the health-promoting properties of *Brassica* vegetables. The impact of domestic cooking on the myrosinase stability, glucosinolates and hydrolysis products in 18 cabbage accession was investigated. Cabbages were steamed, microwaved, and stir-fried before analysis. Cooking significantly affected myrosinase stability and glucosinolate concentrations within and between cabbage morphotypes. Myrosinase was most stable after stir-frying, with up to 65% residual activity. Steaming and microwaving resulted in over 90% loss of myrosinase activity in some accessions. Stir-frying resulted in the greatest decrease in glucosinolate concentration, resulting in up to 70% loss. Steamed cabbages retained the highest glucosinolates after cooking (up to 97%). The profile and abundance of glucosinolate hydrolysis products detected varied across all cooking methods studied. Cooking reduced the amounts of nitriles and epithionitriles formed compared to raw samples. Steaming led to a significant increase in the concentration of beneficial isothiocyanates present in the cabbage and a significantly lower level of nitriles compared to other samples. Microwaving led to a reduction in the concentrations of both nitriles and isothiocyanates when compared to other cooking methods and raw cabbage. The results obtained help provide information on the optimal cooking methods for cabbage, suggesting that steaming may be the best approach to maximising beneficial isothiocyanate production.

Keywords: *Brassica oleracea*; cabbage; myrosinase stability; glucosinolates; glucosinolate hydrolysis products; isothiocyanates; epithionitriles; steaming; microwaving; stir-frying

Citation: Oloyede, O.O.; Wagstaff, C.; Methven, L. The Impact of Domestic Cooking Methods on Myrosinase Stability, Glucosinolates and Their Hydrolysis Products in Different Cabbage (*Brassica oleracea*) Accessions. *Foods* **2021**, *10*, 2908. https://doi.org/10.3390/foods10122908

Academic Editors: Franziska S. Hanschen and Sascha Rohn

Received: 11 October 2021
Accepted: 19 November 2021
Published: 24 November 2021

1. Introduction

Consumption of *Brassica* or cruciferous vegetables such as cabbage (*Brassica oleracea*) is reported to result in chemo-protective effects [1]. This has been attributed to the high amounts of glucosinolates (GSLs) they contain. When plant tissue is damaged as a result of autolysis, plant injury, processing or chewing, GSLs are exposed to, and hydrolysed by, endogenous myrosinase. Upon hydrolysis, glucose and an unstable aglycone (thiohydroxamate-*O*-sulfonate) are produced. The unstable aglycone (thiohydroxamate-*O*-sulfonate) immediately rearranges to form different hydrolysis products such as isothiocyanates (ITCs), thiocyanates, nitriles, epithionitriles (EPTs), and oxazolidine-2-thiones, depending on the conditions of the reaction [2]. Nitriles and EPTs are formed in the presence of epithiospecifier proteins (ESP) instead of the more beneficial ITCs [3]. ITCs and indoles commonly found in cabbages such as sulforaphane (SFP), erucin (ER), allyl ITC (AITC), 2-phenyethyl ITC (PEITC), iberin (IB) and indole-3-carbinol (I3C) are reported to be responsible for some of the health-promoting properties of *Brassicas* [2]. SFP, the most studied of all the ITCs, is reported to possess chemoprotective, antioxidative, antimicrobial and neuroprotective properties [4–9]. AITC has been found to be potent against bladder [10,11], breast [12] and lung [13] cancer cells. I3C is also known to have anti-cancerous activities on reproductive organs, reducing the proliferation of cancer cells in the breast, prostrate, cervical and colon cell lines and preventing tumour development in rodents [14–16]. GSLs and glucosinolate hydrolysis products (GHPs) are also partly responsible for the bitter taste and pungent flavour and aroma of *Brassica* vegetables, which can reduce consumer acceptability of *Brassicas* [17,18]. Cox et al. [19] reported that *Brassica* acceptance was low in adults due to their sensory characteristics.

Where GSL hydrolysis does not occur in the process of preparing *Brassica* for consumption, it can occur as a result of microbial activities in the gastrointestinal tract of humans. However, despite the ability of microorganisms in the human gut to hydrolyse glucosinolates, it has been reported that the conversion is at least three times less efficient when compared to glucosinolate hydrolysis by myrosinase [20]. In a recent in vivo study conducted by Okunade et al. [21], the addition of exogenous myrosinase from brown mustard powder to cooked and pureed broccoli where myrosinase had been inactivated resulted in a four times increase in sulforaphane bioavailability compared to when the pureed broccoli was consumed alone. It is, therefore, important to ensure that myrosinase enzyme remains active during the consumption of *Brassica* vegetables.

Cabbages, like other *Brassica* vegetables, are mostly subjected to some form of thermal processing or domestic cooking before consumption. Cabbages are commonly boiled, steamed, stir-fried, or microwaved prior to consumption. Thermal cooking processes are considered one of the most important factors affecting the stability of myrosinase enzyme and ESP stability and the profiles, concentrations, and bioavailability of GSLs and their hydrolysis products [22]. This, in turn, can influence the health benefits that can be derived from the consumption of these vegetables making it crucial to determine the effect of cooking processes on myrosinase stability, GSLs and GHP formation within plant tissues.

Cooking cabbage can result in total or partial ESP and myrosinase inactivation, which, in turn, influences the type of GHPs formed. The time and temperature of cooking, vegetable matrix and degree of tissue damage all influence the changes observed during cooking [23]. Several studies have shown that myrosinase is inactivated during the thermal processing or domestic cooking of cabbage, leading to a decreased production of beneficial hydrolytic compounds [24–26]. Most of these studies, however, were based on crude myrosinase extracts or cabbage juice [24,26–28]. Myrosinase enzyme, when present within plant tissue, has been shown to have greater thermal stability than its crude extract, with this stability attributed to the rate at which the core temperature increased [24].

Previous studies on GSL concentrations in cooked cabbage showed conflicting results. Some authors have reported an increase in GSL content after microwaving cabbage [24,28]. Rungapamestry et al. [25], Song and Thornalley [29] and Xu et al. [30] reported minimal losses or no change in GSL concentration after steaming and microwaving cabbage. Xu et al. [30] recorded a 77% loss in GSL concentration after stir-frying. GSL losses during cooking have mostly been linked to leaching into cooking water [24,31]. The variation in myrosinase and GSL stability after processing can be attributed to different cooking conditions and the size of cut cabbage pieces, which, in most cases, do not represent standard domestic ways of cooking cabbage. Some of these studies processed the cabbages under much longer time-temperature combinations compared to what would normally be applicable during the domestic cooking of cabbage [24,26]. Furthermore, most of these studies [24,26,28] have focused on closed heart cabbages (mostly red and white cabbage) with limited data available on open leaf varieties such as black kale. There is insufficient evidence to date linking high myrosinase activity and/or GSL accumulation to high stability after processing/cooking; hence, studies considering both the stability of myrosinase and GSLs within plant tissue for various *B. oleracea* species are needed. Available studies have focused on the effect of cooking on specific GSLs and their GHPs, or just ITCs [25,29,32–37]. Some of these studies have been conducted in model systems [36], which do not consider the various reactions occurring within the plant matrix that can influence the GSL-myrosinase system during domestic cooking processes.

In our earlier paper investigating the effect of accession identity and growing conditions on myrosinase activity, GSL and GHP in 18 cabbage accessions, we discussed in detail the results of myrosinase activity, as well as the profile and concentration of GSL and GHP, between the raw accessions [38]. In this paper, the focus is on the effects of cooking on these accessions. This study, therefore, examines the effect of steaming, microwaving, and stir-frying on myrosinase stability, GSL and GHP profiles and concentrations in 18 cabbage accessions. Cooking times were chosen to represent standard domestic practices. It was hypothesised that, through controlled domestic cooking processes, myrosinase and GSL stability would be maximised, thereby increasing the production of ITCs, and improving the health benefits associated with cabbage consumption. It was also hypothesised that the genetic background of the cabbage and its morphotype would impact the observed stability of myrosinase and GSLs, and the production of GHPs.

2. Materials and Methods

2.1. Plant Material

The seeds of the cabbage accessions used for this study were sourced from the University of Warwick Crop Centre Genetic Resources Unit (Wellesbourne, UK), sown in controlled environments, transplanted to pots (2.5 L) and left to grow in the glasshouse for a short time before transplanting to 7 metre beds on the field. A detailed cultivation protocol can be found in Oloyede et al. [38]. Eighteen cabbage accessions from six different cabbage morphotypes (black kale (BK), wild (WD), tronchuda (TC), savoy (SC), red (RC), and white (WC)) were selected based on their head formation (closed heart or open leaf), geographical location and whether they were of hybrid descent (Supplementary Table S1). Cabbages were grown between 7th March and 25th November in the plant growth facilities, Whiteknights campus of the University of Reading, UK. White cabbage accession WC3, did not germinate under the growing condition.

Upon reaching commercial maturity (based on visual inspection), cabbages were harvested, immediately placed on ice in freezer bags and stored in a cold room (4 °C) for 24 h prior to processing. Average weight of each cabbage head per plant was 700 g and 300 g for closed heart and open leaf, respectively. For detailed climatic data and cross section of cultivated cabbages, refer to Supplementary Table S1 in this paper and Figures S1 and S2 in our earlier paper, Oloyede et al. [38].

2.2. Reagents and Chemicals

Sinigrin standard was obtained from Santa Cruz Biotechnology (Heidelberg, Germany) and D-glucose determination kit from R-Biopharm Rhone (Heidelberg, Germany). All other chemicals used were purchased from Sigma–Aldrich (Dorset, UK).

2.3. Cabbage Thermal Processing

In order to achieve a representative sample and remove senescent leaves, the outer leaves and central core of 4–5 cabbage heads (biological replicates) were removed and discarded. Cabbages were chopped into pieces of approximately 1 cm in width using a kitchen knife (representing how cabbages would normally be prepared by consumers), mixed together, washed under running tap water, and drained of excess water using a salad spinner (OXO Good Grips Clear Manual Salad Spinner, Chambersburg, PA, USA). Cabbages were subjected to steaming, microwave, or stir-fry cooking. Unprocessed (raw) cabbage samples were used as controls. Cooking methods were chosen to represent common ways of cooking cabbage.

Time and temperature combinations used for each method were based on a preliminary consumer study with 60 participants to determine consumer acceptability of the samples as steamed, microwaved, and stir-fried cabbage (Supplementary Table S2). These conditions were deemed acceptable with a mean score of between 2.7 and 3.8 on a 5-point degree of cooking scale, where "3" represents 'just about right' for extent of cooking (scale from not cooked enough "1", too much too overcooked "5").

2.3.1. Steaming

The method of Rungapamestry et al. [25] was adopted with slight modifications. A total of 120 g cabbage was placed in the topmost layer of a 3-tier, 18 cm stainless steel steamer (Kitchen craft, Birmingham, UK) containing already-boiling water (in the lowest layer) and allowed to steam for 2 min. Core temperature of cabbage during steaming ranged between 75 and 80 °C and was measured using a temperature probe.

2.3.2. Microwaving

The method of Rungapamestry et al. [25] was adopted. A total of 120 g of cabbage was put into 1-pint Pyrex glass jug, 16 mL water was added, and the jug was covered with a PVC cooking film pierced with 9 holes. Cabbages were microwaved for 3 min. Microwaving was carried out using a 900 W microwave at 60% power output (SANYO microwave oven EM-S355AW/AS, Osaka, Japan). A microwave thermometer was used to measure the core temperature of the cabbage during processing. Core temperature during processing ranged between 88 and 95 °C.

2.3.3. Stir-Frying

Cabbage samples were stir-fried as described by Rungapamestry et al. [39] with modifications. A total of 120 g cabbage was stir-fried in a frying pan for 90 s in 5 mL of preheated olive oil (100 °C) (Asda, Reading, UK) with continuous stirring using a wooden spatula. Core temperature of cabbage during stir-frying ranged between 65 and 70 °C and was measured using a temperature probe.

Samples were put into sterilin tubes immediately after cooking, placed on ice and transferred to a −80 °C freezer. Frozen samples were freeze-dried (Stokes freeze drier, Philadelphia, PA, USA), ground using a tissue grinder (Thomas Wiley® Mini-Mill, Thomas Scientific, Swedesboro, NJ, USA) and stored at −20 °C until further analysis.

2.4. Myrosinase Enzyme Extraction and Assay and Protein Content Analysis

Myrosinase enzyme was extracted using the method described by Ghawi et al. [32] with slight modifications, as described in our previous paper, Oloyede et al. [38]. Myrosinase enzyme was extracted from a 0.1 g sample (at 5 °C) using polyvinylpolypyrrolidone (PVPP) and Tris- HCL buffer. Using a D-glucose determination kit, myrosinase activity was determined following the coupled enzyme method as described by Gatfield and Sand [40] and Wilkinson et al. [41] with some modifications, as outlined in our preceding paper [38]. Myrosinase activity of the samples was calculated using a calibration curve prepared from myrosinase enzyme. One unit of myrosinase activity was defined as the amount of enzyme that produces 1 µmol of glucose per minute from sinigrin substrate at pH 7.5.

Protein content from the enzyme extract was determined using the Bradford method [42]. The method is based on protein complex formation with Brilliant Blue G dye with absorbance read 595 nm in a spectrophotometer (Perkin Elmer, Shelton, CT, USA). A standard curve was constructed using Bovine serum albumin (BSA) and used to calculate protein concentration in the enzyme extracts from which myrosinase-enzyme-specific activity (U/mg protein) was determined.

2.5. Glucosinolate and Glucosinolate Hydrolysis Products Analysis

GSLs and GHPs were extracted following the methods described by Bell et al. [43] and Bell et al. [44], respectively, with modifications as described in our earlier paper Oloyede et al. [38]. GSLs were extracted with 70% methanol, analysed by LC-MS/MS (Agilent, Bracknell, UK), and quantified using sinigrin hydrate standard. Six concentrations of sinigrin hydrate (14–438 µg/mL) were prepared with 70% methanol and used to prepare an external calibration curve (r^2 = 0.996). Compounds were identified using their mass parent ion, characteristic ion fragments and through comparing with ion data from literature (Table 1).

Table 1. Intact glucosinolates identified in cabbage accessions analysed by LC-MS.

Common Name	Chemical Name	Abbreviation	Mass Parent Ion	MS^2 Spectrum Ion (Base Ion in Bold)	Reference
sinigrin	2-propenyl(allyl)GSL	SIN	358	278, 275, **259**, 227, 195, 180, 162	[45,46]
gluconapin	3-butenyl GSL	GPN	372	292, 275, **259**, 195, 194, 176	[45,47]
epi/progoitrin	(R, S)-2-hydroxy-3-butenyl GSL	PROG	388	332, 308, 301, 275, **259**, 210, 195, 146, 136	[45–47]
glucoiberverin	3-(methylthio)propyl GSL	GIBVN	406	326, 275, **259**, 288, 228,195	[43,45,46]
glucoerucin	4-(methylthio)butyl GSL	GER	420	340, 291, 275, **259**, 227, 195, 178, 163	[43,45,46]
glucoiberin	3-(methylsulfinyl)propyl GSL	GIBN	422	407, **358**, 259	[45–47]
glucoraphanin	4-(methylsulfinyl)butyl GSL	GRPN	436	422, **372**, 291, 259, 194	[43,45,46]
glucobrassicin	3-indolylmethyl GSL	GBSN	447	275, **259**, 251, 205	[45–47]
4-hydroxyglucobrassicin	4-hydroxy-3-indolylmethyl GSL	4-HOH	463	383, **285**, 267, 259, 240, 195	[45–47]

Key: GSL = glucosinolate.

GHPs were extracted using dichloromethane and analysed by GC-MS (Agilent, Manchester, UK). Compounds were identified using the literature on ion data (Table 2; see Supplementary Figure S1 for GC-MS chromatograms) and quantified based on an external standard calibration curve. Five concentrations (0.15–0.5 mg/mL) of sulforaphane standard (Sigma Aldrich, UK) were prepared in DCM (r^2 = 0.99). Data analysis was performed using ChemStation for GC-MS (Agilent, Manchester, UK).

2.6. Statistical Analysis

Results are the average of three biological or processing replicates (each replicate consisting of leaves from 4–5 cabbage heads) and two technical replicates (n = 6). All statistical analyses were performed in XLSTAT (version 2019.4.2, Addinsoft, Paris, France). Data obtained were analysed using 2-way ANOVA with both cabbage accession (or morphotype) and processing conditions (raw, steamed, microwaved, and cooked) fitted as treatment effects. Tukey's HSD multiple pairwise comparison test was used to determine significant differences ($p < 0.05$) between samples. Principal component analysis (PCA) and multifactor analysis (MFA) were used to visualise the data in a minimum number of dimensions (two or three). MFA makes it possible to simultaneously analyse several tables of variables, showing relationships and correlations between the observations and variables, which were analysed in such a way that tables that include more variables do not outweigh other tables in the analysis.

Table 2. Glucosinolate hydrolysis products identified in cabbage accessions analysed by GC-MS.

Precursor Glucosinolate	Glucosinolate Hydrolysis Product		Abbreviation	LRI [a]	ID [b]	MS^2 Spectrum Ion (Base Ion in Bold)	Reference
	Common Name	Chemical Name					
sinigrin	allyl thiocyanate	2-propenyl thiocyanate	ATC	871	B	99, 72, 45, 44, **41**, 39	[48]
	allyl-ITC	2-propenyl isothiocyanate	AITC	884	B	**99**, 72, 71, 45, 41, 39	[48,49]
	1-cyano-2,3-epithiopropane	3,4-epithiobutane nitrile	CETP	1004	B	**99**, 72, 66, 59, 45, 41, 39	[48]
gluconapin	3-butenyl-ITC	1-butene, 4-isothiocyanate	3BITC	983	B	113, 85, **72**, 64, 55, 46, 45, 41	[48–50]
	4,5-epithiovaleronitrile	1-cyano-3,4-epithiobutane	EVN	1121	B	**113**, 86, 80, 73, 60, 45	[50]
progoitrin	goitrin	5-vinyloxazolidin-2-thione	GN	1545	B	**129**, 86, 85, 68, 57, 45, 43, 41, 39	[51]
	1-cyano-2-hydroxy-3,4-epithiobutane isomer 1	2-hydroxy-3,4-epithiobutylcyanide diastereomer-1	CHETB-1	1225	B	**129**, 111, 89, 84, 68, 61, 58, 55, 45	
	1-cyano-2-hydroxy-3,4-epithiobutane isomer 2	2-hydroxy-3,4-epithiobutylcyanide diastereomer-2	CHETB-2	1245	B	**129**, 111, 89, 84, 68, 61, 58, 55, 45	
glucoiberverin	iberverin	3-methylthiopropyl-ITC	IBVN	1307	B	147, **101**, 86, 73, 72, 61, 47, 46, 41	[48]
	4-methylthiobutyl nitrile	4-methylthio butanenitrile	4MBN	1085	B	115, 74, 68, **61**, 54, 47, 41	
glucoerucin	erucin	4-(methylthio)-butyl-ITC	ER	1427	B	161, 146, **115**, 85, 72, 61, 55	[48,49]
	erucin nitrile	1-cyano-4-(methylthio) butane	ERN	1200	B	129, 87, 82, **61**, 55, 48, 41, 47	
glucoiberin	iberin	3-methylsulfinylpropyl-ITC	IB	1617	B	163, 130, 116, 102, 100, 86, **72**, 63, 61,41	[48]
	iberin nitrile	4-methylsulfinylbutanenitrile	IBN	1384	B	**131**, 78, 64, 47, 41	
gluconasturtin	2-phenylethyl-ITC	2-isothiocyanatoethyl benzene	PEITC	1458	B	163, 105, **91**, 65, 51, 40	[48]
	benzenepropanenitrile	2-phenylethyl cyanide	BPN	1238	B	131, **91**, 85, 65, 63, 57, 44, 51	[50]
glucoraphanin	sulforaphane	4-methylsulfinylbutyl-ITC	SFP	1757	A	160, 114, 85, **72**, 64, 63, 61, 55. 41, 39	[44,49]
	sulforaphane nitrile	5-(methylsulfinyl) pentanenitrile	SFN	1526	B	145, 128, 82, 64, **55**, 41	
glucobrassiccin	indole-3-carbinol	1H-indole-3-methanol	I3C	1801	B	**144**, 145, 116, 108, 89	[51]
	indoleacetonitrile	1H-indole-3-acetonitrile	1IAN	1796	B	**155**, 145, 144, 130, 116, 89, 101, 63	[52]
pentyl glucosinolate	pentyl-ITC	1-isothiocyanato-pentane	PITC	1165	B	129, 114, 101, 96, 72, 55, **43**, **41**, 39	[53]
indole	1H-indole	Indole (8CI)	1H-I	1290	B	**117**, 90, 89, 63, 58	[54]
glucotropaeolin	benzeneacetonitrile	2-phenylacetonitrile	BAN	1137	A	**117**, 90, 89, 77, 63, 51	

Key: ITC- isothiocyanate. [a] Linear retention index on a HP-5MS non-polar column. [b] A, mass spectrum and LRI agree with those of authentic compound; B, mass spectrum agrees with reference spectrum in the NIST/EPA/NIH mass spectra database and those in the literature.

3. Results and Discussion

3.1. *Effect of Domestic Cooking on Residual Myrosinase Enzyme Activity (Relative Activity) across Cabbage Morphotypes and Accessions*

The myrosinase stability of cabbage accessions after domestic cooking was studied and the results are presented in Figure 1, with relative activity results presented in Supplementary Table S3. Relative activity is defined as the ratio of myrosinase activity of processed (cooked) cabbage to unprocessed (raw) cabbage (A/A_0). Domestic cooking affected the stability of myrosinase enzyme. Myrosinase stability differed significantly ($p < 0.05$) between domestic cooking processes, where there was no difference between steaming and microwaving ($p = 0.912$), but these processes both differed significantly from stir-frying ($p < 0.0001$). Myrosinase was most stable after stir-frying, retaining up to 65% (i.e., $A/A_0 = 0.65$, Supplementary Table S3) of its activity in some studied accessions. Steaming and microwaving resulted in losses of myrosinase activity of up to 98% and 99%, respectively, with the highest stability of 15% and 13%, respectively. Rungapamestry et al. [39], in their study of broccoli florets, reported that stir-frying retained the highest myrosinase activity (17%) compared to boiling (14%).

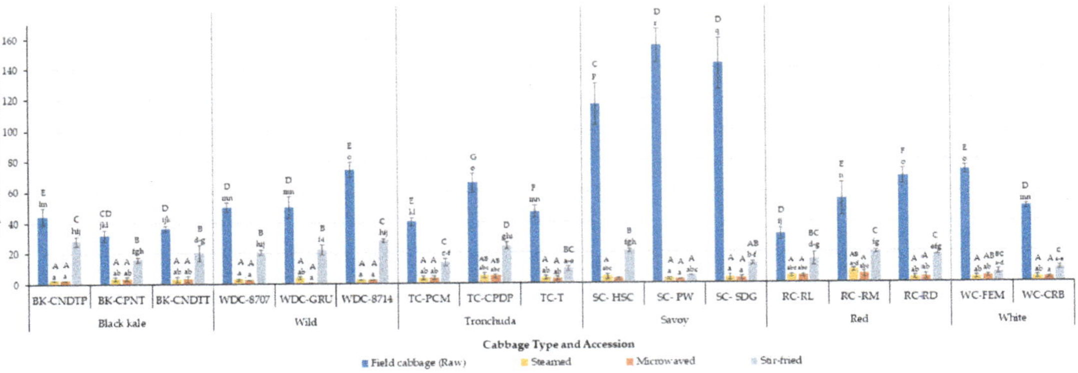

Figure 1. Comparison of the myrosinase activity of raw versus cooked cabbage morphotypes and accessions (U/g DW). Values are means of three biological (raw samples) or processing (cooked samples) replicates (each replicate comprising 4–5 cabbage heads) and two technical replicates ($n = 6$). Error bars represent standard deviation from mean values. Letters "A–G": bars not sharing a common uppercase letter differ significantly ($p < 0.0001$) between accessions and treatments within a cabbage morphotype. Letters "a–r": bars not sharing a common lowercase letter differ significantly ($p < 0.0001$) between cabbage morphotypes, accessions, and treatments. Key: BK-CNDTP: cavolo nero di toscana o senza palla; BK-CPNT: cavolo palmizio; BK-CNDTT: cavolo nero di toscana o senza testa; WD-8707: wild cabbage 8707; WD-GRU: wild cabbage 7338; WD-8714: wild cabbage 8714; TC-PCM: penca mistura; TC-CPDP: penca povoa; TC-T: tronchuda; SC-HSC: hybrid savoy wirosa; SC-PW: pointed winter; SC-SDG: dark green; RC-RL: red langendijker; RC-RM: rocco marner (hybrid); RC-RD: red Danish; WC-FEM: early market; WC-CRB: couve repolho.

The effect of domestic cooking processes on myrosinase stability varied among cabbage morphotypes and accessions and will be discussed in more detail later. The stability of myrosinase in different *Brassica* vegetables under different processing conditions has been discussed by several authors [24–27,39,55]. Differences in myrosinase stability as a result of cooking can be attributed to the maximum core temperature of the vegetable during heating. In our study, stir-frying had the lowest core temperature (65–70 °C) compared to steaming (75–80 °C) and microwaving (88–95 °C). It has previously been reported that, to prevent myrosinase inactivation, the maximum core temperature that cabbage should reach is between 50 and 60 °C, which can be achieved by steaming for 7 min or microwaving (700 W) for 120 s [25]. However, in the study stated, the cabbage samples were cut into wedges, which is not representative of how cabbages are generally prepared before cooking, so the cooking times to reach the same core temperature in cabbage that is more finely chopped would be shorter.

Verkerk and Dekker [24] reported that inactivation of myrosinase enzyme during microwave cooking is affected by the time-energy output combination. Their study showed that a considerable amount of myrosinase activity was retained when red cabbage was microwaved at 180 W for 24 min and 540 W for 8 min, while microwaving for 4.8 min at 900 W resulted in total loss of myrosinase activity even though the total energy output of all three processes was the same (259.2 KJ). The authors explained the resulting effect as a function of the time it takes for the cabbage to reach its maximum core temperature, with a higher energy output and shorter time reaching a high (100 °C) core temperature faster and maintaining that core temperature for the remaining cooking time, while the lower energy output with a longer cooking time resulted in a maximum core temperature of 90 °C at a much slower rate.

In the current study, a physical examination of the cooked cabbage samples showed that the stir-fried cabbage looked firmer than the steamed and microwaved cabbage, which can be a helpful way to assess the severity of the thermal process. The intense heat during stir-frying can lead to drying out of the surface area, thereby resulting in a firmer texture, which reduces the rate of heat penetration as a result of less damage to the cell wall [39,56].

3.2. *Comparison of the Myrosinase Activity of Raw versus Cooked Cabbage Morphotypes and Accessions*

Figure 1 shows the myrosinase activity and subsequent thermal stability of the 17 studied cabbage accessions. Significant differences ($p < 0.0001$) were observed in the myrosinase activity and stability of cabbages as a result of cabbage morphotype, accession, cooking method, and the interactions between these parameters.

There was no relationship between myrosinase activity in raw cabbage and myrosinase stability post-cooking; indeed, some accessions which had high activity in raw cabbage had the lowest stability. Raw savoy cabbage accessions (SC-HSC, SC-PW, and SC-SDG) had the highest myrosinase activity in all studied accessions (116.3, 142.6 and 154.8 U/g DW, respectively) while raw black kale accessions (BK-CNDTP, BK-CPNT, and BK-CNDTT) had the lowest myrosinase activity (31.5, 36.3 and 44.4 U/g DW, respectively). However, black kale, Tronchuda and red cabbage accessions had the highest enzyme stability, while savoy and white cabbage accessions, which had the highest myrosinase activity, were the least stable after domestic processing (Figure 1). As discussed earlier, steaming and microwaving resulted in lower myrosinase stability overall, with up to 99% inactivation occurring in some cases. However, a critical look at the stability of myrosinase in steamed and microwaved cabbages (Figure 1) shows that some accessions had relatively higher myrosinase stability compared to others. Red cabbage accessions RC-RM and RC-RL were the most stable, retaining up to 15% after steaming (RC-RM) and 13% after microwaving (RC-RL). This result is in agreement with the results of Yen and Wei [27], who stated that red cabbage myrosinase was more stable than white cabbage myrosinase after thermal processing.

A possible reason for the difference in myrosinase stability across accessions might be due to differences in the myrosinase isoenzymes found in each accession, with the red cabbage accessions having more thermally stable myrosinase isoenzyme. Red cabbage contains anthocyanins, which, in addition to being bioactive compounds with health promoting properties, are also pigments that offer effective protection to plants under stress [57,58]. Therefore, red cabbage is more adapted to stressful conditions, and it stands to reason that the myrosinase isoenzyme in red cabbage may also be adapted to operating under heat stress.

Different types of myrosinase isoenzymes have been identified and they vary between *Brassica* vegetables. They can differ to some extent in characteristics and activity, with distribution in plants appearing to be both organ- and species-specific [27,59]. Rask et al. [60] reported that some of the myrosinase isoforms form complexes by interacting with myrosinase-binding proteins, which may enhance their stability during processing.

The myrosinase activity values obtained in this study were higher in most cases than those reported by other authors [25,61], except in the case of white cabbage accessions, where values were similar to those obtained by Penas et al. [62]. This might be because, in most previous studies, cabbages were obtained from supermarkets, while in this study and the study conducted by Penas et al. [62], the cabbages were grown for the experiment and transferred into cold conditions immediately after harvest. These minimal transfer and storage times reduce the postharvest effects experienced by the supermarket samples.

There was no relationship found between accession origin, physical characteristics (open-leaf or heart-forming) and whether cabbages were hybrid on the myrosinase activity and stability of the accessions studied.

3.3. Protein Content and Specific Activity of Raw and Cooked Cabbages

The protein content and specific activity of cabbage myrosinase before and after cooking is presented in Table 3. There were significant ($p < 0.05$) differences in the protein content and specific activity of all accessions for both raw and cooked samples. Protein content decreased with cooking, with the rate of reduction corresponding to the severity of the cooking process. Stir-fried samples had significantly higher protein contents than steamed and microwave samples. Black kale and red accessions, with the highest protein content, also retained the most protein after stir-frying (up to 87% in BK-CNDTP) but the lowest after steaming and microwaving (up to 67% in steamed BK-CPNT). This can be attributed to the denaturation of protein into free amino acids during cooking.

Cooking led to a significant reduction ($p < 0.05$) in the specific activity of cabbage samples. The specific activity of the cabbages followed a similar trend to myrosinase activity and protein content, where specific myrosinase activity decreased with the severity of the cooking method. The result shows a correlation between myrosinase activity and specific activity, implying that denaturation of the protein is equal to denaturation of the enzyme. Stir-fried cabbages had the most stable specific activity and differed significantly ($p < 0.001$) from steamed and microwaved samples between accessions for all studied morphotypes, with the exception of savoy morphotype, where no significant difference was observed in specific activity between accessions for the studied cooking methods (Table 3). It is worth mentioning that the results obtained for savoy accessions were mostly due to the significantly higher specific activity of the raw samples, instead of a comparable stability across the cooking methods. Similar to our earlier discussion on myrosinase stability, samples with the highest specific activity were not always the most stable after cooking. For example, Savoy cabbage accessions (SC-HSC, SC-PW, and SC-SDG), which had the highest specific activity (4.7, 6.4 and 5.8 U/mg soluble protein, respectively) had the lowest stability after cooking, with an up 97% loss in specific activity after steaming and microwave cooking observed in SC-PW accession. On the other hand, black kale accessions, with some of the lowest specific activity in raw samples, retained the most specific activity after cooking, with up to 80% specific activity observed in BK-CNDTT accession. As expected, the result obtained is in agreement with myrosinase activity results discussed earlier, where black kale accessions with the least myrosinase activity were the most stable after domestic cooking. The differences observed in specific activity can be attributed to variations in myrosinase isoenzyme stability for the different morphotypes and accessions, as discussed in Sections 3.1 and 3.2.

Table 3. Protein content ((mg/g ± SD) DW) and specific activity ((U/mg soluble protein ± SD) DW) of cabbage accessions before and after domestic processing.

Cabbage Morphotype [a]/Accession	Protein Content (mg/g ± SD) DW				Specific Activity (U/mg Soluble Protein ± SD) DW			
	Raw	Steamed	Microwaved	Stir-Fried	Raw	Steamed	Microwaved	Stir-Fried
Black Kale								
BK-CNDTP	33.7 ± 0.6 $^{no, E}$	11.0 ± 0.3 $^{ab, A}$	11.2 ± 0.4 $^{ab, A}$	29.0 ± 0.7 $^{kl, D}$	1.3 ± 0.2 $^{d–k, D}$	0.2 ± 0.0 $^{a, A}$	0.2 ± 0.0 $^{a, A}$	1.0 ± 0.1 $^{a–j, C}$
BK-CPNT	35.4 ± 1.0 $^{op, EF}$	11.7 ± 0.6 $^{b, A}$	11.9 ± 1.4 $^{b, A}$	21.6 ± 1.9 $^{hi, B}$	0.9 ± 0.1 $^{a–i, BC}$	0.3 ± 0.1 $^{abc, A}$	0.3 ± 0.1 $^{abc, A}$	0.7 ± 0.0 $^{a–h, B}$
BK-CNDTT	36.7 ± 0.7 $^{p, F}$	12.7 ± 0.1 $^{bc, A}$	12.5 ± 0.1 $^{bc, A}$	24.9 ± 1.6 $^{j, C}$	1.0 ± 0.0 $^{a–j, C}$	0.2 ± 0.1 $^{a, A}$	0.3 ± 0.1 $^{ab, A}$	0.8 ± 0.1 $^{a–h, BC}$
Wild								
WD-8707	31.4 ± 0.1.2 $^{lmn, E}$	11.1 ± 0.1 $^{ab, A}$	10.9 ± 0.1 $^{ab, A}$	19.1 ± 0.4 $^{fgh, C}$	1.6 ± 0.1 $^{g–l, C}$	0.2 ± 0.1 $^{a A}$	0.2 ± 0.0 $^{a, A}$	1.1 ± 0.1 $^{a–j, B}$
WD-GRU	29.9 ± 0.6 $^{kl, D}$	10.7 ± 0.4 $^{ab, A}$	10.6 ± 0.1 $^{ab, A}$	18.1 ± 1.1 $^{efg, C}$	1.7 ± 0.2 $^{h–l, C}$	0.3 ± 0.1 $^{a–c A}$	0.2 ± 0.0 $^{a, A}$	1.2 ± 0.2 $^{c–k, B}$
WD-8714	30.6 ± 0.8 $^{lm, DE}$	10.9 ± 0.1 $^{ab, A}$	11.0 ± 0.2 $^{ab, A}$	16.9 ± 0.5 $^{def, B}$	2.4 ± 0.2 $^{l, D}$	0.2 ± 0.0 $^{a A}$	0.2 ± 0.0 $^{a, A}$	1.7 ± 0.1 $^{h–l, C}$
Tronchuda								
TC-PCM	33.6 ± 0.2 $^{no, F}$	11.1 ± 0.3 $^{ab, A}$	11.1 ± 0.1 $^{ab, A}$	19.9 ± 1.47 $^{gh, D}$	1.2 ± 0.1 $^{b–k, D}$	0.3 ± 0.1 $^{a–c A}$	0.3 ± 0.1 $^{a–c, A}$	0.7 ± 0.1 $^{a–g, C}$
TC-CPDP	27.8 ± 0.6 $^{k, E}$	11.0 ± 0.3 $^{ab, A}$	11.0 ± 0.3 $^{ab, A}$	18.1 ± 0.8 $^{efg, C}$	2.4 ± 0.3 $^{l, E}$	0.4 ± 0.2 $^{a–e ABC}$	0.4 ± 0.1 $^{a–e, AB}$	1.4 ± 0.2 $^{e–k, D}$
TC-T	33.1 ± 0.8 $^{mno, F}$	10.9 ± 0.2 $^{ab, A}$	10.8 ± 0.2 $^{ab, A}$	15.7 ± 0.9 $^{de, B}$	1.4 ± 0.1 $^{f–k, D}$	0.3 ± 0.1 $^{abc A}$	0.3 ± 0.1 $^{abc, A}$	0.6 ± 0.1 $^{a–f, BC}$
Savoy								
SC-HSC	24.6 ± 1.43 $^{j, D}$	10.7 ± 0.4 $^{ab, AB}$	10.6 ± 0.3 $^{ab, AB}$	12.0 ± 1.1 $^{b, B}$	4.7 ± 0.3 $^{n, B}$	0.4 ± 0.1 $^{a–d A}$	0.2 ± 0.0 $^{ab, A}$	1.8 ± 0.2 $^{i–l, A}$
SC-PW	24.3 ± 0.3 $^{j, D}$	12.0 ± 1.2 $^{b, B}$	10.1 ± 0.2 $^{b, AB}$	14.8 ± 0.4 $^{cd, C}$	6.4 ± 0.5 $^{o, B}$	0.2 ± 0.0 $^{a A}$	0.2 ± 0.0 $^{a, A}$	0.3 ± 0.0 $^{abc, A}$
SC-SDG	24.4 ± 0.5 $^{j, D}$	10.3 ± 0.4 $^{ab, AB}$	8.9 ± 0.2 $^{a, A}$	11.4 ± 0.3 $^{ab, AB}$	5.8 ± 0.7 $^{o, B}$	0.3 ± 0.1 $^{abc A}$	1.5 ± 0.1 $^{f–l, A}$	1.1 ± 0.1 $^{a–j, A}$
Red								
RC-RL	33.6 ± 0.6 $^{no, E}$	11.0 ± 0.3 $^{ab, A}$	11.2 ± 0.4 $^{ab, A}$	29.0 ± 0.7 $^{kl, D}$	0.9 ± 0.1 $^{a–j, D}$	0.4 ± 0.1 $^{a–d A}$	0.4 ± 0.1 $^{a–d, A}$	0.5 ± 0.1 $^{a–f, ABC}$
RC-RM	35.4 ± 1.0 $^{op, EF}$	11.7 ± 0.6 $^{b, A}$	11.9 ± 1.4 $^{b, A}$	21.6 ± 1.9 $^{hi, B}$	1.5 ± 0.3 $^{g–l, E}$	0.7 ± 0.1 $^{a–g BCD}$	0.4 ± 0.1 $^{a–e, AB}$	0.9 ± 0.1 $^{a–i, D}$
RC-RD	36.7 ± 0.7 $^{p, F}$	12.7 ± 0.1 $^{bc, A}$	12.5 ± 0.1 $^{bc, A}$	24.9 ± 3.9 $^{j, C}$	1.9 ± 0.1 $^{jkl, F}$	0.2 ± 0.1 $^{a A}$	0.3 ± 0.1 $^{ab, A}$	0.7 ± 0.1 $^{a–h, CD}$
White								
WC-FEM	21.3 ± 0.4 $^{hi, C}$	10.1 ± 0.3 $^{ab, A}$	10.1 ± 0.1 $^{ab, A}$	10.9 ± 0.2 $^{ab, A}$	3.4 ± 0.2 $^{m, E}$	0.3 ± 0.1 $^{abc, A}$	0.4 ± 0.1 $^{a–d, AB}$	0.6 ± 0.2 $^{a–f, BC}$
WC-CRB	23.0 ± 1.2 $^{ij, D}$	10.2 ± 0.1 $^{ab, A}$	10.2 ± 0.1 $^{ab, A}$	12.1 ± 0.7 $^{b, B}$	2.1 ± 0.1 $^{kl, D}$	0.3 ± 0.1 $^{abc, A}$	0.2 ± 0.1 $^{ab, A}$	0.7 ± 0.2 $^{a–h, C}$

[a] Names in bold refer to cabbage morphotype. Values are means of three processing replicates and two technical replicates ($n = 6 \pm$ SD). SD: standard deviation from mean. Letters "A–F": mean values not sharing a common uppercase letter differ significantly ($p < 0.05$) between accessions and treatments within a cabbage type for each parameter (i.e., protein content and specific activity). Letters "a–p": mean values not sharing a common lowercase letter differ significantly ($p < 0.05$) between cabbage types, accessions, and treatments for each parameter (i.e., protein content and specific activity). Key: BK-CNDTP: cavolo nero di toscana o senza palla BK-CPNT: cavolo palmizio; BK-CNDTT: cavolo nero di toscana o senza testa; WD-8707: wild cabbage 8707; WD-GRU: wild cabbage 7338; WD-8714: wild cabbage 8714; TC-PCM: penca mistura; TC-CPDP: penca povoa; TC-T: tronchuda; SC-HSC: hybrid savoy wirosa; SC-PW: pointed winter; SC-SDG: dark green; RC-RL: red langendijker; RC-RM: rocco marner (Hybrid); RC-RD: red Danish; WC-FEM: early market; WC-CRB: couve repolho.

3.4. Effect of Domestic Cooking on GSL Profile and Concentration of Cabbage Accessions

GSL profile and concentrations for all samples before and after cooking are presented in Figure 2, with significant differences within and between cabbage morphotypes presented in Supplementary Table S4.

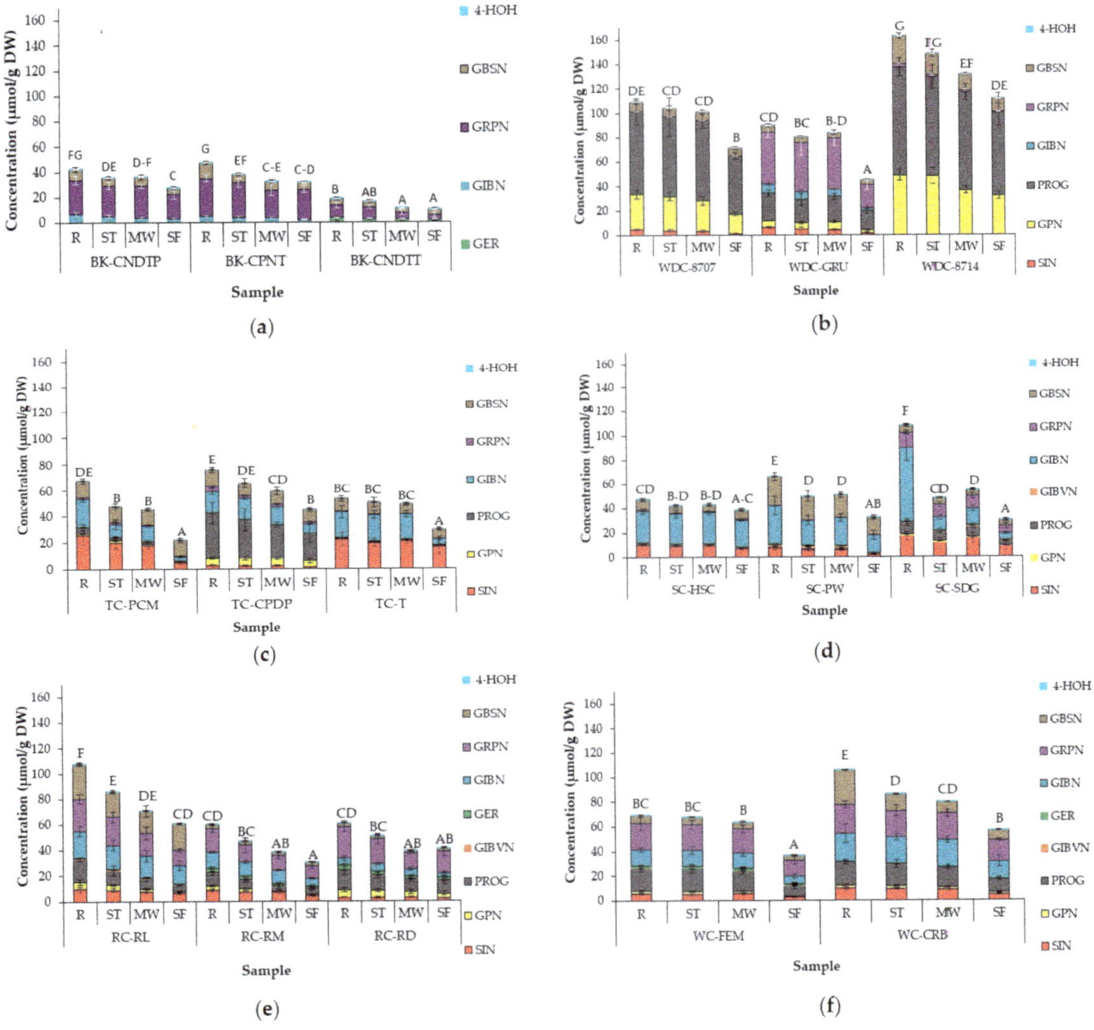

Figure 2. Glucosinolate concentrations (µmol/g DW) in different accessions of (**a**) Black kale; (**b**) Wild cabbage (**c**) Tronchuda cabbage; (**d**) Savoy cabbage; (**e**) Red cabbage and (**f**) White cabbage before and after domestic processing. Error bars represent standard deviation from mean values. Letters above bars refer to differences in total GSL concentration. Letters 'A–G': bars not sharing a common uppercase letter differ significantly ($p < 0.05$) between accession and cooking conditions within a cabbage morphotype (i.e., within each separate graph). For significant differences between cabbage morphotypes, accessions, and cooking methods (i.e., across the separate cabbage morphotype graphs) see Supplementary Table S4. Abbreviations: R = raw, ST = steamed; MW = microwaved; SF = stir-fried; BK-CPNT: cavolo palmizio; BK-CNDTT: cavolo nero di toscana o senza testa; WD-8707: wild cabbage 8707; WD-GRU: wild cabbage 7338; WD-8714: wild cabbage 8714; TC-PCM: penca mistura; TC-CPDP: penca povoa; TC-T: tronchuda; SC-HSC: hybrid savoy wirosa; SC-PW: pointed winter; SC-SDG: dark green; RC-RL: red langendijker; RC-RM: rocco marner (Hybrid); RC-RD: red Danish; WC-FEM: early market; WC-CRB: couve repolho. For abbreviations of compounds, see Table 1 (GSLs).

GSL profile and concentrations varied across accessions within and between cabbage morphotypes. From five to nine individual GSLs were identified within all cabbages studied; seven aliphatic GSLs, namely sinigrin (SIN), gluconapin (GPN) and epi/progoitrin (PROG), Glucoibeverin (GIBVN), glucoerucin (GER), glucoiberin (GIBN) and glucoraphanin (GRPN) and two indole GSLs: glucobrassicin (GBSN) and 4-hydroxyglucobrassicin (4-HOH) (Table 1). Black kale accessions had the lowest number of identified GSLs (five), while nine GSLs were identified in red and white cabbages. GBSN and 4-HOH were the only GSLs identified in all studied accessions. Total GSLs differed significantly between accessions ($p < 0.0001$), post different cooking methods ($p < 0.0001$) and in the interaction between these two factors ($p < 0.0001$). Aliphatic GSLs were the most abundant GSLs in all accessions, making up about 95% of total GSLs.

Cooking significantly reduced GSL concentration in all cabbage samples. GSL stability varied across the accessions and cooking methods studied. GIBN was the least stable GSL resulting in an average loss of 59% across all accessions. However, GIBN loss varied largely between accessions, with tronchuda accession TC-PCM recording a loss of up to 83%, while the loss in savoy SC-HSC was as low as 14%. The obtained results agree with those reported by Oerlemans et al. [28] and Dekker et al. [63], who report variations in GSL stability between GSLs and variations in the stability of the same GSL across different *Brassica* vegetables. In a previous study, concentrations of GIBN (aliphatic GSL) and GBSN (indole GSL) in white cabbage were found to significantly decrease during cooking due to their high potential to leach into the cooking water [64,65].

Total GSLs in steamed cabbage ranged between 16.5 μmol/g DW (BK-CNDTT) and 148.8 μmol/g DW (WD-8714). There was a significant difference in GSL concentrations of steamed cabbages across accessions and between accessions of the same cabbage morphotype. The observed differences were mostly due to the initial GSL concentration of the raw samples rather than the steaming process. In relation to the residual GSL content of cabbage samples after steaming, steamed WC-FEM had the most stable total GSL, retaining up to 97% GSL concentration, while the biggest loss of total GSL was in steamed SC-SDG, where up to 56% loss was recorded. The differences observed in GSL stability may be due to variations in leaf thickness between the accessions, which would impact the rate of heat transfer within the leaves. Thicker leaves would lead to a slower heat transfer rate within the leaves, resulting in reduced GSL degradation and better stability compared to thinner leaves. In some accessions, steaming did not affect the concentrations of some individual GSLs, e.g., SIN and PROG in WD-8714 and WC-FEM, respectively. There was a significant ($p < 0.0001$) reduction in individual and total GSL content for all samples after cooking, except for GPN, which did not differ significantly from raw to cooked samples within each accession for most of the studied accessions. The stability of individual GSLs varied greatly between accessions, within and between cabbage morphotypes. For example, after steaming, in BK samples, the loss of GRPN did not differ between the three accessions (8–10%), while in TC samples, it led to a between 44% (TC-CPDP) and <1% (TC-PCM) loss of GRPN content.

Previous studies reported no loss [25,29,35,66,67] or minimal losses [30,31,35,68] of GSL in broccoli, turnip, and cabbages after domestic processing. Xu et al. [30] reported a loss of about 15% in steamed red cabbage; however, the large sample size (3 cm cubes) may have caused lower losses in comparison to the present study, as this would have had an impact on the core temperature of the samples during cooking. Similar to the current study, Vallejo et al. [31] reported losses in some individual GSL (GRPN) and no loss in others (GIBN) after steaming for 3.5 min. In kale samples steamed for 15 min, SIN degraded more rapidly than GIBN, with more than 80% and 40% loss recorded for SIN and GIBN, respectively [33]. The minimal GSL losses reported in steamed samples were due to the low levels of leaching into cooking water compared to that normally reported under boiling conditions [23].

In microwaved samples, total GSL varied between 11.2 μmol/g DW (BK-CNDTT) and 131.6 μmol/g DW (WD-8714). Microwaving significantly affected the amount of GSLs in cabbage samples, with reductions up to 76% of GRPN in TC1 and residual total GSL varying between 50% and 93%. Microwaving led to significantly lower GSL concentrations when compared to raw cabbages. As in steamed samples, the effect of microwaving differed between accessions and individual GSLs. Some GSLs were more stable than others in certain accessions within and between cabbage morphotypes. As discussed in Section 3.1, high core temperatures (85–95 °C) of microwaved samples led to myrosinase enzyme inactivation, which could have prevented GSL hydrolysis during the microwave process and can account for the high retention of GSL concentrations in some microwaved cabbages.

There are several conflicting reports on the effect of microwaving on GSL content in *Brassica* vegetables. In a recent study, no significant difference was observed in broccoli and red cabbage samples microwaved under different time and power combinations while retaining the same final energy (1080 kJ) [35]. Song and Thornalley [29] and Xu et al. [30] also reported no significant difference in GSL concentration after microwaving green and red cabbage samples for three and five minutes, respectively. The authors stated that the stability of GSL might be due to myrosinase inactivation, and that the absence of water during microwaving prevented GSL leaching into cooking water. The large size of the shredded cabbage pieces in the two studies may also have reduced the loss of GSLs. A study on broccoli resulted in a 74% decrease in total GSL content after microwaving and was attributed to leaching in water and more intense microwave conditions (150 g broccoli to 150 g water and microwaving for 5 min at 1000 W power) [31]. However, a contrary result was observed by Verkerk and Dekker [24] and Oerlemans et al. [28], who reported an increase of up to 78% and 35%, respectively in GSL concentrations after microwaving red cabbage, though the increase was not significant in the Oerlemans et al. [28] study due to the large sample variability. The authors attributed the increase to the enhanced extractability of GSL after microwaving, which could be more of an analytical artefact than an actual increase in GSL concentration.

Stir-frying led to a significant decrease in the total and individual GSL content of cabbages. Total GSL ranged between 10.3 µmol/g DW (BK-CNDTT) and 111.7 µmol/g DW (WD-8714). There was a significant difference in GSL concentrations between accessions, within and between cabbage morphotypes. Residual total GSL varied between 28% (SC-SDG) and 81% (SC-HSC). The highest loss of aliphatic individual GSL concentration was recorded in stir-fried TC-PCM accession, where there was a decrease of between 79 to 83%. Indole GSLs, GBSN and 4-HOH were the most stable GSLs in stir-fried cabbages. The relative thermostability of individual GSLs (if under the same myrosinase level and stability) can be influenced by their chemical structure and has been reported to vary with heating temperature [28,69]. Among all the studied cooking methods, stir-frying resulted in significantly greater losses of GSL than steaming or microwaving, which agrees with previous reports. A study on the effect of different types of cooking oil on GSLs in stir-fried broccoli resulted in up to 49% losses, irrespective of the cooking oil used [70]. Xu et al. [30] also reported a 77% loss in GSL concentration after stir-frying red cabbage while there was no significant loss in GSL content when green cabbage was stir-fried for 5 min [29]. The difference in leaf structure may have influenced GSL stability in green cabbage. Green cabbage can have thicker leaves with a more uneven surface texture, which may create a microclimates around the leaf during the cooking process [29]. It was observed in the present study that green cabbage tended to have thicker leaves than other morphotypes based on visual observation. It is hypothesized that losses due to stir-frying can be attributed to substantial moisture evaporation. During stir-frying, cabbage loses moisture and GSLs are leached into the moisture, which evaporates during the cooking process. A study conducted by Adler-Nissen [56] showed that when carrot cubes were stir-fried, despite temperatures only reaching 70 °C, a high evaporation loss was observed. Another possible reason for the lower GSL amounts in stir-fried cabbages can be attributed to GSL hydrolysis by myrosinase and ESP during the cooking process. As mentioned previously, (see Sections 3.1 and 3.2), the low core temperatures (65–70 °C) of stir-fried cabbages resulted in higher myrosinase stability of the samples when compared to steamed and microwaved cabbages. Contrary to the findings of this study, however, some recent studies showed that stir-frying preserved total and most individual GSL contents of various *Brassica* vegetables, with the authors attributing this to quick myrosinase inactivation and no leaching losses into cooking water [33,71,72].

The relative stability of individual and total aliphatic GSLs to indoles varied between accessions and cooking methods, however, generally, indole GSLs were more stable than aliphatics. Previous studies have largely reported indole GSLs to be more heat-labile under domestic cooking conditions than aliphatic GSLs [31,33,64,73–75]. The higher stability of indoles in the present study may be due to the type of *Brassica* and cooking methods that were investigated. The nature of the vegetable matrix and heat treatment have been shown to have an effect on the stability of individual indole GSLs. For example, in the absence of leaching and enzymatic degradation during thermal treatment, 4HOH has been reported to be more thermolabile than GBSN and neoglucobrassicin (NEO) [28,75,76]. NEO has also been found to be more thermostable than GIBVN after roasting broccoli sprouts for 15 min [75]. It is important to mention that the GSL concentrations obtained in the current study are much higher than those reported for mature cabbage in the literature [62,65,77–80]. However, up to 80 µmol/g DW [81] and 111 µmol/g DW [82] have been reported in mature rocket leaves, which is similar to the concentrations reported for most accessions in this study. We hypothesis that the reason for the high GSL accumulation in the accessions studied may be because they are gene bank accessions, which means they have not been thoroughly characterized for their phytochemical content. It is common for wild Brassicas to have much higher concentrations than cultivated varieties, and many of the high GSL genotypes that have been bred came from crosses with gene bank material, an example of which is the "*Beneforte*" broccoli [83].

In summary, WD-8707 accession had the most stable individual and total GSL, while GSLs of SC-SDG were the most thermolabile across all cooking methods, despite having one of the highest GSL concentrations in the raw sample. Different accessions of the same cabbage morphotype can vary in their GSL stability during cooking, resulting in large differences in GSL loss between species. The rate and extent of loss is dependent on the morphotype of the cabbage, sample cut size, cooking time and temperature, amount of moisture, and initial concentration of GSL [65,66]. The variation in residual GSL in the cabbages will have an impact on the amounts of GHPs produced.

3.5. *Effect of Domestic Cooking on GHP Profile and Concentration in Cabbage Accessions*

The profile and concentration of GHPs resulting from cooking cabbage are presented in Figure 3, with significant differences between and within the cabbage morphotypes presented in Supplementary Table S5. Twenty-three (23) different GHPs were detected as a result of GSL hydrolysis during cooking. Accession, cooking method and interaction between the two significantly ($p < 0.0001$) influenced GHP profile and concentrations. Total GHPs across all accessions and cooking methods ranged between 0.33 µmol/g DW (microwaved TC-T) and 18.66 µmol/g DW (raw WD-8707). In raw samples, GSL hydrolysis led to the production of majorly nitriles and epithionitriles. Matusheski and Jeffery [84] and Mithen et al. [85], in their studies of fresh and freeze-dried raw broccoli, found that GRPN hydrolysis primarily led to the formation of SFN rather than its ITC, SFP. In most studied accessions, raw and stir-fried cabbages had the highest total GHPs in all samples, apart from red and white cabbage accessions, where the highest total GHPs was recorded in steamed cabbages. Black kale samples had the lowest identified GHPs, which could be related to the lower number of individual GSL present in the accession.

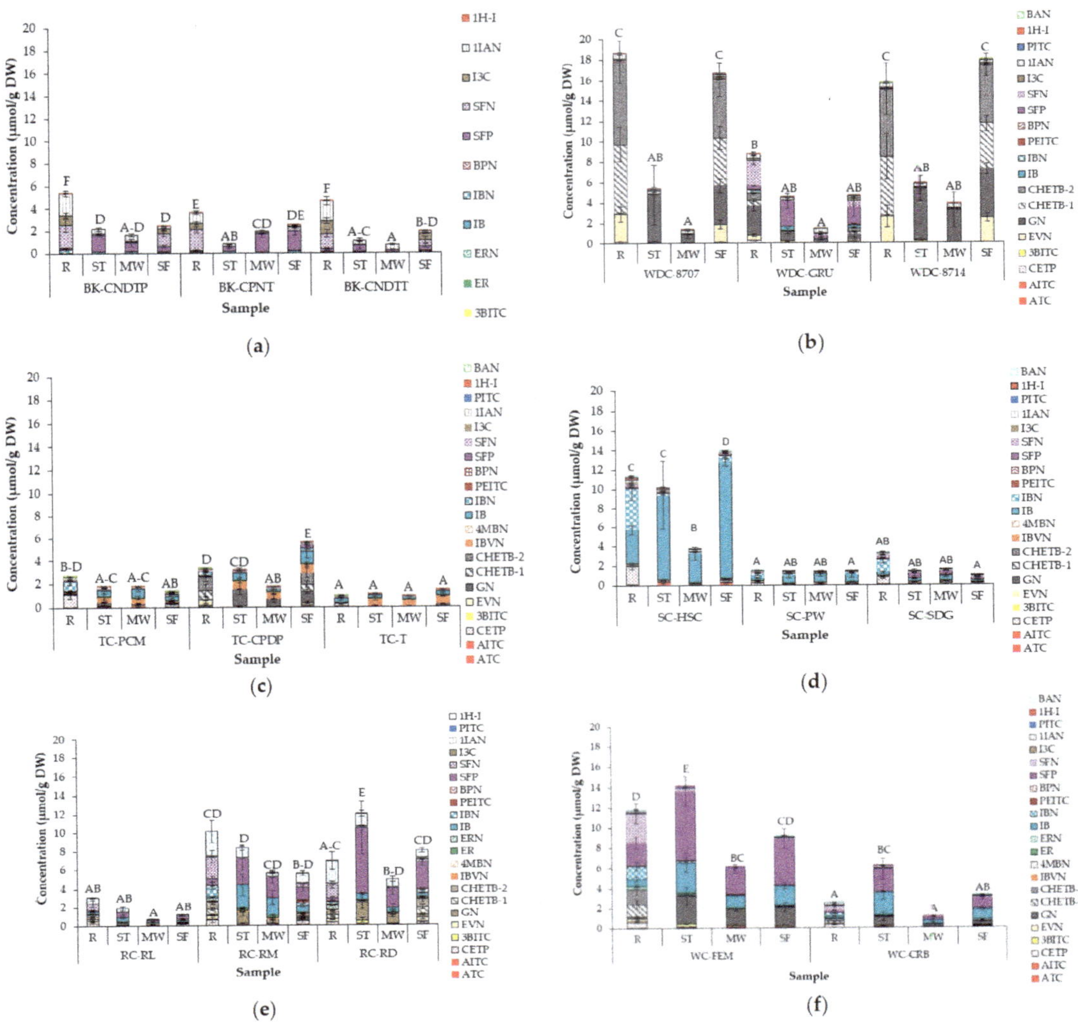

Figure 3. Glucosinolate hydrolysis products (GHPs) (µmol/g DW) in different accessions of (**a**) black kale; (**b**) wild cabbage (**c**) tronchuda cabbage; (**d**) Savoy cabbage; (**e**) Red cabbage and (**f**) white cabbage before and after domestic cooking. Results are expressed as sulforaphane equivalents. Error bars represent standard deviation from mean values. Letters above bars refer to differences in total GHP concentration. Letters 'A–F': bars not sharing a common uppercase letter differ significantly ($p < 0.05$) between accessions and cooking methods within a cabbage morphotype (i.e., within each separate graph). For significant differences between cabbage morphotypes, accessions, and cooking methods (i.e., across the separate cabbage morphotype graphs), see Supplementary Table S5. Compounds with similar colour shades are GHPs (with nitriles in pattern fill) of corresponding GSL, presented in Figure 2. Abbreviations: R = raw, ST = steamed; MW = microwaved and SF = stir-fried; BK-CPNT: cavolo palmizio; BK-CNDTT: cavolo nero di toscana o senza testa; WD-8707: wild cabbage 8707; WD-GRU: wild cabbage 7338; WD-8714: wild cabbage 8714; TC-PCM: penca mistura; TC-CPDP: penca povoa; TC-T: tronchuda; SC-HSC: hybrid savoy wirosa; SC-PW: pointed winter; SC-SDG: dark green; RC-RL: red langendijker; RC-RM: rocco marner (Hybrid); RC-RD: red Danish; WC-FEM: early market; WC-CRB: couve repolho. For abbreviations of compounds, see Table 2 (GHPs).

However, some GHPs were identified where intact GSL was not detected, and this occurred across all tested accessions. In black kale and savoy cabbage accessions, 3-butenyl-

ITC (3BITC) was detected in cooked samples though intact GPN was not present. A similar trend was noticed by Bell et al. [44], who found 3BITC in rocket samples, in the absence of GPN. The presence of 3BITC might be the result of SFP degradation. A study conducted on broccoli showed that standard SFP solution was degraded to 3BITC under thermal conditions [86]. PEITC and benzenepropanenitrile (BPN), hydrolysis products of gluconasturtiin, were detected in low amounts across all accessions, although intact gluconasturtiin was not detected in samples. The small amounts detected suggest that the GSL was present in low amounts in the sample and may have been hydrolysed during sample preparation, or the amounts present were below the limit of detection for the LC-MS.

Other studies have reported the presence of GHPs where their precursor, GSL, was not detected in different *Brassicas* [82,87,88]. Bell et al. [82] suggested that this discrepancies may be due to the degradation of other hydrolysis products during the analytical process, the inaccurate identification of GSLs and GHPs, very low amounts of GSLs being present in the plant, which were below the limit of detection (LOD) of the analytical method, or a yet-to-be identified mechanism by which GHPs are modified after hydrolysis.

Cooking significantly reduced the nitriles and EPTs that were formed and increased the number of ITCs compared to raw cabbage. Goitrin (GN), iberin (IB) and SFP were the major GHPs in cooked cabbages. Of all the studied cooking methods, microwaved samples had the lowest levels of GHPs; few or no nitriles and EPTs were detected, while very low amounts of ITCs were formed. However, in most cases, more ITCs were formed in microwaved samples when compared to raw samples. The highest concentrations of ITCs were formed in steamed samples across all studied accessions, with up to 23-fold increases in ITCs, compared to those observed in raw samples (SFP in steamed RC-RD), where few or no nitriles were present. In most samples, total and individual GHPs did not significantly differ between stir-fried and raw samples, although higher numbers of ITCs were formed in stir-fried samples. The pattern of GHP formation did not differ across accessions.

Differences in the severity of cooking methods, which may have influenced residual myrosinase activity in relation to ESP activity, can account for the difference in the types and concentration of GHPs present. ESP promotes the formation of nitriles and EPTs from GSL hydrolysis instead of ITCs from myrosinase [3]. The stir-fry cooking temperature was the least severe, leading to the formation of EPTs, nitriles and ITCs, as ESP and myrosinase would have still been active in the samples. The lower amounts of GSL detected in stir-fried cabbages did not seem to affect total GHPs but might have been partly responsible for the higher amounts of nitriles formed, as GSL was hydrolysed by the ESP present in the samples during the stir-frying process. Microwave cooking was the most severe cooking method employed, which was responsible for the negligible amounts of nitriles and low amounts of ITCs. The high core temperatures during microwaving (85–95 °C) would have led to the complete denaturation of ESP and almost total myrosinase inactivation, as highlighted earlier (Section 3.1). However, the steaming temperature would have been enough to denature ESP whilst still retaining substantial myrosinase activity (see Sections 3.1 and 3.2). The nitriles detected in both microwaved and steamed samples may have been formed with the residual ESP present during the cooking process, while the ITCs present in microwaved samples could be the result of residual myrosinase activity. In cooked broccoli, ESP was found to be denatured at temperatures above 50 °C, with a corresponding reduction in SFN production [3]. Rungapamestry et al. [25], in their study of SIN hydrolysis products in cooked cabbage, found that microwaving for 120 secs resulted in a reduction in nitriles, allyl cyanide and CEP (about 87%), with an increase in AITC formation (about 88%). The authors found that steaming cabbages for seven minutes resulted in an increase in AITC of up to 578%. The authors also found that AITC was formed in cabbages with no residual myrosinase activity and attributed this to formation during the hydrolysis and cooking process, which may have been bound to the cell membranes but released during processing. In a recent study, the steaming and stir-frying of broccolini and kale for

15 min resulted in significantly lower amounts of SFP and IB, respectively, when compared to the uncooked sample. The lower concentrations reported were probably due to the longer cooking time used in the study, which would have resulted in total myrosinase inactivation, preventing the conversion of GSLs to GHPs, although myrosinase activity was not measured in the study [33]. In another study, steaming broccoli resulted in increased SFP content after 5 min with a decrease observed beyond that, while microwaving broccoli and red cabbage for a minute led to a 5-fold increase in SFP with a decline reported beyond this time [35]. This suggests that there is an optimum cooking time to achieve maximum ITC formation, beyond which beneficial ITCs are lost. The low conversion of GSLs to their hydrolysis products in some of the studied samples is underscored. The results show that GHP recovery reduced with increases in the severity of the cooking procedure, with much lower concentrations (about 1% in microwaved TC-T) observed in microwaved samples, which was the most severe treatment employed. The low recovery of hydrolysis products observed in the microwaved samples is not necessarily a surprise, given that most of the ESP and myrosinase enzyme in the samples were already inactivated as a result of the cooking temperature (88–95 °C), as discussed in Sections 3.1 and 3.2 (see Figure 1 and Supplementary Table S3), which suggests that most of the GSL present in the sample remained unhydrolysed. However, the GSL concentration of the hydrolysed sample was not measured in the current study. In raw accessions with a low GHP content, we hypothesise that this may be due to the environmental responses of the plant, which, unfortunately, are not very well understood in the context of myrosinase activity and GSL hydrolysis. The low conversion of GSLs to GHPs has also been reported in some other *Brassica* species [44,81].

The results obtained in this study are similar to those observed by several authors during the thermal processing of *Brassica* vegetables [25,29,32,35,55,89]. This study adds to the findings of previous researchers; however, the study is particularly conclusive as it demonstrates similar findings across cabbage morphotypes and accessions. To improve the health benefits derived from cabbage consumption, this paper concludes that steaming is the optimum preparation method, due to the resulting increase in the ITCs formed.

3.6. *Principal Component Analysis (PCA) and Multifactor Analysis (MFA) of GSLs and GHPs in Raw and Cooked Cabbage*

To differentiate samples based on their GSLs and GHPs content, PCA analysis was conducted, as shown in Figure 4. Figure 4a shows the biplot for GSL distribution in samples, where dimensions 1 and 2 account for 56.4% of the observed variation. The plot shows TC2 (TC-CPDP), and wild cabbage accessions were characterized by high PROG and GPN contents, while black kale and most red cabbages, except for RC1 (RC-RL), had a higher tendency to accumulate GRPN and GER. Savoy cabbages, RC1 (RC-RL), TC1 (TC-PCM) and TC3 (TC-T) correlated positively with one another and were characterized by the amounts of SIN, GIBVN and GIBN they accumulated. Samples were separated based on cabbage morphotype and accession rather than cooking methods, suggesting that cabbage accession had a higher influence on GSL concentration than the tested cooking methods. However, the PCA biplot for GHPs (Figure 4b) shows differentiations in samples based on cooking. F1 and F2 explain only 39.6% of the variations; however, other dimensions did not provide any new information. Steamed and stir-fried cabbages correlated positively with ITCs, while nitriles mostly correlated with raw cabbages. There was no correlation observed in microwaved samples with GHPs, due to the low amounts of nitriles and ITCs present in the samples. Samples were separated based on their GHP profile and concentrations.

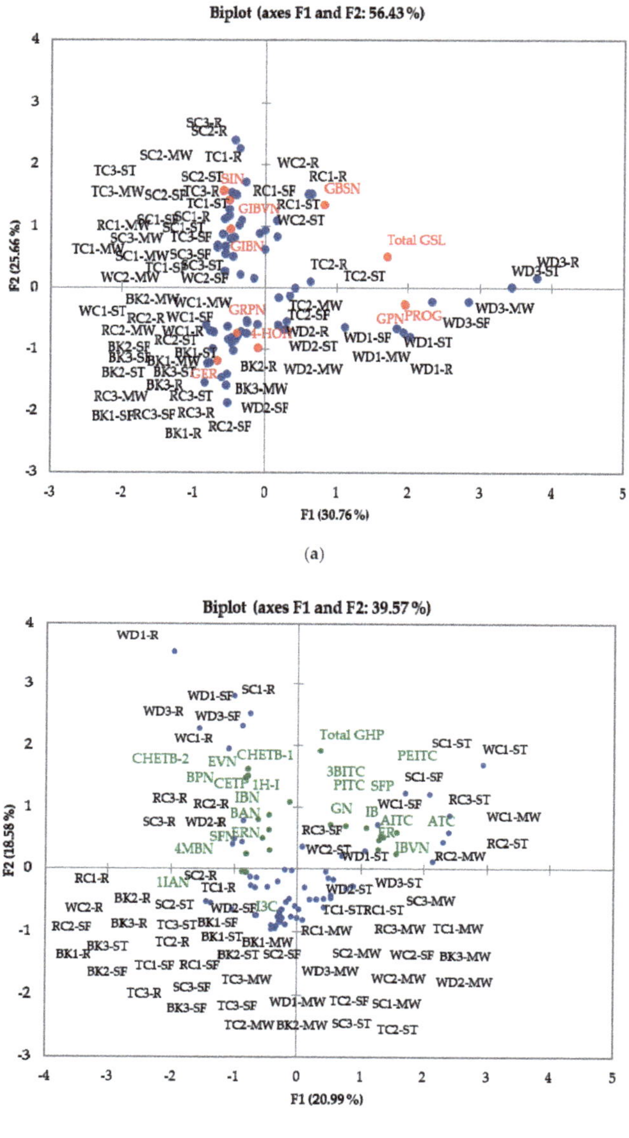

(a)

(b)

Figure 4. (**a**) PCA plot for tested samples and their relative distributions in relation to GSL concentrations. (**b**) PCA plot for tested samples and their relative distributions in relation to GHP concentrations. Abbreviations: R = raw, ST = steamed; MW = microwaved; SF = stir-fried; BK1: cavolo nero di toscana o senza palla (BK-CNDTP); BK2: cavolo palmizio (BK-CPNT); BK3: cavolo nero di toscana o senza testa (BK-CNDTT); WD1: wild cabbage 8707 (WD-8707); WD2: wild cabbage 7338 (WD-GRU); WD3: wild cabbage 8714 (WD-8714); TC1: penca mistura (TC-PCM); TC2: penca povoa (TC-CPDP); TC3 tronchuda (TC-T); SC1: hybrid savoy wirosa (SC-HSC); SC2: pointed winter (SC-PW); SC3: dark green (SC-SDG); RC1: red langendijker (RC-RL); RC2: rocco marner (Hybrid) (RC-RM); RC3: red Danish (RC-RD); WC1: early market (WC-FEM); WC2: couve repolho (WC-CRB). Red coloured compounds = GSLs; Green coloured compounds = GHPs; Blue dots = Samples. For full names of compounds, see Tables 1 and 2.

To better understand the results, MFA was performed on the accessions in relation to their GSL and GHP concentrations, as shown in Figure 5. Dimensions 1 and 2 (F1 and F2) represent only 34.3% of the variations, but other dimensions did not provide additional information. The observed results are similar to those observed in the biplot of GSL. Samples were separated in the same pattern as GSLs, based on cabbage morphotype and accession rather than cooking method. Individual GSLs correlated with their corresponding GHPs. The results observed from the MFA analysis confirm the results obtained from the PCA analysis, confirming the robustness of the findings. The results show that cooking has a greater effect on GHPs than GSLs but, when combined, samples were differentiated on their GSL content and the type of GHP present.

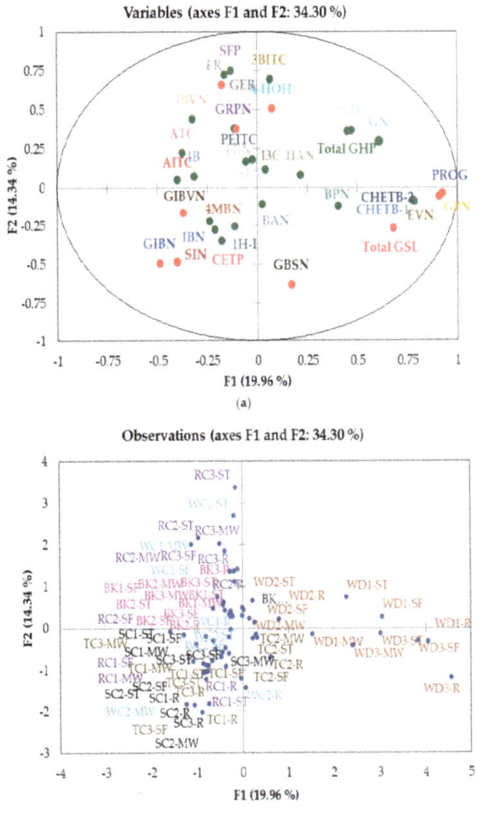

Figure 5. MFA map of glucosinolates and glucosinolate hydrolysis products (**a**) distribution of variables and (**b**) sample distribution. Abbreviations: R = raw, ST = steamed; MW = microwaved; SF = stir-fried; BK1: cavolo nero di toscana o senza palla (BK-CNDTP); BK2: cavolo palmizio (BK-CPNT); BK3: cavolo nero di toscana o senza testa (BK-CNDTT); WD1: wild cabbage 8707 (WD-8707); WD2: wild cabbage 7338 (WD-GRU); WD3: wild cabbage 8714 (WD-8714); TC1: penca mistura (TC-PCM); TC2: penca povoa (TC-CPDP); TC3 tronchuda (TC-T); SC1: hybrid savoy wirosa (SC-HSC); SC2: pointed winter (SC-PW); SC3: dark green (SC-SDG); RC1: red langendijker (RC-RL); RC2: rocco marner (Hybrid) (RC-RM); RC3: red Danish (RC-RD); WC1: early market (WC-FEM); WC2: couve repolho (WC-CRB). Colour codes: Pink = Black kale; Brown = Wild cabbage; Dijon yellow = Tronchuda cabbage; Black = Savoy cabbage; Purple = Red cabbage; Blue = White cabbage. Compounds with different shades of the same colour in Figure 5a refer to the GSL and corresponding GHPs. For compound codes on plot, refer to Tables 1 and 2.

This study is not without its limitations. Although precautions were followed to ensure accuracy during sample preparation, the size of cut cabbages, stirring during stir-frying and general reproducibility of the cooking processes across all samples may have slightly differed. In addition, some very volatile GHPs may have been lost during the cooking and analytical processes due to the long extraction method used, which may have affected the results. However, given that standard measures were employed to limit variations due to the above, it is unlikely that possible variations within the samples could have had a significant influence on the results, as only small variations were observed in the biological and technical replicates.

4. Conclusions

The results of this study confirm that domestic cooking has an effect on myrosinase stability, GSL concentration and GHP profiles, and concentration. Domestic cooking resulted in a significant loss of myrosinase activity, with stir-frying having the highest residual activity compared to the other two cooking methods that were investigated. Microwave cooking was the most severe heat treatment, resulting in the highest loss of myrosinase activity, reaching up to 99% in some cases. The study showed that mild cooking prevents the complete inactivation of myrosinase enzyme. Myrosinase enzyme stability differed significantly between cabbage accessions and morphotypes. Black kale myrosinase was the most stable after stir-frying, while red cabbage accessions were most stable after steaming and microwaving. No correlation was found between myrosinase activity and stability, as the accessions with the highest myrosinase activity did not have the most stable myrosinase after domestic processing.

Cooking led to a reduction in GSL concentrations compared to raw cabbage, with stir-frying leading to the greatest loss compared to the other two cooking methods and mild steaming enabling the greatest retention of GSL compounds. Considering that cabbages are usually consumed cooked, it important for breeders to work alongside nutritionists to select accessions with more thermally stable GSL and myrosinase for breeding, to ensure that the health benefits from cabbage consumption are not lost. The study found a relationship between cabbage core temperature during cooking, myrosinase stability and final GHPs profile. GHPs of raw cabbages were mainly nitriles and EPTs, probably due to the presence of active ESP in the samples. Cooking led to a reduction in the number of nitriles and EPTs formed, with levels differing between cooking methods. Optimal cooking conditions led to the degradation of ESP but retention of active myrosinase. Microwaving resulted in significantly lower amounts of nitriles, EPT and ITC formed, while steaming cabbages led to the production of significantly higher amounts of ITCs. However, the study showed that low residual myrosinase activity can still result in ITC formation.

The study concludes that consumption of raw or severely heat-treated cabbage can reduce possible health benefits, while mild cooking of cabbages, such as mild steaming, maximises beneficial isothiocyanate formation. This was especially true for IB and SFP in the studied cabbages and could provide information to guide consumers on how to improve the possible health benefits derived from cabbage consumption.

Supplementary Materials: The following are available online at https://www.mdpi.com/article/10.3390/foods10122908/s1, Figure S1: Examples of GC-MS chromatograms for raw and cooked samples for each morphotype of cabbage studied (a) black kale; (b) wild cabbage; (c) tronchuda cabbage; (d) savoy cabbage; (e) red cabbage and (f) white cabbage. Table S1: Origin, botanical and common names of planted cabbage accessions. Table S2. Consumption intent and cooking time scores from preliminary consumer study. Table S3. Relative activity ($A/A0 \pm SD$) of myrosinase enzyme after domestic cooking of cabbage. Table S4: Glucosinolate concentration of raw and cooked cabbage (μmol/g DW). Table S5: Glucosinolate hydrolysis products concentration in raw and cooked cabbage (μmol/g DW sulforaphane equivalent).

Author Contributions: Conceptualization, O.O.O., C.W. and L.M.; methodology, O.O.O., C.W. and L.M.; software, O.O.O. and L.M.; validation, O.O.O., C.W. and L.M.; formal analysis, O.O.O.; investigation, O.O.O.; resources, O.O.O.; data curation, O.O.O.; writing—original draft preparation,

O.O.O.; writing—review and editing, O.O.O., C.W. and L.M.; visualization, O.O.O., C.W. and L.M.; supervision, C.W. and L.M.; project administration, O.O.O., C.W. and L.M.; funding acquisition, O.O.O. All authors have read and agreed to the published version of the manuscript.

Funding: This research was funded by the Commonwealth Scholarship commission (CSC), UK as part of the doctoral research of the first author (O.O.O.), scholar ID: NGCS-2013-363.

Institutional Review Board Statement: Not applicable.

Informed Consent Statement: Not applicable.

Data Availability Statement: The data presented in this study are available on request from the corresponding author.

Acknowledgments: We would like to specially thank Warwick Genetic Bank for providing the cabbage seeds used for the study, Chelsea Snell for her advice on the cabbage-growing conditions and Valerie A. Jasper, Tobias James Lane and Matthew J. Richardson of the plant-growth unit, the University of Reading for their help with growing the cabbages. A big thank you to Denise Macdonald, Bindukala Radha, Chris Bussey, Josh Stapleford and Charwin Piyapinyo for their help with sample preparation. Our thanks go to Sameer Khalil Ghawi and Olukayode Okunade for support and guidance with myrosinase extraction and assay; Luke Bell, Nicholas Michael, Stella Lignou, Hanis Nadia Yahya and Rashed Alarfaj for support and guidance with glucosinolate extraction and LC-MS analysis and finally, Salah Abukhabta and Stephen Elmore for help and guidance with glucosinolate hydrolysis product extraction and the GC-MS analysis, respectively.

Conflicts of Interest: The authors declare no conflict of interest. The funders had no role in the design of the study; in the collection, analyses, or interpretation of data; in the writing of the manuscript, or in the decision to publish the results.

References

1. Herr, I.; Buchler, M.W. Dietary constituents of broccoli and other cruciferous vegetables: Implications for prevention and therapy of cancer. *Cancer Treat. Rev.* **2010**, *36*, 377–383. [CrossRef]
2. Mithen, R.F.; Dekker, M.; Verkerk, R.; Rabot, S.; Johnson, I.T. The nutritional significance, biosynthesis and bioavailability of glucosinolates in human foods. *J. Sci. Food Agric.* **2000**, *80*, 967–984. [CrossRef]
3. Matusheski, N.V.; Swarup, R.; Juvik, J.A.; Mithen, R.; Bennett, M.; Jeffery, E.H. Epithiospecifier protein from broccoli (*Brassica oleracea* L. ssp. *italica*) inhibits formation of the anticancer agent sulforaphane. *J. Agric. Food Chem.* **2006**, *54*, 2069–2076. [CrossRef]
4. Qazi, A.; Pal, J.; Maitah, M.I.; Fulciniti, M.; Pelluru, D.; Nanjappa, P.; Lee, S.; Batchu, R.B.; Prasad, M.; Bryant, C.S.; et al. Anticancer Activity of a Broccoli Derivative, Sulforaphane, in Barrett Adenocarcinoma: Potential Use in Chemoprevention and as Adjuvant in Chemotherapy. *Transl. Oncol.* **2010**, *3*, 389–399. [CrossRef]
5. Li, Y.; Karagöz, G.E.; Seo, Y.H.; Zhang, T.; Jiang, Y.; Yu, Y.; Duarte, A.M.S.; Schwartz, S.J.; Boelens, R.; Carroll, K.; et al. Sulforaphane inhibits pancreatic cancer through disrupting Hsp90–p50Cdc37 complex and direct interactions with amino acids residues of Hsp90. *J. Nutr. Biochem.* **2012**, *23*, 1617–1626. [CrossRef]
6. Guerrero-Beltran, C.E.; Calderon-Oliver, M.; Pedraza-Chaverri, J.; Chirino, Y.I. Protective effect of sulforaphane against oxidative stress: Recent advances. *Exp. Toxicol. Pathol.* **2012**, *64*, 503–508. [CrossRef]
7. Tarozzi, A.; Angeloni, C.; Malaguti, M.; Morroni, F.; Hrelia, S.; Hrelia, P. Sulforaphane as a potential protective phytochemical against neurodegenerative diseases. *Oxidative Med. Cell. Longev.* **2013**, *2013*, 1–10. [CrossRef]
8. Fahey, J.W.; Zalcmann, A.T.; Talalay, P. The chemical diversity and distribution of glucosinolates and isothiocyanates among plants. *Phytochemistry* **2001**, *56*, 5–51. [CrossRef]
9. Hanschen, F.S.; Lamy, E.; Schreiner, M.; Rohn, S. Reactivity and stability of glucosinolates and their breakdown products in foods. *Angew. Chem. Int. Ed.* **2014**, *53*, 11430–11450. [CrossRef]
10. Bhattacharya, A.; Tang, L.; Li, Y.; Geng, F.; Paonessa, J.D.; Chen, S.C.; Wong, M.K.K.; Zhang, Y. Inhibition of bladder cancer development by allyl isothiocyanate. *Carcinogenesis* **2010**, *31*, 281–286. [CrossRef]
11. Geng, F.; Tang, L.; Li, Y.; Yang, L.; Choi, K.-S.; Kazim, A.L.; Zhang, Y. Allyl isothiocyanate arrests cancer cells in mitosis, and mitotic arrest in turn leads to apoptosis via Bcl-2 protein phosphorylation. *J. Biol. Chem.* **2011**, *286*, 32259–32267. [CrossRef]
12. Tsai, S.-C.; Huang, W.-W.; Huang, W.-C.; Lu, C.-C.; Chiang, J.-H.; Peng, S.-F.; Chung, J.-G.; Lin, Y.-H.; Hsu, Y.-M.; Amagaya, S. ERK-modulated intrinsic signaling and G2/M phase arrest contribute to the induction of apoptotic death by allyl isothiocyanate in MDA-MB-468 human breast adenocarcinoma cells. *Int. J. Oncol.* **2012**, *41*, 2065–2072. [CrossRef]
13. Tripathi, K.; Hussein, U.K.; Anupalli, R.; Barnett, R.; Bachaboina, L.; Scalici, J.; Rocconi, R.P.; Owen, L.B.; Piazza, G.A.; Palle, K. Allyl isothiocyanate induces replication-associated DNA damage response in NSCLC cells and sensitizes to ionizing radiation. *Oncotarget* **2015**, *6*, 5237. [CrossRef]

14. Cashman, J.R.; Xiong, Y.; Lin, J.; Verhagen, H.; van Poppel, G.; van Bladeren, P.J.; Larsen-Su, S.; Williams, D.E. In vitro and in vivo inhibition of human flavin-containing monooxygenase form 3 (FMO3) in the presence of dietary indoles. *Biochem. Pharmacol.* **1999**, *58*, 1047–1055. [CrossRef]
15. Bonnesen, C.; Eggleston, I.M.; Hayes, J.D. Dietary indoles and isothiocyanates that are generated from cruciferous vegetables can both stimulate apoptosis and confer protection against DNA damage in human colon cell lines. *Cancer Res.* **2001**, *61*, 6120–6130.
16. Kim, D.J.; Han, B.S.; Ahn, B.; Hasegawa, R.; Shirai, T.; Ito, N.; Tsuda, H. Enhancement by indole-3-carbinol of liver and thyroid gland neoplastic development in a rat medium-term multiorgan carcinogenesis model. *Carcinogenesis* **1997**, *18*, 377–381. [CrossRef]
17. Baik, H.Y.; Juvik, J.; Jeffery, E.H.; Wallig, M.A.; Kushad, M.; Klein, B.P. Relating glucosinolate content and flavor of broccoli cultivars. *J. Food Sci.* **2003**, *68*, 1043–1050. [CrossRef]
18. Kubec, R.; Drhová, V.; Velíšek, J. Thermal Degradation of S-Methylcysteine and Its SulfoxidesImportant Flavor Precursors of Brassica and Allium Vegetables. *J. Agric. Food Chem.* **1998**, *46*, 4334–4340. [CrossRef]
19. Cox, D.N.; Melo, L.; Zabaras, D.; Delahunty, C.M. Acceptance of health-promoting Brassica vegetables: The influence of taste perception, information and attitudes. *Public Health Nutr.* **2012**, *15*, 1474–1482. [CrossRef]
20. Conaway, C.C.; Getahun, S.M.; Liebes, L.L.; Pusateri, D.J.; Topham, D.K.W.; Botero-Omary, M.; Chung, F.-L. Disposition of glucosinolates and sulforaphane in humans after ingestion of steamed and fresh broccoli. *Nutr. Cancer* **2000**, *38*, 168–178. [CrossRef]
21. Okunade, O.; Niranjan, K.; Ghawi, S.K.; Kuhnle, G.; Methven, L. Supplementation of the diet by exogenous myrosinase via mustard seeds to increase the bioavailability of sulforaphane in healthy human subjects after the consumption of cooked broccoli. *Mol. Nutr. Food Res.* **2018**, *62*, 1700980. [CrossRef] [PubMed]
22. Wu, X.; Huang, H.; Childs, H.; Wu, Y.; Yu, L.; Pehrsson, P.R. Glucosinolates in Brassica Vegetables: Characterization and Factors That Influence Distribution, Content, and Intake. *Annu. Rev. Food Sci. Technol.* **2021**, *12*, 485–511. [CrossRef]
23. Dekker, M.; Verkerk, R.; Jongen, W.M.F. Predictive modelling of health aspects in the food production chain: A case study on glucosinolates in cabbage. *Trends Food Sci. Technol.* **2000**, *11*, 174–181. [CrossRef]
24. Verkerk, R.; Dekker, M. Glucosinolates and myrosinase activity in red cabbage (*Brassica oleracea* L. Var. *capitata* f. *rubra* DC.) after various microwave treatments. *J. Agric. Food Chem.* **2004**, *52*, 7318–7323. [CrossRef]
25. Rungapamestry, V.; Duncan, A.J.; Fuller, Z.; Ratcliffe, B. Changes in glucosinolate concentrations, myrosinase activity, and production of metabolites of glucosinolates in cabbage (*Brassica oleracea* var. *capitata*) cooked for different durations. *J. Agric. Food Chem.* **2006**, *54*, 7628–7634. [CrossRef] [PubMed]
26. Ghawi, S.K.; Methven, L.; Rastall, R.A.; Niranjan, K. Thermal and high hydrostatic pressure inactivation of myrosinase from green cabbage: A kinetic study. *Food Chem.* **2012**, *131*, 1240–1247. [CrossRef]
27. Yen, G.-C.; Wei, Q.-K. Myrosinase activity and total glucosinolate content of cruciferous vegetables, and some properties of cabbage myrosinase in Taiwan. *J. Sci. Food Agric.* **1993**, *61*, 471–475. [CrossRef]
28. Oerlemans, K.; Barrett, D.M.; Suades, C.B.; Verkerk, R.; Dekker, M. Thermal degradation of glucosinolates in red cabbage. *Food Chem.* **2006**, *95*, 19–29. [CrossRef]
29. Song, L.; Thornalley, P.J. Effect of storage, processing and cooking on glucosinolate content of *Brassica* vegetables. *Food Chem. Toxicol.* **2007**, *45*, 216–224. [CrossRef]
30. Xu, F.; Zheng, Y.; Yang, Z.; Cao, S.; Shao, X.; Wang, H. Domestic cooking methods affect the nutritional quality of red cabbage. *Food Chem.* **2014**, *161*, 162–167. [CrossRef]
31. Vallejo, F.; Tomás-Barberán, F.; García-Viguera, C. Glucosinolates and vitamin C content in edible parts of broccoli florets after domestic cooking. *Eur. Food Res. Technol.* **2002**, *215*, 310–316. [CrossRef]
32. Ghawi, S.K.; Methven, L.; Niranjan, K. The potential to intensify sulforaphane formation in cooked broccoli (*Brassica oleracea* var. *italica*) using mustard seeds (*Sinapis alba*). *Food Chem.* **2013**, *138*, 1734–1741. [CrossRef]
33. Baenas, N.; Marhuenda, J.; García-Viguera, C.; Zafrilla, P.; Moreno, D. Influence of cooking methods on glucosinolates and isothiocyanates content in novel cruciferous foods. *Foods* **2019**, *8*, 257. [CrossRef] [PubMed]
34. Hanschen, F.S.; Kühn, C.; Nickel, M.; Rohn, S.; Dekker, M. Leaching and degradation kinetics of glucosinolates during boiling of *Brassica oleracea* vegetables and the formation of their breakdown products. *Food Chem.* **2018**, *263*, 240–250. [CrossRef]
35. Tabart, J.; Pincemail, J.; Kevers, C.; Defraigne, J.-O.; Dommes, J. Processing effects on antioxidant, glucosinolate, and sulforaphane contents in broccoli and red cabbage. *Eur. Food Res. Technol.* **2018**, *244*, 2085–2094. [CrossRef]
36. Hanschen, F.S.; Kaufmann, M.; Kupke, F.; Hackl, T.; Kroh, L.W.; Rohn, S.; Schreiner, M. *Brassica* vegetables as sources of epithionitriles: Novel secondary products formed during cooking. *Food Chem.* **2018**, *245*, 564–569. [CrossRef] [PubMed]
37. Hanschen, F.S. Domestic boiling and salad preparation habits affect glucosinolate degradation in red cabbage (*Brassica oleracea* var. *capitata f. rubra*). *Food Chem.* **2020**, *321*, 126694. [CrossRef]
38. Oloyede, O.O.; Wagstaff, C.; Methven, L. Influence of cabbage (*Brassica oleracea*) accession and growing conditions on myrosinase activity, glucosinolates and their hydrolysis products. *Foods* **2021**, accepted.
39. Rungapamestry, V.; Duncan, A.J.; Fuller, Z.; Ratcliffe, B. Influence of blanching and freezing broccoli (*Brassica oleracea* var. *italica*) prior to storage and cooking on glucosinolate concentrations and myrosinase activity. *Eur. Food Res. Technol.* **2008**, *227*, 37. [CrossRef]

40. Gatfield, I.L.; Sand, T. A coupled enzymatic procedure for the determination of myrosinase activity. *Lebensm.-Wiss. Technol.* **1983**, *16*, 73–75.
41. Wilkinson, A.P.; Rhodes, M.J.C.; Fenwick, R.G. Myrosinase Activity of Cruciferous Vegetables. *J. Sci. Food Agric.* **1984**, *35*, 543–552. [CrossRef]
42. Bradford, M.M. A rapid and sensitive method for the quantitation of microgram quantities of protein utilizing the principle of protein-dye binding. *Anal. Biochem.* **1976**, *72*, 248–254. [CrossRef]
43. Bell, L.; Oruna-Concha, M.J.; Wagstaff, C. Identification and quantification of glucosinolate and flavonol compounds in rocket salad (*Eruca sativa*, *Eruca vesicaria* and *Diplotaxis tenuifolia*) by LC–MS: Highlighting the potential for improving nutritional value of rocket crops. *Food Chem.* **2015**, *172*, 852–861. [CrossRef] [PubMed]
44. Bell, L.; Yahya, H.N.; Oloyede, O.O.; Methven, L.; Wagstaff, C. Changes in rocket salad phytochemicals within the commercial supply chain: Glucosinolates, isothiocyanates, amino acids and bacterial load increase significantly after processing. *Food Chem.* **2017**, *221*, 521–534. [CrossRef] [PubMed]
45. Rochfort, S.J.; Trenerry, V.C.; Imsic, M.; Panozzo, J.; Jones, R. Class targeted metabolomics: ESI ion trap screening methods for glucosinolates based on MSn fragmentation. *Phytochemistry* **2008**, *69*, 1671–1679. [CrossRef]
46. Lelario, F.; Bianco, G.; Bufo, S.A.; Cataldi, T.R.I. Establishing the occurrence of major and minor glucosinolates in Brassicaceae by LC–ESI-hybrid linear ion-trap and Fourier-transform ion cyclotron resonance mass spectrometry. *Phytochemistry* **2012**, *73*, 74–83. [CrossRef] [PubMed]
47. Bennett, R.N.; Mellon, F.A.; Kroon, P.A. Screening crucifer seeds as sources of specific intact glucosinolates using ion-pair high-performance liquid chromatography negative ion electrospray mass spectrometry. *J. Agric. Food Chem.* **2004**, *52*, 428–438. [CrossRef]
48. Al-Gendy, A.A.; Lockwood, G.B. GC-MS analysis of volatile hydrolysis products from glucosinolates in Farsetia aegyptia var. ovalis. *Flavour Fragr. J.* **2003**, *18*, 148–152. [CrossRef]
49. Arora, R.; Sharma, D.; Kumar, R.; Singh, B.; Vig, A.P.; Arora, S. Evaluating extraction conditions of glucosinolate hydrolytic products from seeds of *Eruca sativa* (Mill.) Thell. using GC-MS. *J. Food Sci.* **2014**, *79*, C1964–C1969. [CrossRef]
50. Hong, E.; Kim, G.-H. GC-MS analysis of the extracts from Korean cabbage (*Brassica campestris* L. ssp. *pekinensis*) and its seed. *Prev. Nutr. Food Sci.* **2013**, *18*, 218–221. [CrossRef]
51. Spencer, G.F.; Daxenbichler, M.E. Gas chromatography-mass spectrometry of nitriles, isothiocyanates and oxazolidinethiones derived from cruciferous glucosinolates. *J. Sci. Food Agric.* **1980**, *31*, 359–367. [CrossRef]
52. Hanschen, F.S.; Klopsch, R.; Oliviero, T.; Schreiner, M.; Verkerk, R.; Dekker, M. Optimizing isothiocyanate formation during enzymatic glucosinolate breakdown by adjusting pH value, temperature and dilution in *Brassica* vegetables and *Arabidopsis thaliana*. *Sci. Rep.* **2017**, *7*, 40807. [CrossRef]
53. de Pinho, P.G.; Valentão, P.; Gonçalves, R.F.; Sousa, C.; Andrade, P.B. Volatile composition of *Brassica oleracea* L. var. *costata* DC leaves using solid-phase microextraction and gas chromatography/ion trap mass spectrometry. *Rapid Commun. Mass Spectrom.* **2009**, *23*, 2292–2300. [CrossRef] [PubMed]
54. Vaughn, S.F.; Palmquist, D.E.; Duval, S.M.; Berhow, M.A. Herbicidal activity of glucosinolate-containing seedmeals. *Weed Sci.* **2017**, *54*, 743–748. [CrossRef]
55. Matusheski, N.V.; Juvik, J.A.; Jeffery, E.H. Heating decreases epithiospecifier protein activity and increases sulforaphane formation in broccoli. *Phytochemistry* **2004**, *65*, 1273–1281. [CrossRef] [PubMed]
56. Adler-Nissen, J. The continuous work—A new unit operation in industiral food processes. *J. Food Process Eng.* **2002**, *25*, 435–453. [CrossRef]
57. Gould, K.S. Nature's Swiss Army Knife: The Diverse Protective Roles of Anthocyanins in Leaves. *J. Biomed. Biotechnol.* **2004**, *2004*, 314–320. [CrossRef]
58. Park, S.; Arasu, M.V.; Lee, M.-K.; Chun, J.-H.; Seo, J.M.; Al-Dhabi, N.A.; Kim, S.-J. Analysis and metabolite profiling of glucosinolates, anthocyanins and free amino acids in inbred lines of green and red cabbage (*Brassica oleracea* L.). *LWT-Food Sci. Technol.* **2014**, *58*, 203–213. [CrossRef]
59. Bones, A.M.; Rossiter, J.T. The myrosinase-glucosinolate system, its organisation and biochemistry. *Physiol. Plant.* **1996**, *97*, 194–208. [CrossRef]
60. Rask, L.; Andréasson, E.; Ekbom, B.; Eriksson, S.; Pontoppidan, B.; Meijer, J.; Andreasson, E. Myrosinase: Gene family evolution and herbivore defense in *Brassicaceae*. *Plant Mol. Biol.* **2000**, *42*, 93–113. [CrossRef]
61. Charron, C.S.; Sams, C.E. Glucosinolate content and myrosinase activity in rapid-cycling *Brassica oleracea* grown in a controlled environment. *J. Am. Soc. Hortic. Sci.* **2004**, *129*, 321–330. [CrossRef]
62. Penas, E.; Frias, J.; Martinez-Villaluenga, C.; Vidal-Valverde, C. Bioactive compounds, myrosinase activity, and antioxidant capacity of white cabbages grown in different locations of Spain. *J. Agric. Food Chem.* **2011**, *59*, 3772–3779. [CrossRef] [PubMed]
63. Dekker, M.; Hennig, K.; Verkerk, R. Differences in thermal stability of glucosinolates in five *Brassica* vegetables. *Czech J. Food Sci.* **2009**, *27*, S85–S88. [CrossRef]
64. Rosa, E.A.S.; Heaney, R.K. The effect of cooking and processing on the glucosinolate content: Studies on four varieties of portuguese cabbage and hybrid white cabbage. *J. Sci. Food Agric.* **1993**, *62*, 259–265. [CrossRef]
65. Ciska, E.; Kozłowska, H. The effect of cooking on the glucosinolates content in white cabbage. *Eur. Food Res. Technol.* **2001**, *212*, 582–587. [CrossRef]

66. Jones, R.B.; Faragher, J.D.; Winkler, S. A review of the influence of postharvest treatments on quality and glucosinolate content in broccoli (Brassica oleracea var. italica) heads. *Postharvest Biol. Technol.* **2006**, *41*, 1–8. [CrossRef]
67. Gliszczyńska-Świgło, A.; Ciska, E.; Pawlak-Lemańska, K.; Chmielewski, J.; Borkowski, T.; Tyrakowska, B. Changes in the content of health-promoting compounds and antioxidant activity of broccoli after domestic processing. *Food Addit. Contam.* **2006**, *23*, 1088–1098. [CrossRef]
68. Francisco, M.; Velasco, P.; Moreno, D.A.; García-Viguera, C.; Cartea, M.E. Cooking methods of *Brassica rapa* affect the preservation of glucosinolates, phenolics and vitamin C. *Food Res. Int.* **2010**, *43*, 1455–1463. [CrossRef]
69. Wathelet, J.-P.; Mabon, N.; Foucart, M.; Marlier, M. Influence of blanching on the quality of Brussels sprouts (*Brassica oleracea* L cv *gemmifera*). *Sci. Aliment.* **1996**, *16*, 393–402.
70. Moreno, D.A.; López-Berenguer, C.; García-Viguera, C. Effects of Stir-Fry Cooking with Different Edible Oils on the Phytochemical Composition of Broccoli. *J. Food Sci.* **2007**, *72*, S064–S068. [CrossRef]
71. Nugrahedi, P.Y.; Oliviero, T.; Heising, J.K.; Dekker, M.; Verkerk, R. Stir-frying of chinese cabbage and pakchoi retains health-promoting glucosinolates. *Plant Foods Hum. Nutr.* **2017**, *72*, 439–444. [CrossRef]
72. Wu, Y.; Shen, Y.; Wu, X.; Zhu, Y.; Mupunga, J.; Bao, W.; Huang, J.; Mao, J.; Liu, S.; You, Y. Hydrolysis before stir-frying increases the isothiocyanate content of broccoli. *J. Agric. Food Chem.* **2018**, *66*, 1509–1515. [CrossRef]
73. Yuan, G.-F.; Sun, B.; Yuan, J.; Wang, Q.-M. Effects of different cooking methods on health-promoting compounds of broccoli. *J. Zhejiang Univ. Sci. B* **2009**, *10*, 580–588. [CrossRef] [PubMed]
74. Palermo, M.; Pellegrini, N.; Fogliano, V. The effect of cooking on the phytochemical content of vegetables. *J. Sci. Food Agric.* **2014**, *94*, 1057–1070. [CrossRef]
75. Hanschen, F.S.; Rohn, S.; Mewis, I.; Schreiner, M.; Kroh, L.W. Influence of the chemical structure on the thermal degradation of the glucosinolates in broccoli sprouts. *Food Chem.* **2012**, *130*, 1–8. [CrossRef]
76. Jensen, S.K.; Liu, Y.-G.; Eggum, B.O. The effect of heat treatment on glucosinolates and nutritional value of rapeseed meal in rats. *Anim. Feed Sci. Technol.* **1995**, *53*, 17–28. [CrossRef]
77. Kushad, M.M.; Brown, A.F.; Kurilich, A.C.; Juvik, J.A.; Klein, B.P.; Wallig, M.A.; Jeffery, E.H. Variation of glucosinolates in vegetable crops of *Brassica oleracea*. *J. Agric. Food Chem.* **1999**, *47*, 1541–1548. [CrossRef] [PubMed]
78. Cartea, M.E.; Velasco, P.; Obregon, S.; Padilla, G.; de Haro, A. Seasonal variation in glucosinolate content in Brassica oleracea crops grown in northwestern Spain. *Phytochemistry* **2008**, *69*, 403–410. [CrossRef]
79. Choi, S.-H.; Park, S.; Lim, Y.P.; Kim, S.-J.; Park, J.-T.; An, G. Metabolite profiles of glucosinolates in cabbage varieties (Brassica oleracea var. capitata) by season, color, and tissue position. *Hortic. Environ. Biotechnol.* **2014**, *55*, 237–247. [CrossRef]
80. Park, S.; Valan Arasu, M.; Lee, M.-K.; Chun, J.-H.; Seo, J.M.; Lee, S.-W.; Al-Dhabi, N.A.; Kim, S.-J. Quantification of glucosinolates, anthocyanins, free amino acids, and vitamin C in inbred lines of cabbage (*Brassica oleracea* L.). *Food Chem.* **2014**, *145*, 77–85. [CrossRef]
81. Jasper, J.; Wagstaff, C.; Bell, L. Growth temperature influences postharvest glucosinolate concentrations and hydrolysis product formation in first and second cuts of rocket salad. *Postharvest Biol. Technol.* **2020**, *163*, 111157. [CrossRef]
82. Bell, L.; Kitsopanou, E.; Oloyede, O.O.; Lignou, S. Important odorants of four *Brassicaceae* species, and discrepancies between glucosinolate profiles and observed hydrolysis products. *Foods* **2021**, *10*, 1055. [CrossRef] [PubMed]
83. Traka, M.H.; Saha, S.; Huseby, S.; Kopriva, S.; Walley, P.G.; Barker, G.C.; Moore, J.; Mero, G.; van den Bosch, F.; Constant, H.; et al. Genetic regulation of glucoraphanin accumulation in Beneforté broccoli. *New Phytol.* **2013**, *198*, 1085–1095. [CrossRef] [PubMed]
84. Matusheski, N.V.; Jeffery, E.H. Comparison of the bioactivity of two glucoraphanin hydrolysis products found in broccoli, sulforaphane and sulforaphane nitrile. *J. Agric. Food Chem.* **2001**, *49*, 5743–5749. [CrossRef] [PubMed]
85. Mithen, R.; Faulkner, K.; Magrath, R.; Rose, P.; Williamson, G.; Marquez, J. Development of isothiocyanate-enriched broccoli, and its enhanced ability to induce phase 2 detoxification enzymes in mammalian cells. *Theor. Appl. Genet.* **2003**, *106*, 727–734. [CrossRef]
86. Chiang, W.C.K.; Pusateri, D.J.; Leitz, R.E.A. Gas chromatography/mass spectrometry method for the determination of sulforaphane and sulforaphane nitrile in broccoli. *J. Agric. Food Chem.* **1998**, *46*, 1018–1021. [CrossRef]
87. Nor, N.D.M.; Lignou, S.; Bell, L.; Houston-Price, C.; Harvey, K.; Methven, L. The relationship between glucosinolates and the sensory characteristics of steamed-pureed turnip (*Brassica rapa* subsp. *Rapa* L.). *Foods* **2020**, *9*, 1719. [CrossRef] [PubMed]
88. Klopsch, R.; Witzel, K.; Börner, A.; Schreiner, M.; Hanschen, F.S. Metabolic profiling of glucosinolates and their hydrolysis products in a germplasm collection of *Brassica rapa* turnips. *Food Res. Int.* **2017**, *100*, 392–403. [CrossRef]
89. Jones, R.B.; Frisina, C.L.; Winkler, S.; Imsic, M.; Tomkins, R.B. Cooking method significantly effects glucosinolate content and sulforaphane production in broccoli florets. *Food Chem.* **2010**, *123*, 237–242. [CrossRef]

foods

MDPI

Article

Determination of Isothiocyanate-Protein Conjugates in a Vegetable-Enriched Bread

Mareike Krell [1], Lina Cvancar [1], Michael Poloczek [1], Franziska S. Hanschen [2] and Sascha Rohn [1,3,4,*]

[1] Institute of Food Chemistry, Hamburg School of Food Science, University of Hamburg, Grindelallee 117, 20146 Hamburg, Germany; mareike.krell@chemie.uni-hamburg.de (M.K.); Lina.Cvancar@gmx.de (L.C.); m-poloczek@gmx.de (M.P.)
[2] Leibniz Institute of Vegetable and Ornamental Crops (IGZ) e.V., Theodor-Echtermeyer-Weg 1, 14979 Großbeeren, Germany; hanschen@igzev.de
[3] Institute of Food Technology and Food Chemistry, Technische Universität Berlin, Gustav-Meyer-Allee 25, 13355 Berlin, Germany
[4] Institute for Food and Environmental Research (ILU) e.V., Papendorfer Weg 3, 14806 Bad Belzig, Germany
* Correspondence: rohn@tu-berlin.de; Tel.: +49-030-314-72583

Abstract: Vegetables of the plant order Brassicales are believed to have health-promoting properties, as they provide high contents of glucosinolates (GLS) and deriving from these, enzymatically and heat-induced breakdown products, such as isothiocyanates (ITC). Besides their positive physiological effects, ITC are electrophilic and can undergo reactions with food components such as proteins. Following the trend of improving traditional food products with GLS-rich ingredients, interactions of ITC with proteins can diminish the properties of both components—protein's value and functionality as well as ITC's bioactivity. In vegetable-enriched bread, where cresses (*Lepidium sativum* L. or *Tropaeolum majus* L.) are added to the initial dough, together with benzyl cyanide, benzyl isothiocyanate (BITC) is formed during the baking process. The aim of the present study was to investigate the possible migration behavior of the GLS breakdown products and the formation of ITC-wheat protein conjugates. After the baking process, the breads' proteins were enzymatically hydrolyzed, and the ITC-amino acid conjugates analyzed using a LC-ESI-MS/MS methodology. In all samples, BITC-protein conjugates were detected as thiourea derivatives, while formation of dithiocarbamates could not be detected. The study showed that GLS and their breakdown products such as ITC migrate into the surrounding food matrix and undergo reactions with proteins, which could in turn lead to modified protein properties and reduce the bioavailability of ITC and lysine.

Keywords: glucosinolates; benzyl isothiocyanate; protein conjugates; functional foods; nasturtium; garden cress; thiourea

Citation: Krell, M.; Cvancar, L.; Poloczek, M.; Hanschen, F.S.; Rohn, S. Determination of Isothiocyanate-Protein Conjugates in a Vegetable-Enriched Bread. *Foods* **2021**, *10*, 1300. https://doi.org/10.3390/foods10061300

Academic Editor: Montserrat Dueñas Paton

Received: 17 May 2021
Accepted: 3 June 2021
Published: 5 June 2021

1. Introduction

Plants possess a wide variation of bioactive compounds derived from plant secondary metabolism. When consumed these can provide pharmacological properties and beneficial effects on human health [1,2]. Consequently, the *World Health Organisation* (WHO) recommends an intake of at least 400 g of fruits and vegetables per day, which is still not being reached sufficiently [3,4]. In order to further increase the supply of vegetables with their health-promoting secondary plant metabolites (SPM), different pasta and bread recipes have already been adapted, applying raw materials rich in SPM [5–7]. Bread seems to be a particularly suitable product for a fortification with SPM, as it is consumed quite frequently in Western countries, regardless of age and gender [8].

When vegetables of the plant order Brassicales are used as ingredients in foods, glucosinolates (GLS) are in many cases the dominating SPM [9]. Besides antibacterial and anti-inflammatory effects, some studies even suggest a reduced risk of suffering from certain types of cancer, when vegetables containing GLS are regularly consumed [10,11].

However, the health-beneficial effects seem to be not directly associated with the native GLS, but rather linked to their breakdown products, especially isothiocyanates (ITC) [12]. In addition to ITC, nitriles, thiocyanates, and epithionitriles can be formed after the damage of the plant tissue, as occurring during chewing or cutting. During food processing nitriles are formed via thermally-induced degradation [13,14]. The tissue damage causes the enzyme myrosinase to come into contact with the GLS and hydrolyze them into unstable aglycone intermediates. These are then further converted by a Lossen-like rearrangement or specific proteins into the different breakdown products [15,16]. The profile of the breakdown products formed is dependent on the pH value, the presence of modifier proteins, and the initial composition of the GLS [17].

Due to their electrophilic nature, ITC can react with nucleophiles such as the amino or thiol groups of amino acids, peptides, or proteins. The reaction of ITC with the side chains leads to the formation of thiourea and dithiocarbamate derivatives, respectively, presenting different stabilities and subsequent reaction pathways [18–20]. Conjugates between ITC and different food proteins such as myoglobin, egg white proteins, and whey proteins have already been studied. It was found that the ITC-protein conjugation does not only lead to altered physicochemical properties, but also to a structural change of the proteins [18,21,22]. In food, ITC-protein conjugates were already identified when garden cress (*Lepidium sativum*) was mixed with curd (as in the typical German spread 'Kräuterquark') with the result that 28% of the original GLS were conjugated to lysine and cysteine [23].

In a recent study, bread was enriched with pak choi (*Brassica rapa* subsp. *chinensis*) and kale (*Brassica oleracea* var. *sabellica*) for increasing the intake of GLS and their breakdown products [24]. During breadmaking, the ingredients undergo a heating process, reaching temperatures around 100 °C. Consequently, mainly nitriles rather than ITC were detected in those experiments, as higher temperatures conditions preferably lead to the formation of nitriles, which have a higher thermostability [24,25]. The desired health-promoting ITC were only found in bread enriched with pak choi (*Brassica rapa* subsp. *chinensis*) microgreens in a concentration below 0.005 µmol/g fresh weight. Further reasons for the low ITC concentration could have been based on the chemical structure of the GLS, the water content of the bread, and the plant tissue being only marginally destroyed [24]. The behavior of ITC in a processed food with regard to interactions with nucleophilic amino acid side chains was not yet investigated in detail. However, reactions between ITC from the added vegetables and proteins in a more highly processed product such as bread, still need characterization, as fate of the GLS and ITC during breadmaking is not yet completely understood. Such interactions can affect the physiological as well as the technofunctional properties of both components.

It is hypothesized that the thermally-induced breakdown products of GLS can migrate into the bread crumb and form ITC-wheat protein conjugates to a certain extent, away from their place of formation. Consequently, the aim of this study was to investigate the formation of ITC, their migration behavior in a model bread matrix, and the possible conjugation with wheat proteins. Based on a recent experiment, wheat dough was used, mixed with fresh garden cress (*Lepidium sativum* L.) or freeze-dried nasturtium leaf-powder (*Tropaeolum majus* L.) [7]. Microgreens and homogenized material from the different cress genera were used for studying the factors that can influence the formation of ITC conjugates such as different BITC sources and the form of plant-material addition.

2. Materials and Methods

2.1. Chemicals and Materials

Boric acid solution (4%), hydrochloric acid (0.1 M), hydrochloric acid (32%), formic acid (FA; 98%), sodium hydroxide (100%), trichloroacetic acid (≥99%), Kjeldahl tablets, sulfuric acid (96–98%, p.a.), methanol (LC-MS grade), water (LC-MS grade), methylene blue, methyl red and methylene chloride (GC Ultra Grade, solvent) were purchased from Carl Roth GmbH & Co. KG (Karlsruhe, Germany). Pronase E (from *Streptomyces griseus*), pepsin (from porcine gastric mucosa), pancreatin (from porcine pancreas), β-glucuronidase (from *Helix pomatia*), DEAE Sephadex® A-25, 4-hydroxybenzyl glucosinolate potassium salt, 2-propenyl glucosinolate potassium salt, benzyl cyanide (≥98%), benzonitrile (≥99.9%), and benzyl isothiocyanate (≥98.5%) were obtained from Merck KGaA (Darmstadt, Germany). 1*H*-Imidazole was purchased from AppliChem GmbH (Darmstadt, Germany), boric acid (powder, pure) from Honeywell International Inc. (Seelze, Germany) and 4-(methylsulfinyl)butyl glucosinolate from PhytoLab GmbH & Co. KG (Vestenbergsgreuth, Germany). Sodium sulfate anhydrous (≥99%) was obtained from VWR International GmbH (Darmstadt, Germany). C18ec solid phase extraction cartridges (3 mL, 200 mg) were obtained from Machery-Nagel GmbH & Co. KG (Düren, Germany). The BITC-lysine and BITC-cysteine standards used were synthesized by Kühn et al. [23].

2.2. Plant Material

Garden cress seeds (*Lepidium sativum* L.) were purchased from Dürr Samen (Reutlingen, Germany). The seeds were cultivated for seven days at room temperature on cotton wool with natural light on a windowsill and sprayed with water every day. The sprouts were harvested and divided into two parts. One part was used for the analysis of the GLS content of the non-processed material, the other part was directly applied as ingredient in the bread dough.

Freeze-dried nasturtium (*Tropaeolum majus* L.) leaves were obtained from the Leibniz Institute of Vegetable and Ornamental Crops (IGZ) e.V. (Großbeeren, Germany). The cultivation was done according to standard procedures for *Brassica* vegetables, including fertilization, irrigation, and plant protection. After 10 weeks, leaves were harvested and immediately frozen at −50 °C, freeze-dried, and milled to a fine powder [26]. A part of the homogenized material was analysed for the GLS content and the other part applied into the bread dough.

2.3. Bread Baking and Material for Analysis

Three bread types with different cress applications and percentage addition relative to the dough weight were prepared: freeze-dried, powdered nasturtium leaves (*Tropaeolum majus* L., 4%) mixed homogenously into the dough (application No. I); parts of fresh garden cress microgreens (*Lepidium sativum* L., 1.5%) mixed into the dough (application No. II), and freeze-dried nasturtium leaves (*Tropaeolum majus* L., 1.5%) placed centrally in the dough (application No. III).

The different bread types were prepared in triplicate. The basis was a wheat dough with 65% wheat flour (Type 450, EDEKA Zentrale AG & Co. KG, Hamburg, Germany), 33% water, 1% salt, 0.5% sugar, and 1% yeast (EDEKA Zentrale AG & Co. KG). The dough was prepared by adding the water to the dry ingredients and kneaded by hand for 5 min. The dough was divided, and the plain dough was used as control. The second part of the dough was mixed with freeze-dried powdered nasturtium leaves (application No. I) or respectively with parts of fresh garden cress microgreens (application No. II). The resulting levels of benzyl glucosinolate (BG) were around 50 µmol/100 g bread dough for application No. I and 20 µmol/100 g bread dough for application No. II. For the rise and the baking, a commercially available breadmaking device (UNOLD Backmeister® extra Modell 65811, UNOLD AG, Hockenheim, Germany) was used. The program sequence is listed in Table 1 and the temperature profile is given in Figure 1. After cooling down the whole bread, the crust was removed, and the crumb was frozen at −80 °C. For the analysis of GLS breakdown products the material was grounded using liquid nitrogen to prevent thawing and stored frozen at −80 °C. To investigate ITC-protein conjugates lyophilized powder was used, also stored at −80 °C.

Table 1. Baking program of the UNOLD Backmeister® extra Modell 65811.

Section Number	Section Name	Time [min]	Temperatures Ø [°C]
1	tempering	17	22.4
2	slow stirring	3	25.6
3	kneading clockwise	2	25.6
4	kneading anti-clockwise/clockwise	13	27.8
5	1st rising without heat	45	28.0
6	smoothing dough	1	28.0
7	2nd rising without heat	18	26.8
8	smoothing dough	1	26.8
9	3rd rising with heat	45	33.6
10	baking	55	87.8
	total	200	

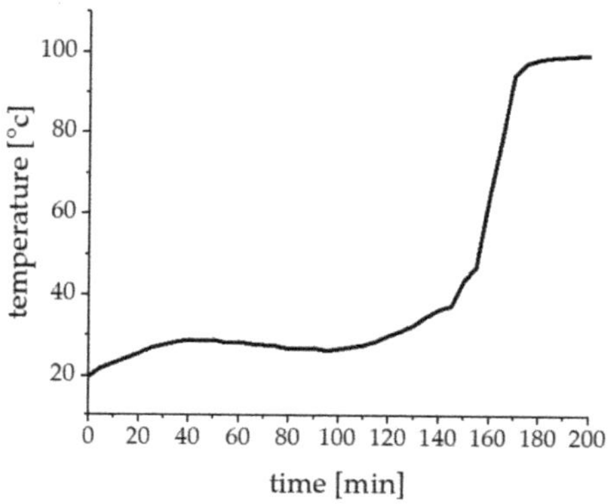

Figure 1. Temperature profile of the rise and baking in the bread maker (UNOLD Backmeister® extra Modell 65811, UNOLD AG, Hockenheim, Germany).

For the bread with the freeze-dried nasturtium bolus in the center of the dough (application No. III), the plain dough was treated according to program section number 8. Then, the additional plant material was placed into the dough and the treatment was continued with program sections number 9 and number 10. The BG amount was around 17 µmol/100 g bread dough. After cooling down, the crust and the plant material were removed. The crumb was separated into three further fractions as shown in Figure 2: Crumb material in direct and close contact to the freeze-dried material was the core fraction (fr-1). With increased distance of the original bolus, an intermediate fraction (fr-2), and an outer fraction (fr-3) were prepared.

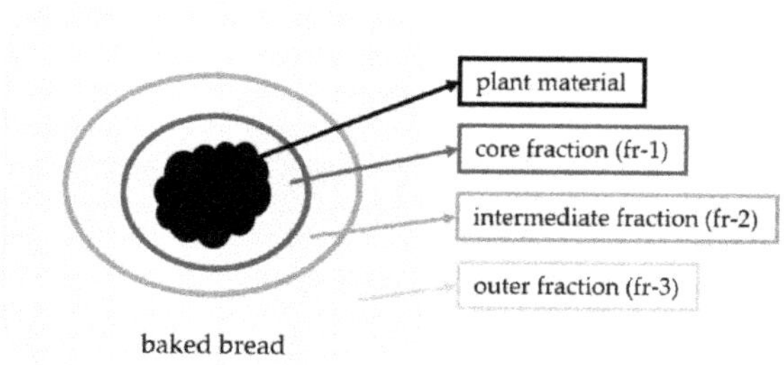

Figure 2. Scheme of the vegetable-enriched model bread application No III (bolus application). The crumb is separated into three fractions: core fraction (fr-1), intermediate fraction (fr-2), and outer fraction (fr-3).

The separated garden cress material from application No. II and nasturtium from application No. III were also used for a GLS analysis to verify the loss of GLS and/or breakdown product formation resulting from the baking process or further transformations of the plant material. This further enables one to evaluate indirectly the leaching and migration into the dough. Additionally, an experiment was performed to investigate a possible heat-induced breakdown of GLS during the baking process of application No. III. For this, homogenized nasturtium was boiled in water for 10 min to deactivate the myrosinase and freeze-dried afterwards. The samples were then parted: one part was used for GLS analysis. The other part was baked into the bread, freeze-dried and also analyzed for the GLS content.

2.4. Protein Content

For the determination of the protein content the control breads were used and a traditional Kjeldahl protocol was performed. Freeze-dried bread (2.5 g) was mixed with 20 mL sulfuric acid (96–98%, p.a.) and a catalyst tablet. The reaction was heated for 4 to 5.5 h, till the solution was clear. For the distillation, a distillation unit (Büchi Labortechnik GmbH, Essen, Germany) was used. To the cooled down reaction mixture, 75 mL of water and 75 mL of aqueous sodium hydroxide (33%, *w*/*v*) were added, and steam distilled for 5 min at 75% steam pressure into 50 mL aqueous boric acid (4%). Finally, the distillates were titrated with 0.1 M hydrochloric acid until a color change from green to grey/colorless of the Tashiro indicator was observed. The calculation of the protein content was based on the nitrogen content and a protein conversion factor of 5.7 [27].

2.5. Enzymatic Hydrolysis of the Samples

For the extraction of ITC-amino acid conjugates, a combination of different enzymatic digestion protocols were used. First, the protocol of Pasini et al. was applied, in which an in vitro digestion was imitated to hydrolyze the proteins into peptides [28]. Therefore, 50 mg of freeze-dried bread was mixed with 4 mL of a pepsin solution (0.05 mg/mL in 0.2 M hydrochloric acid, enzyme/protein ratio: 1:30) and incubated for 30 min at 37 °C and 400 rpm in a thermoshaker (Hettich Benelux B.V., Geldermalsen, the Netherlands). Afterwards, 1.15 mL of a pancreatin solution (0.25 mg/mL in 1 M boric acid and 0.5 M sodium hydroxide solution, pH 6.8, enzyme/protein ratio 1:21) were added to the samples with a resulting pH value of 7.6. The reaction mixture was incubated again (37 °C, 400 rpm) for 150 min and the enzymatic digestion was finally stopped by the addition of 1 mL 20% (*w*/*v*) trichloroacetic acid (TCA). The reaction mixture was allowed to precipitate for 60 min, centrifuged for 10 min at 2576× *g*, and the supernatant was freeze-dried. The latter was diluted in 2 mL PBS-buffer and for further protein digestion, the protocol described by Kühn et al. was used with some modifications [23]. A pronase E solution (60 µL, 7 U/mg, 10 mg/mL in PBS buffer, enzyme/protein ratio 1:100) were added and the mixture was incubated for 18 h at 37 °C in a thermoshaker (400 rpm). The enzymatic digestion was stopped by adding 100 µL 20% (*w*/*v*) TCA and the extracts were centrifuged for 10 min at 24,104× *g*.

The supernatants were purified using solid phase extraction (SPE). For the SPE, a C18ec column was used, which was conditioned with methanol (3 mL) and equilibrated with formic acid (FA, 3 mL, 0.1% in water, *v*/*v*). Subsequently, the samples were applied and washed twice with aqueous FA (3 mL, 0.1% in water, *v*/*v*). The amino acid conjugates were eluted with methanolic FA (3 mL, 0.1% in methanol, *v*/*v*). The extracts were evaporated to dryness under a continuous stream of nitrogen and finally concentrated in 100 µL FA (0.1% in methanol/water, 80:20, *v*/*v*).

2.6. HPLC-ESI-MS/MS Analysis of Protein Conjugates

For LC-ESI-MS/MS analysis of the ITC protein conjugates, the protocol described by Kühn et al. was used with some modifications [20]. An aliquot (4 µL) was injected into the LC-ESI-MS/MS-system. For quantification, an external calibration with BITC-lysine (BITC-Lys) conjugate synthesized by Kühn et al. was used in a concentration range from 0.01 to 4 μmol/L [23]. The LC-ESI-MS/MS system consisted of a 5500 QTrap triple quadrupole MS/MS system (AB Sciex Germany GmbH, Darmstadt, Germany) combined with an Agilent 1260 Infinity II HPLC-system (Agilent Technologies Deutschland GmbH, Waldbronn, Germany). For data acquisition and analysis, the software Analyst 1.7.0 (AB Sciex Germany GmbH) was used.

The separation of the analytes was performed on a Kinetex® C18 column (5 μm, 100 Å, 150 mm × 2.1 mm; Phenomenex Ltd., Aschaffenburg, Germany). The autosampler temperature was set to 4 °C and the column oven to 20 °C. The separation was performed with a flow rate of 300 μL/min of the mobile phase, consisting of 0.1% FA in water (eluent A) and 0.1% FA in MeOH (eluent B). At the beginning, the gradient consisted of 90% eluent A for the first minute. Subsequently, the concentration of eluent B increased linearly to 90% within 8 min and remained constant for 1 min. Afterwards, the start composition of eluent A and B was reached within 1 min. Finally, the column was re-equillibrated for 4 min with a composition of 90% eluent A.

The MS system was set to positive ionization mode with an entrance potential of 10 V. The ion spray voltage consisted of 5.5 kV, the desolvation gas temperature was 550 °C, the ion source gas pressure 70 psi for gas 1 and 55 psi for gas 2, and the curtain gas pressure was 40 psi.

2.7. HPLC-UV Analysis of Desulfo-GLS

The extraction of GLS and a conversion into desulfo-GLS were performed as previously described by Wiesner et al. with some modifications [29]. For extracting GLS from 10 mg freeze-dried plant material, 750 μL methanol (70% in water, *v*/*v*, 70 °C) were added as well as 25 μL of a *p*-hydroxybenzyl GLS solution (sinalbin, 1 mM in water) as internal standard. The mixture was heated at 70 °C for 10 min and afterwards centrifuged at 7748× *g* for 5 min at room temperature. The residue was re-extracted twice with 500 μL methanol (70% in water, *v*/*v*, 70 °C) and an incubation time of 5 min. All supernatants were combined.

To clean-up the extracts, small glass columns containing 500 μL of DEAE-Sephadex A-25 were used. The sorbent was activated with 6 M imidazole (2 mL, in FA 30% in water) and washed two times with water (1 mL), before sample-extracts were applied. Afterwards, 75 μL of purified β-glucuronidase (from *Helix pomatia*) were added and incubated overnight. Then, desulfo-GLS were eluted with 2 × 500 μL of water, transferred to Corning® Costar® Spin-X® centrifuge tube filters (Merck KGaA) and centrifuged for 2 min at 5165× *g*. The sample solutions passed through the filters were used for HPLC-UV analysis.

An Agilent 1260 Infinity II LC system with an UV detector, equipped with a Poroshell 120 EC-C18 column (2.7 μm, 2.1 mm × 100 mm), was used for the separation. The autosampler temperature was set to 4 °C and the column oven to 23 °C. The mobile phase consisted of water (eluent A) and acetonitrile (eluent B) with a flow rate of 400 μL/min. In the beginning, eluent A was set to 99.8% for 2 min. Subsequently, the concentration of eluent B was linearly increased to 19.8% within 10 min and hold for 2 min. After that, the concentration of eluent B was further increased to 50% within 1 min and was kept constant for 1 min. Re-equilibration of the column was finally performed for 2 min with 99.8% of eluent A.

The identification of the desulfo-GLS based on a comparison of the retention time with reference GLS such as BG, 4-(methylsulfinyl)butyl glucosinolate and 2-propenyl glucosinolate at 229 nm. The quantification was done with *p*-hydroxybenzyl glucosinolate as internal standard and the specific response factor for each compound [30].

2.8. GC-MS Analysis of GLS Breakdown Products

Freshly frozen and ground bread (500 mg) was weighed frozen into a solvent resistant vessel. The extraction started with frozen bread powder that was allowed to heat to room temperature after adding the first round of methylene chloride. GLS breakdown products were extracted and analyzed with gas chromatography-mass spectrometry (GC–MS) using the extraction protocol described by Wermter et al. and the GC-MS protocol described by Hanschen et al., except that the transfer line temperature was set to 250 °C [31,32].

2.9. Statistical Analysis

For statistical analysis Statistica 64 (Version 13, Dell Inc., Tulsa, OK, USA) was used. The differences between the results of the two different applications No. I and No. II, and the non-processed and baked plant material was analyzed using the *t*-test. A statistical confidence level of 95% ($p \leq 0.05$) was defined.

3. Results

3.1. Quantification of GLS in Nasturtium and Garden Cress before and after the Baking Process

In this experiment, the thermally-induced degradation of GLS during the baking process was investigated with a main focus on glucotropaeolin (BG) and its breakdown product BITC. For a comparison, the GLS content of the plant materials (*Tropaeolum majus* L., *Lepidium sativum* L.) was analyzed before and after the baking process. The results are shown in Table 2.

Table 2. GLS profiles and contents in nasturtium (*Tropaeolum majus* L.) and garden cress (*Lepidium sativum* L.) material, either non-processed (before baking) or after the baking process. Results are presented in µmol/g dry weight (DW) ± standard deviation. Because of the use of fresh garden cress the result of benzyl glucosinolate is additionally expressed in fresh weight (FW) for this material. Abbreviations: n.d.: not detected, BG: benzyl glucosinolate, IMG: indol-3-ylmethyl glucosinolate, 2-PE: 2-phenylethyl glucosinolate, 1-MeO-IMG: 1-methoxy-indol-3-ylmethyl glucosinolate.

		Nasturtium		Garden Cress	
		Non-Processed	After Baking [1]	Non-Processed	After Baking [1]
BG	DW	11.51 ± 0.42	n.d.	48.80 ± 3.78	27.28 ± 2.92
	FW	-	-	12.14 ± 0.94	6.79 ± 0.73
IMG	DW	-	-	0.36 ± 0.15-	0.54 ±0.06-
2-PE	DW	-	-	0.44 ± 0.07	0.29 ± 0.13
1-MeO-IMG	DW	-	-	0.08 ± 0.01	n.d.

[1] Results of three treatment replicates analyzed in technical duplicates.

In the non-processed nasturtium (*Tropaeolum majus* L.), the GLS profile consisted of only BG with a content of 11.5 µmol/g freeze-dried material, which results around 0.5 µmol BG/g bread before the baking process for application No. I and 0.18 µmol BG/g bread for application No. III. In the separated material after the baking process of the nasturtium in application No. III, no BG could be detected, which corresponds to a release and possible degradation of 100% BG during the baking process.

The GLS profile of the fresh garden cress (*Lepidium sativum* L.) microgreens consisted mainly of BG with an amount of 48.80 µmol/g in freeze-dried material, which corresponds to an amount of 12.14 µmol/g in fresh material and around 0.2 µmol BG/g bread in application No. II. It also contained 0.36 µmol/g indol-3-ylmethyl glucosinolate (glucobrassicin, IMG), 0.45 µmol/g 2-phenylethyl glucosinolate (gluconasturtiin, 2-PE), and 0.08 µmol/g 1-methoxy-indol-3-ylmethyl glucosinolate (neoglucobrassicin, 1-MeO-IMG) in freeze-dried material. After the baking process, the BG content in garden cress microgreens decreased to 27.28 µmol/g freeze-dried material, which corresponds to a decline of 44.0%. The reduction of the other GLS in the garden cress differed from 17% to 100% in the following order: 1-MeO-IMG > 2-PE > BG > IMG (Figure 3).

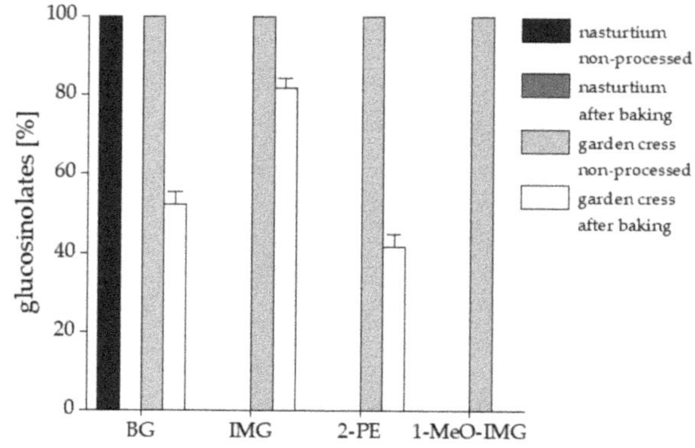

Figure 3. Thermal degradation of GLS in nasturtium and garden cress during the baking process. Results are expressed in % of residual glucosinolates of three treatment replicates analyzed in technical duplicates. Abbreviations: BG: benzyl glucosinolate, IMG: indolyl-3-methyl glucosinolate, 2-PE: 2-phenylethyl glucosinolate, 1-MeO-IMG: 1-methoxy-3-indolylmethyl glucosinolate.

With regard to the BG degradation and release of BITC, more BG was degraded when using the homogenized nasturtium leaves instead of the fresh garden cress microgreens, which could lead to a higher concentration of BITC in bread with freeze-dried material.

For the additional experiment where the myrosinase was deactivated by boiling the freeze-dried nasturtium material for 10 min, Figure 4 shows that even by adding the homogenized material to boiling water for 10 min BG is degraded by 75% from 11.51 μmol/g to 2.79 μmol/g in freeze-dried material. After the baking process no BG could be detected. Therefore, GLS in freeze-dried homogenized material can be strongly degraded by heat.

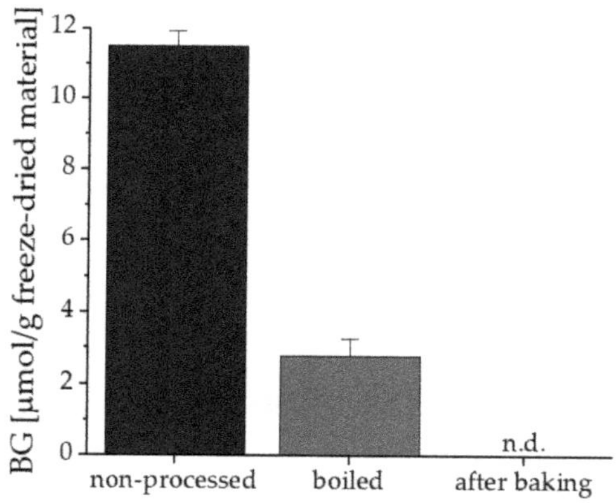

Figure 4. Thermal degradation of GLS in nasturtium after boiling in water for 10 min to deactivate the myrosinase and further baked into bread. The experiment and analysis were performed in duplicate. Abbreviations: BG: benzyl glucosinolates, n.d.: not detected.

3.2. *ITC-Protein Conjugates in Bread with Different Cress Genera*

To investigate possible reactions between BITC and wheat proteins, homogenized plant material of nasturtium (*Tropaeolum majus* L) was used for creating a large reaction surface and compared to the addition of garden cress (*Lepidium sativum* L.) microgreens, being comparatively more compact in structure.

The baked breads showed a difference in appearance, because of the differently applied plant materials. When adding a freeze-dried, powdered material, a homogenous green color of the crumb was obvious (Figure 5a), while the addition of fresh material was more heterogeneously distributed. So, individual green particles were noticed and spread all over the crumb (Figure 5b); the reference bread showed no further particularities (Figure 5c).

(a) (b) (c)

Figure 5. Pictures of the different cress-enriched model breads. (**a**) freeze-dried nasturtium (*Tropaeolum majus* L, 4%) mixed into the dough (application No. I); (**b**) fresh garden cress (*Lepidium sativum* L., 1.5%) mixed into the dough (application No. II); (**c**) reference bread without plant material.

The LC-ESI-MS/MS method was developed for analyzing modifications with amino and thiol groups analyzed as ITC-lysine and ITC-cysteine conjugates. At hand of synthesized standards, the thiourea derivative BITC-lysine (BITC-Lys) and the dithiocarbamate BITC-cystein (BITC-Cys) can be quantified [20].

As shown in Figure 6, in the bread with freeze-dried powdered nasturtium leaves (*Tropaeolum majus* L., application No. I), 20.9 nmol BITC-Lys/g protein corresponding to 3.36 nmol BITC-Lys/g bread were determined, while in the bread with the fresh garden cress (*Lepidium sativum* L.) microgreens (application No. II), 17.2 nmol BITC-Lys/g protein (3.07 nmol BITC-Lys/g bread) were found.

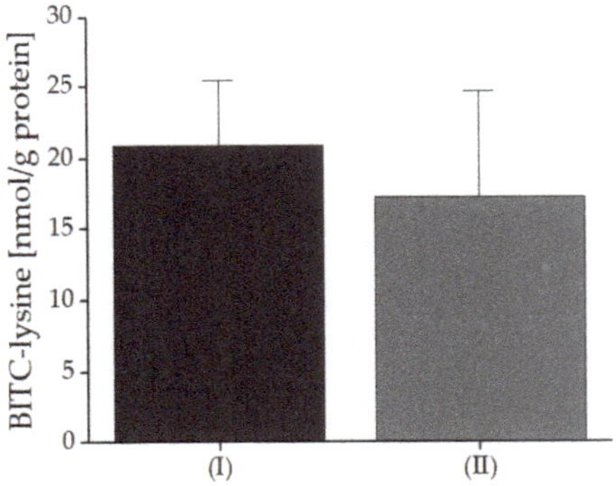

Figure 6. Formed BITC-Lys conjugates in bread with freeze-dried nasturtium (application No. I) and garden cress microgreens (application No. II) mixed into the dough. The breads were prepared and analyzed in triplicate. The initial amount of BG in the breads differs around 40%, but the amounts of BITC-Lys conjugates differed in a range of 10% per gram bread. Therefore, the addition of 4% homogenized nasturtium leaves resulted only in a 10% higher concentration of BITC-Lys conjugation compared to the addition of 1.5% fresh garden cress.

In comparison to BITC-Lys, the dithiocarbamate BITC-Cys was not detected in the samples.

3.3. *Quantification of BITC and Benzyl Cyanide in Bread with Different Cress Genera*

Next to the identification and quantification of BITC-protein conjugates, the breakdown products of BG, BITC and the corresponding nitrile benzyl cyanide (BC), were investigated. This analysis was performed to obtain an indication of whether free BITC is detectable or whether the GLS has been transformed to the corresponding nitrile during the baking process. This would be visible in a low concentration of BITC and a high concentration of BC. Again, it was suspected that there might be a difference between the homogenized nasturtium leaves, and the fresh material of garden cress, due to the different degree of destruction of the plant material prior to and during the baking process.

The results in Figure 7 show only a slight difference (ratio of 1:1.38) in the concentration of BITC (5.93 nmol/g bread) and BC (4.30 nmol/g bread) for application No. I. In the bread application No. II, only 1.51 nmol BITC/g bread and 42.74 nmol BC/g bread were analyzed, resulting in a higher ratio of 1:28 BITC to BC. Comparing the different applications, more BITC was detected in application No. I, but 10 times more BC was analyzed in application No. II.

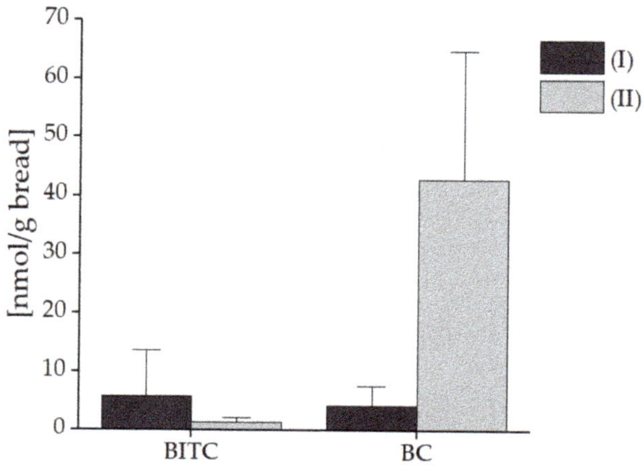

Figure 7. Formed benzyl glucosinolate breakdown products BITC and BC in bread with freeze-dried nasturtium (application No. I) and garden cress microgreens (application No. II) mixed into the dough. The breads were prepared in triplicate. Abbreviations: BITC: benzyl isothiocyanate, BC: benzyl cyanide.

3.4. ITC-Protein Conjugates in Bread with a Centrally Placed Nasturtium Bolus

The migration behavior of ITC into the food matrix and the distribution of reaction products during food processing were further studied in a model bread, where a bolus of freeze-dried nasturtium material was placed centrally (application No. III). Following the baking process, the bolus was removed and the GLS content analyzed. The bread was divided into a core, an intermediate, and an outer fraction (Figure 2). The appearance of the crumb did not show any differences compared to the reference bread (Figure 8a,b).

Figure 8. Picture of the cress-enriched model bread. (**a**) with freeze-dried nasturtium (*Tropaeolum majus* L, 1.5%) placed as centrally bolus (application No. III); (**b**) Picture of the reference bread without plant material.

BITC-Lys was detectable in the crumb. As anticipated, the highest concentration with 9.02 nmol BITC-Lys/g protein (1.38 nmol BITC-Lys/g bread) was found in the core fraction, while the analysis of the intermediate fraction showed a lower concentration of only 0.89 nmol BITC-Lys/g protein (0.12 nmol BITC-Lys/g bread) (Figure 9). In the outer fraction BITC-Lys could still be detected with a concentration of 0.26 nmol BITC-Lys/g protein (0.02 nmol BITC-Lys/g bread). Again, the BITC-Cys conjugate could not be detected in any of the fractions.

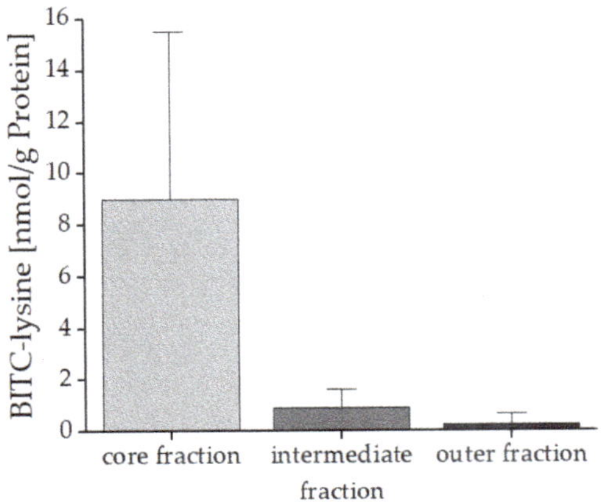

Figure 9. Amounts of the BITC-Lys conjugates in the core, intermediate, and outer fraction of the bread with freeze-dried nasturtium added selectively (application No. III). Three different breads were prepared and analyzed in triplicate.

3.5. Quantification of BITC and Benzyl Cyanide in Bread with a Centrally Placed Nasturtium Bolus

It was also of interest to investigate the migration behavior of the BG breakdown products BITC and BC in the different fractions of the bread. As shown in Figure 10a, it was found that only in the core fraction BITC could be detected with a concentration of 0.52 nmol/g bread. BC was found in all three fractions with decreasing concentration from the core fraction to the outer fraction. In detail the results in Figure 10b show that in the core fraction 653.02 nmol BC/g bread was found, decreased by 64% to the intermediate fraction (235.25 nmol BC/g bread). The difference from the intermediate to the outer fraction (66.98 nmol BC/g bread) was around 70%.

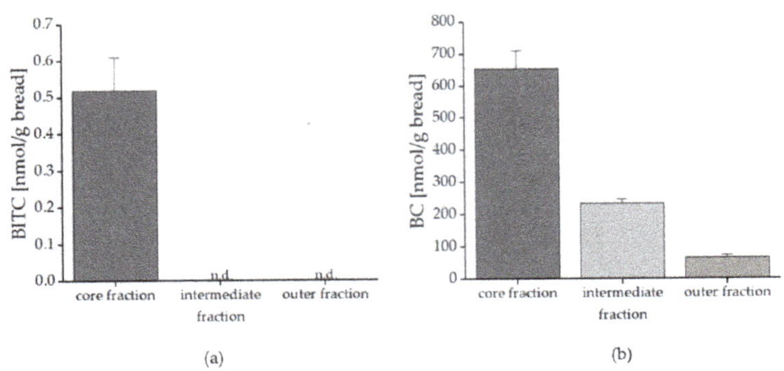

Figure 10. Formed benzyl glucosinolate breakdown products BITC (**a**) and BC (**b**) in bread with freeze-dried nasturtium added as a bolus in the middle of the bread (application No. III). The breads were prepared in triplicate. Abbreviations: BITC: benzyl isothiocyanate, BC: benzyl cyanide.

4. Discussion

In the present study, the degradation of GLS, the migration of their main breakdown product ITC, the corresponding nitrile, and the formation of ITC-protein conjugates were exemplarily investigated in a vegetable-enriched bread.

For these studies, bread was baked with parts of fresh garden cress (*Lepidium sativum* L.) microgreens and homogenized freeze-dried powdered nasturtium leaves (*Tropaeolum majus* L.). The different cress genera mainly contained the GLS BG, which can be degraded to BITC and the corresponding nitrile. This allowed a direct conclusion on the used material (homogenized or microgreens), independent of different chemical structures and therefore, different degradation rates of GLS during a heat treatment [33]. Because of the different percentage of addition the bread with nasturtium contained 2.5 times more BG before the baking process, therefore it was assumed that there is a more intense release of BITC during the baking process and consequently to an enhanced conjugation with proteins.

With regard to the GLS profile before and after the baking process, 100% BG in the homogenized freeze-dried nasturtium were degraded. In comparison, only 44% BG were released during the process from the garden cress microgreens into the crumb. Therefore, in application No. I around 50 µmol/100 g bread dough were degraded and in application No. II around 9 µmol BG/100 g bread dough. A former study assumed that due to a low destruction of the plant matrix, the GLS remain longer intact in the plant material [24]. The results of the present study underlined this assumption. When using fresh, non-homogenously material, the migration of released BITC into the surrounding food matrix might be hampered by the intact plant structure and the reduced reaction surface in comparison to a freeze-dried material (Figure 5b). The plant tissue of freeze-dried material is already more or less disintegrated. By adding the material to the moist bread dough, the GLS could already be degraded by the enzyme myrosinase before the baking process, which is a difference to the use of garden cress microgreens. During the baking process, further heat-induced degradation can occur. The more homogeneous, pulverized material also leads to easier migration of the GLS and their breakdown products into the surrounding matrix before and during the baking process because the homogenized material ensures a broader distribution in the crumb (Figure 5a).

For these reasons and the higher BG concentration it was assumed that there would be significantly more protein conjugates in application No. I. The results confirmed that there is indeed a higher concentration of BITC-Lys, but it is only 10% higher than in application No. II even though the BITC levels likely formed were lower, because only 44% BG is released. To explain these results, a correlation can be made to the analyzed breakdown products. The breakdown product results show that there is only a slight difference in the amount of unreacted BITC in the different breads. However, the BC results show a 10-times higher concentration in the bread with fresh garden cress compared to the bread with freeze-dried, homogenized nasturtium. Thus, overall, a higher amount of breakdown products is detectable in application No. II.

The difference can be explained by the different states of the plant materials. In application No. I, BG might be already enzymatically degraded to BITC before the baking process began. This BITC might be degraded by heat action at an early stage of the baking process and thus might be no longer available for follow-up reactions. In addition, there appears to be volatilization of BITC and BC, so that overall fewer breakdown products were detectable in the baked bread of application No. I. When fresh material is used, the degradation of BG starts probably mainly during the baking process, so that initially the released BITC is available for reactions with proteins directly during the baking process. BC very likely is formed in high amounts due to thermal degradation of BG, as nitriles have been found to be the main degradation products from GLS in heated vegetables [13,25]. Therefore, despite the lower initial concentration of BG in application No. II, a larger amount of BITC may have been available for the formation of protein conjugates, as shown in the results of the ITC conjugates.

In the present study, 0.01% of BG in application No. I and 0.02% in application No. II were transformed into BITC-Lys conjugates, resulting in around 0.01% modified lysine in both breads. In an experiment described by Kühn et al., curd was mixed with cress, resulting in a typical German spread ('Kräuterquark') and analyzed for possible ITC protein conjugates. The results showed a higher conjugation of BG to BITC-Lys of up to 2–4% [23]. Considering the pH value, which not only influences the formation of ITC conjugates, but also their stability, more lysine conjugates should be detectable in wheat bread (pH 5.9) than in curd (pH 4.5), since a pH value close to the neutral range leads to an increased formation and stabilization of thioureas as shown by Platz et al. [34]. Therefore, the different results could be explained by the far different processing approaches. Baking produces temperatures of up to 95 °C for around 30 min in the bread matrix (Figure 1), whereas curd with herbs is only mixed at room temperature.

In regard to the enzymatic degradation of GLS selected studies showed that a short heat exposure of white cabbage (*Brassica oleracea var. capitata f. alba* cv. Minicole), red cabbage (*Brassica oleracea var. capitata f. rubra* cv. Primero), and kohlrabi (*Brassica oleracea var. gongylodes cv.* Kolibri,) for 10 min at around 60 °C led to an increased ITC formation due to inactivation of specifier proteins that enhance enzymatically nitrile and epithionitrile formation. In contrast, longer heat exposure durations (of up to 120 min), especially in the range of 100 °C, led to a nitrile formation due to thermal GLS degradation. Nitriles then accumulate in the matrix, because of their heat stability. Compared to the nitriles, the concentration of ITC decreases with prolonged heating, which results in less ITC available for follow-up reactions with proteins [13]. During the baking process, ITC could decompose to a wide range of further breakdown products, depending on their chemical structure. As an example for an aromatic ITC, a study applying phenethyl isothiocyanate (PEITC) in aqueous solutions showed that PEITC can be thermally decomposed to phenethylamine and can further react to *N,N′*-diphenethylthiourea [35]. This hydrolysis reaction of ITC results in less protein conjugates in heated food products because less ITC is available for follow-up reactions with proteins.

To understand not only the formation of ITC-protein conjugates, but also the migration of GLS, ITC, and possible reaction products, another bread with a centrally placed freeze-dried nasturtium bolus was prepared. This experiment showed that BITC migrated up to the outer fraction and the concentration of BITC must have been still high enough to detect BITC-protein conjugates. It could be demonstrated that the concentration of the conjugates was reduced by 90% from the core fraction to the intermediate fraction (Figure 7). However, ITC-protein conjugates are still detectable in the outer fraction reduced by 97% to the core fraction.

The analysis of the BG breakdown products BITC and BC showed, that BITC was only detectable in the core fraction, whereas BC, which was highly abundant, could be quantified in all three fractions reduced by 90% from the core fraction to the outer fraction. This result shows that BITC migrates to the outer fraction during the baking process and reacts with proteins. Furthermore, not only the BITC migrates from the core fraction to the outer fraction, as proven by the protein adducts, but also formed BC.

In comparison to application No. I with homogenized nasturtium leaves mixed into the dough, there might also be a difference of degradation of BG. Whereas BG might be mainly degraded enzymatically in application No. I due to the contact with the dough, the BG in freeze-dried material added as a bolus might be degraded as well by the heat during the baking process, because not all the material is in contact with the moisture of the bread dough. When adding homogenized nasturtium to boiling water and treating it for 10 min at 100 °C, already 75% of BG were degraded. Therefore, BG seems to be very heat labile, and this might explain the total degradation of BG that was observed.

In this study, 1.5% to 4% homogenized nasturtium was applied to the bread doughs. In a former study, bread has been enriched with vegetable powder in a range of 1–3%, which lies within the range of quantity used [36]. Studies using fresh diced vegetable ingredients, such as kale, beetroot, and spinach, enriched the breads in a range of 10 to 40% [6,24,37].

Here, the prepared breads with fresh garden cress microgreens were enriched with 1.5%. With regard to a higher enrichment around 10 to 40% the concentration of BG and thus, resulting BITC likely would further increase significantly. This again can lead to a more intense release and migration of ITC and more follow-up reactions with proteins.

In all experiments, only BITC-Lys conjugates were detected, although the calculated difference in the concentrations of lysine (2.3% in wheat proteins) and cysteine (1.6% in wheat proteins) in the initial wheat proteins was only 0.7%. In another study, where garden cress was implemented in curd, higher concentrations of BITC-Cys conjugates were detected, despite the lower concentration of cysteine (0.3%) to lysine (8.2%) of the milk proteins [23]. Because of these results it was assumed that more cysteine conjugates would be formed in the present study, which could not be confirmed. These differences can be explained by the pH value and the heat treatment during the baking process (Figure 1). The pH value of the food matrix not only influences the formation of BITC protein conjugates but also their stability. The pH of the bread (5.9) should lead to a higher formation of BITC-Lys conjugates [34]. Additionally, BITC-Cys conjugation is reversible and shows only little stability with increasing temperature, whereas reaction between the amino group of lysine and ITC are stable and temperature has only a minor influence on BITC-Lys conjugates [20,38,39].

Next to a potentially lower biological value, protein conjugations have further consequences. In former studies, an influence on typical digestive enzymes has already been observed: While trypsin and chymotrypsin were not able to fully hydrolyze ITC-modified proteins, pepsin activity was not hindered. These differences are assigned to the typical specificities of those enzymes to split near selected amino acids. Trypsin, for example, hydrolyzes peptide linkages containing lysine or arginine, whereas pepsin is a non-specific protease [19,40–42]. Therefore, the digestibility by trypsin can be more affected due to the BITC-Lys conjugation. Next to the influence on the digestibility, another experiment showed that ITC modifications also lead to a decrease of enzyme activity as shown with tyrosine phosphatase [43]. In food production, a reduced water solubility of proteins due to a conjugation with the hydrophobic BITC could be relevant, as well as, altered heat aggregation, foaming and emulsifying properties which was demonstrated at hand of AITC modified β-lactoglobulin [41,44].

In addition to GLS, Brassicales vegetables also contain other SPM such as flavonoids and carotenoids. The heat exposure during the baking might also affect these SPM, which could also lead to follow-up reactions [24,45]. It has already been shown that the antioxidant activity of phenolic compounds in vegetables is often not diminished, even when there is a thermally-induced breakdown to smaller compounds, but there is also a formation of complexes with proteins during the baking process [45].

5. Conclusions

In foods containing proteins and GLS-rich vegetables, reactions between ITC and proteins may occur and the reaction of BITC with lysine can lead to a lower biological value of the protein [46]. However, in comparison to non-heated foods, less protein conjugates are formed in heated food products, because the formation of ITC during the baking process is diminished due to evaporation and hydrolysis reactions. Therefore, a lower bioavailability of lysine due to reactions with ITC is less of a concern in heated processed foods, but there is also no health benefit from more available ITC.

In addition to the ITC-amino acid conjugates with cysteine and lysine investigated herein, there could be further reactions with other amino acids, such as methionine (1.1% in the bread) and tyrosine (1.5% in the bread). Such reactions have not yet been investigated and could lead to a higher ITC-protein conjugation than previously anticipated [47,48]. It could also be interesting to study how the concentration of conjugates change over a longer period of time, because bread is typically consumed over several days.

Author Contributions: Conceptualization, M.K. and S.R.; Data curation, M.K.; Formal analysis, M.K., M.P. and F.S.H.; Investigation, M.K. and F.S.H.; Methodology, M.K., L.C. and F.S.H.; Resources,

S.R.; Supervision, S.R.; Visualization, M.K.; Writing—original draft, M.K., F.S.H. and S.R.; Writing—review & editing, M.K., F.S.H. and S.R. All authors have read and agreed to the published version of the manuscript.

Funding: Franziska S. Hanschen is funded by the Leibniz-Association (Leibniz-Junior Research Group OPTIGLUP; J16/2017).

Data Availability Statement: The data sets presented in this study are available on request from the corresponding author.

Acknowledgments: We would like to thank Maria Riedner and Gaby Graack for providing excellent support with the QTRAP 5500 mass spectrometric measurements and Andrea Jankowsky for her support regarding the quantification of the glucosinolates. We acknowledge support by the German Research Foundation and the Open Access Publication Fund of the TU Berlin.

Conflicts of Interest: The authors declare no conflict of interest.

References

1. Raskin, I.; Ribnicky, D.M.; Komarnytsky, S.; Ilic, N.; Poulev, A.; Borisjuk, N.; Brinker, A.; Moreno, D.A.; Ripoll, C.; Yakoby, N. Plants and human health in the twenty-first century. *Trends Biotechnol.* **2002**, *20*, 522–531. [CrossRef]
2. Bernhoft, A. A brief review on bioactive compounds in plants. *Bioact. Compd. Plants-Benefits Risks Man Anim.* **2010**, *50*, 11–17.
3. de Ridder, D.; Kroese, F.; Evers, C.; Adriaanse, M.; Gillebaart, M. Healthy diet: Health impact, prevalence, correlates, and interventions. *Psychol. Health* **2017**, *32*, 907–941. [CrossRef] [PubMed]
4. WHO. Obesity and Overweight. 2020. Available online: https://www.who.int/news-room/fact-sheets/detail/obesity-and-overweight (accessed on 4 January 2021).
5. Oliviero, T.; Fogliano, V. Food design strategies to increase vegetable intake: The case of vegetable enriched pasta. *Trends Food Sci. Technol.* **2016**, *51*, 58–64. [CrossRef]
6. Hobbs, D.; Ashouri, A.; George, T.; Lovegrove, J.; Methven, L. The consumer acceptance of novel vegetable-enriched bread products as a potential vehicle to increase vegetable consumption. *Food Res. Int.* **2014**, *58*, 15–22. [CrossRef]
7. Klopsch, R.; Baldermann, S.; Voss, A.; Rohn, S.; Schreiner, M.; Neugart, S. Bread enriched with legume microgreens and leaves—Ontogenetic and baking-driven changes in the profile of secondary plant metabolites. *Front. Chem.* **2018**, *6*, 1–19. [CrossRef] [PubMed]
8. Belderok, B. Developments in bread-making processes. *Plant Foods Hum. Nutr.* **2000**, *55*, 1–14. [CrossRef] [PubMed]
9. Mithen, R.; Lewis, B.; Fenwick, G. In vitro activity of glucosinolates and their products against *Leptosphaeria maculans*. *Trans. Br. Mycol. Soc.* **1986**, *87*, 433–440. [CrossRef]
10. Palliyaguru, D.L.; Yuan, J.M.; Kensler, T.W.; Fahey, J.W. Isothiocyanates: Translating the power of plants to people. *Mol. Nutr. Food Res.* **2018**, *62*, 1700965. [CrossRef]
11. Veeranki, O.L.; Bhattacharya, A.; Tang, L.; Marshall, J.R.; Zhang, Y. Cruciferous vegetables, isothiocyanates, and prevention of bladder cancer. *Curr. Pharmacol. Rep.* **2015**, *1*, 272–282. [CrossRef]
12. Gupta, P.; Wright, S.E.; Kim, S.-H.; Srivastava, S.K. Phenethyl isothiocyanate: A comprehensive review of anti-cancer mechanisms. *Biochim. Biophys. Acta (BBA) Rev. Cancer* **2014**, *1846*, 405–424. [CrossRef]
13. Hanschen, F.S.; Kühn, C.; Nickel, M.; Rohn, S.; Dekker, M. Leaching and degradation kinetics of glucosinolates during boiling of *Brassica oleracea* vegetables and the formation of their breakdown products. *Food Chem.* **2018**, *263*, 240–250. [CrossRef] [PubMed]
14. Bones, A.M.; Rossiter, J.T. The enzymic and chemically induced decomposition of glucosinolates. *Phytochemistry* **2006**, *67*, 1053–1067. [CrossRef]
15. Kliebenstein, D.J.; Kroymann, J.; Mitchell-Olds, T. The glucosinolate–myrosinase system in an ecological and evolutionary context. *Curr. Opin. Plant Biol.* **2005**, *8*, 264–271. [CrossRef]
16. Mocniak, L.E.; Elkin, K.; Bollinger, J.M., Jr. Lifetimes of the aglycone substrates of specifier proteins, the autonomous iron enzymes that dictate the products of the glucosinolate-myrosinase defense system in Brassica plants. *Biochemistry* **2020**, *59*, 2432–2441. [CrossRef]
17. Wittstock, U.; Kurzbach, E.; Herfurth, A.-M.; Stauber, E. Glucosinolate breakdown. *Adv. Bot. Res.* **2016**, *80*, 125–169.
18. Kroll, J.; Rawel, H.; Kröck, R.; Proll, J.; Schnaak, W. Interactions of isothiocyanates with egg white proteins. *Food/Nahrung* **1994**, *38*, 53–60. [CrossRef]
19. Kawakishi, S.; Kaneko, T. Interaction of proteins with allyl isothiocyanate. *J. Agric. Food Chem.* **1987**, *35*, 85–88. [CrossRef]
20. Kühn, C.; von Oesen, T.; Herz, C.; Schreiner, M.; Hanschen, F.S.; Lamy, E.; Rohn, S. In vitro determination of protein conjugates in human cells by LC-ESI-MS/MS after benzyl isothiocyanate exposure. *J. Agric. Food Chem.* **2018**, *66*, 6727–6733. [CrossRef]
21. Kroll, J.; Rawel, H. Chemical reactions of benzyl isothiocyanate with myoglobin. *J. Sci. Food Agric.* **1996**, *72*, 376–384. [CrossRef]
22. Keppler, J.K.; Koudelka, T.; Palani, K.; Stuhldreier, M.C.; Temps, F.; Tholey, A.; Schwarz, K. Characterization of the covalent binding of allyl isothiocyanate to β-lactoglobulin by fluorescence quenching, equilibrium measurement, and mass spectrometry. *J. Biomol. Struct. Dyn.* **2014**, *32*, 1103–1117. [CrossRef]

23. Kühn, C.; von Oesen, T.; Hanschen, F.S.; Rohn, S. Determination of isothiocyanate-protein conjugates in milk and curd after adding garden cress (*Lepidium sativum* L.). *Food Res. Int.* **2018**, *108*, 621–627. [CrossRef] [PubMed]
24. Klopsch, R.; Baldermann, S.; Hanschen, F.S.; Voss, A.; Rohn, S.; Schreiner, M.; Neugart, S. Brassica-enriched wheat bread: Unraveling the impact of ontogeny and breadmaking on bioactive secondary plant metabolites of pak choi and kale. *Food Chem.* **2019**, *295*, 412–422. [CrossRef] [PubMed]
25. Hanschen, F.S.; Platz, S.; Mewis, I.; Schreiner, M.; Rohn, S.; Kroh, L.W. Thermally induced degradation of sulfur-containing aliphatic glucosinolates in broccoli sprouts (*Brassica oleracea* var. *italica*) and model systems. *J. Agric. Food Chem.* **2012**, *60*, 2231–2241. [CrossRef] [PubMed]
26. Kühn, C.; Kupke, F.; Baldermann, S.; Klopsch, R.; Lamy, E.; Hornemann, S.; Pfeiffer, A.F.; Schreiner, M.; Hanschen, F.S.; Rohn, S. Diverse excretion pathways of benzyl glucosinolate in humans after consumption of nasturtium (*Tropaeolum majus* L.)—A pilot study. *Mol. Nutr. Food Res.* **2018**, *62*, e1800588. [CrossRef]
27. Williams, P.C. The use of titanium dioxide as a catalyst for large-scale Kjeldahl determination of the total nitrogen content of cereal grains. *J. Sci. Food Agric.* **1973**, *24*, 343–348. [CrossRef]
28. Pasini, G.; Simonato, B.; Giannattasio, M.; Peruffo, A.D.; Curioni, A. Modifications of wheat flour proteins during in vitro digestion of bread dough, crumb, and crust: An electrophoretic and immunological study. *J. Agric. Food Chem.* **2001**, *49*, 2254–2261. [CrossRef] [PubMed]
29. Wiesner, M.; Zrenner, R.; Krumbein, A.; Glatt, H.; Schreiner, M. Genotypic variation of the glucosinolate profile in Pak Choi (*Brassica rapa* ssp. *chinensis*). *J. Agric. Food Chem.* **2013**, *61*, 1943–1953. [CrossRef]
30. EN ISO 9167-1:1995. *Rapssamen–Bestimmung des Glucosinolatgehaltes—Teil 1: HPLC-Verfahren (ISO 9167-1: 1992)*; DIN Deutsches Institut für Normung e.V: Berlin, Germany, 1995; ICS 67.200.20; pp. 1–8.
31. Wermter, N.S.; Rohn, S.; Hanschen, F.S. Seasonal variation of glucosinolate hydrolysis products in commercial white and red cabbages (*Brassica oleracea var. capitata*). *Foods* **2020**, *9*, 1682. [CrossRef] [PubMed]
32. Hanschen, F.S.; Schreiner, M. Isothiocyanates, nitriles, and epithionitriles from glucosinolates are affected by genotype and developmental stage in *Brassica oleracea* varieties. *Front. Plant Sci.* **2017**, *8*, 1095. [CrossRef]
33. Hanschen, F.S.; Rohn, S.; Mewis, I.; Schreiner, M.; Kroh, L.W. Influence of the chemical structure on the thermal degradation of the glucosinolates in broccoli sprouts. *Food Chem.* **2012**, *130*, 1–8. [CrossRef]
34. Platz, S.; Kühn, C.; Schiess, S.; Schreiner, M.; Mewis, I.; Kemper, M.; Pfeiffer, A.; Rohn, S. Determination of benzyl isothiocyanate metabolites in human plasma and urine by LC-ESI-MS/MS after ingestion of nasturtium (*Tropaeolum majus* L.). *Anal. Bioanal. Chem.* **2013**, *405*, 7427–7436. [CrossRef]
35. Chen, C.-W.; Rosen, R.T.; Ho, C.-T. Analysis of thermal degradation products of allyl isothiocyanate and phenethyl isothiocyanate. *ACS Symp. Ser.* **1998**, *705*, 152–166.
36. Odunlade, T.; Famuwagun, A.; Taiwo, K.; Gbadamosi, S.; Oyedele, D.; Adebooye, O. Chemical composition and quality characteristics of wheat bread supplemented with leafy vegetable powders. *J. Food Qual.* **2017**, *2017*, 1–7. [CrossRef]
37. López-Nicolás, R.; Frontela-Saseta, C.; González-Abellán, R.; Barado-Piqueras, A.; Perez-Conesa, D.; Ros-Berruezo, G. Folate fortification of white and whole-grain bread by adding Swiss chard and spinach. Acceptability by consumers. *LWT-Food Sci. Technol.* **2014**, *59*, 263–269. [CrossRef]
38. Brown, K.K.; Hampton, M.B. Biological targets of isothiocyanates. *Biochim. Biophys. Acta* **2011**, *1810*, 888–894. [CrossRef] [PubMed]
39. Nakamura, T.; Kawai, Y.; Kitamoto, N.; Osawa, T.; Kato, Y. Covalent modification of lysine residues by allyl isothiocyanate in physiological conditions: Plausible transformation of isothiocyanate from thiol to amine. *Chem. Res. Toxicol.* **2009**, *22*, 536–542. [CrossRef]
40. Rawel, H.; Kroll, J.; Haebel, S.; Peter, M.G. Reactions of a glucosinolate breakdown product (benzyl isothiocyanate) with myoglobin. *Phytochemistry* **1998**, *48*, 1305–1311. [CrossRef]
41. Rawel, H.; Kroll, J.; Schröder, I. Reactions of isothiocyanates with food proteins: Influence on enzyme activity and tryptical degradation. *Food/Nahrung* **1998**, *42*, 197–199. [CrossRef]
42. Rawel, H.M.; Kroll, J.; Schröder, I. In vitro enzymatic digestion of benzyl-and phenylisothiocyanate-derivatized food proteins. *J. Agric. Food Chem.* **1998**, *46*, 5103–5109. [CrossRef]
43. Lewis, S.M.; Li, Y.; Catalano, M.J.; Laciak, A.R.; Singh, H.; Seiner, D.R.; Reilly, T.J.; Tanner, J.J.; Gates, K.S. Inactivation of protein tyrosine phosphatases by dietary isothiocyanates. *Bioorganic Med. Chem. Lett.* **2015**, *25*, 4549–4552. [CrossRef] [PubMed]
44. Rade-Kukic, K.; Schmitt, C.; Rawel, H.M. Formation of conjugates between β-lactoglobulin and allyl isothiocyanate: Effect on protein heat aggregation, foaming and emulsifying properties. *Food Hydrocoll.* **2011**, *25*, 694–706. [CrossRef]
45. Świeca, M.; Gawlik-Dziki, U.; Dziki, D.; Baraniak, B.; Czyż, J. The influence of protein–flavonoid interactions on protein digestibility in vitro and the antioxidant quality of breads enriched with onion skin. *Food Chem.* **2013**, *141*, 451–458. [CrossRef]
46. Hernández-Triana, M.; Kroll, J.; Proll, J.; Noack, J.; Petzke, K.J. Benzyl-isothiocyanate (BITC) decreases quality of egg white proteins in rats. *J. Nutr. Biochem.* **1996**, *7*, 322–326. [CrossRef]
47. McDermott, E.; Pace, J. The content of amino-acids in white flour and bread. *Br. J. Nutr.* **1957**, *11*, 446–452. [CrossRef]
48. Murthy, N.K.K.; Rao, M.N. Interaction of allyl isothiocyanate with mustard 12S protein. *J. Agric. Food Chem.* **1986**, *34*, 448–452. [CrossRef]

foods

MDPI

Article

Low pH Enhances the Glucosinolate-Mediated Yellowing of Takuan-zuke under Low Salt Conditions

Taito Kobayashi [1], Kei Kumakura [1], Asaka Takahashi [2] and Hiroki Matsuoka [1,*]

1 Department of Health and Nutrition, Takasaki University of Health and Welfare, 37-1 Nakaorui-machi, Takasaki-shi, Gunma 370-0033, Japan; kobayashi-tai@takasaki-u.ac.jp (T.K.); kumakura@takasaki-u.ac.jp (K.K.)

2 Department of Nutritional Sciences, Tohto University, 4-2-7 Kamishima-cho Nishi, Fukaya-shi, Saitama 366-0052, Japan; asaka.takahashi@tohto.ac.jp

* Correspondence: matsuoka@takasaki-u.ac.jp; Tel.: +81-27-388-8390

Received: 9 September 2020; Accepted: 20 October 2020; Published: 23 October 2020

Abstract: This study was performed to clarify the enhancement of the 4-methylthio-3-butenyl isothiocyanate induced yellowing of salted radish root (takuan-zuke) by low pH during short-term salt-aging at low temperature and low salinity. We used two different methods to prepare the dehydrated daikon prior to salt-aging: air-drying outdoors (hoshi takuan-zuke) or salting with a stone press (shio-oshi takuan-zuke). Low salt-aging at low temperature was carried out under pH control with citrate-phosphate buffer. The yellowing of both types of takuan-zuke was accelerated below pH 5, and the color of air-dried takuan-zuke was deeper than that of salt-pressed takuan-zuke. To elucidate this phenomenon, several previously reported yellowing-related compounds were analyzed by high-performance liquid chromatography. The result showed that the production of the primary pigment, 2-[3-(2-thioxopyrrolidin-3-ylidene)methyl]-tryptophan, was low compared with that in previous reports. Therefore, we suggest that an unknown pigment was generated through a previously unreported pathway.

Keywords: pickled vegetables; yellowing salted radish root; glucosinolate–myrosinase system; tryptophan biosynthesis; isothiocyanates

1. Introduction

Takuan-zuke (salted radish root) is a popular and traditional Japanese processed food. Japanese radish roots (*Raphanus sativus* L.; daikon) are dehydrated by either air-drying outdoors (hoshi takuan-zuke) or salting with a stone press (shio-oshi takuan-zuke) before pickling. The dehydrated daikon radishes are then pickled in salt or salty rice bran for several months. The color of takuan-zuke is transformed from the white color of daikon to bright yellow during the long salt-aging process at ambient temperature [1]. However, the yellow color of takuan-zuke is easily photobleached by visible light and often fades when it is displayed in stores [2]. Recently, the color of commercial takuan-zuke, which is pre-pickled at low temperature and low salt, has been noted to be a dull yellow. Commercial takuan-zuke is prepared using yellow coloring agents such as tartrazine and gardenia pigment for consistency. However, modern Japanese consumers prefer white takuan-zuke to yellow takuan-zuke, possibly due to the misunderstanding of why white radish is intentionally dyed yellow. The need for adding colorants can be avoided if the yellowing reaction is controlled.

In our previous reports, we elucidated the detailed mechanisms of the yellowing reaction. Figure 1 shows the production process for the yellow pigment contained in takuan-zuke. Briefly, the starting compound for the yellowing reaction is 4-methylthio-3-butenyl isothiocyanate (MTB-ITC; raphasatin), which is the main pungent compound of radish. MTB-ITC is generated enzymatically from 4-methylthio-3-butenyl glucosinolate (MTB-GLS; glucoraphasatin), which is induced by cell damage during the dehydration process. Because MTB-ITC is readily degraded in the aqueous phase, it plays an essential role in the taste and flavor of processed radish [3]. MTB-ITC is converted to 2-thioxo-3-pyrrolidinecarbaldehyde (TPC) and 3-(methylthio) methylene-2-thioxopyrolidine (MeSTP), gaining a pyrrolidine ring by intramolecular cyclization and elimination of the methylthio group [4,5]. Furthermore, TPC, which has an aldehyde group, reacts with tryptophan via the Pictet–Spengler reaction to form 1-(2-thioxopyrrolidin-3-yl)-1,2,3,4-tetrahydro-β-carboline-3-carboxylic acid (TPCC) as a precursor to the yellow pigment [1,6]. 2-[3-(2-Thioxopyrrolidin-3-ylidene)methyl]-tryptophan (TPMT) from TPCC is the main yellow pigment in long-term salt-aged takuan-zuke [7]. The proportion of geometric isomers in TPMT is an important factor for yellow brightness and intensity, as the color of the (*Z*)-isomer at 100 ppm is significantly yellow with Δb^* 10.5 compared to (*E*)-isomer. The *Z* to *E*-isomerization under visible light irradiation causes a chain reaction, making the preservation of the yellow color difficult without a light-shielding film [2]. We previously reported that TPMT formation from TPCC is a rate-limiting step that is pH- and temperature-dependent [7,8]. Therefore, it is difficult to use the characteristic natural color in the modern manufacturing method.

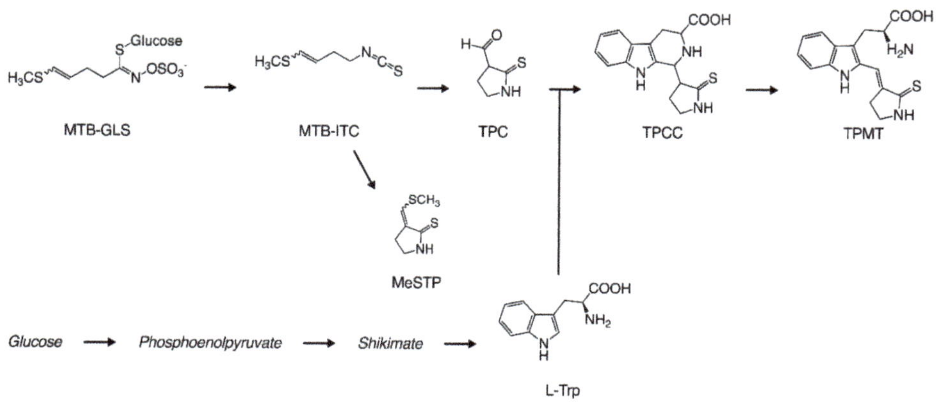

Figure 1. Yellow pigmentation process in takuan-zuke. Abbreviation. 4-methylthio-3-butenyl glucosinolate (MTB-GLS); 4-methylthio-3-butenyl isothiocyanate (MTB-ITC); 2-thioxo-3-pyrrolidinecarbaldehyde (TPC); 3-(methylthio) methylene-2-thioxopyrolidine (MeSTP); 1-(2-thioxopyrrolidin-3-yl)-1,2,3,4-tetrahydro-β-carboline-3-carboxylic acid (TPCC); 2-[3-(2-Thioxopyrrolidin-3-ylidene)methyl]-tryptophan (TPMT).

In our preliminary experiments, takuan-zuke was prepared with 10 mmol/kg buffering agent (pH 4, 5, and 6) at low temperature, to clarify the influence of pH on the yellowing reaction during the salt-aging process. We found that the yellow coloring during short-term salt-aging at low temperature was promoted under acidic conditions. Importantly, it is expected that microorganisms can be suppressed by pickling under acidic conditions [9]. In the present study, we aimed to evaluate the levels of known yellowing-related substances in takuan-zuke, and to understand the influence of acidic pH condition on the yellowing reaction for takuan-zuke.

2. Materials and Methods

2.1. Preparation of Yellowing-Related Substances from MTB-ITC

The yellowing-related substances, MeSTP, TPCC, and TPMT, from MTB-ITC were synthesized using previously reported methods [4,7,8,10]. MTB-ITC was synthesized by the reaction of a crude MTB-GLS solution extracted from radish sprouts, with myrosinase extracted from radish. TPC was synthesized by first dissolving MTB-ITC in acetone, adding acetic acid, and then sonicating the mixture.

For synthesizing MeSTP, methanol and phosphoric acid were added to the MTB-ITC solution, and the mixture was heated under reflux (70 °C) overnight to obtain a precursor fraction. The precursor fraction was trans-solved in acetone, hydrochloric acid was added, and the mixture was synthesized by heating under reflux (60 °C). TPCC was synthesized by adding tryptophan and water to MTB-ITC, adjusting the pH to 2 or less with phosphoric acid, and sonicating. TPMT was synthesized by adjusting the crude TPCC solution to pH 7 and heating under reflux (37 °C). The synthesized compound was purified using column chromatography.

For preparing the internal standard for TPCC and TPMT analysis, 1-ethyl-1,2,3,4-tetrahydro-β-carboline-3-carboxylic acid (ETCA) was synthesized from L-tryptophan and propionaldehyde [11]. L-tryptophan (612 mg, 3.0 mmol) and propionaldehyde (191 mg, 3.3 mmol) were dissolved in 0.025 mol/L H_2SO_4 (26 mL) and allowed to react at 40 °C overnight. Next, a solid precipitate was collected. Subsequently, *(1S, 3S)*- and *(1R, 3S)*-ETCA were separated by preparative Octa Decyl Silyl (ODS) column using middle-pressure liquid chromatography (MPLC) (Smart Flash EPCLC AI-580S, YAMAZEN Co., Yodogawa-ku, Osaka, Japan) equipped with a Biotage® SNAP Ultra C18 column (Filling amount 60 g; Biotage, Sweden).

To prepare the internal standard for MeSTP analysis, (*R*)-3-((*S*)-ethoxy(methylthio)methyl)-2-thioxopyrrolidine (EtOTP) was synthesized from MTB-ITC. MTB-ITC (2.0 mmol) was dissolved in the mixture of ethanol (80 mL) and 1 M phosphoric acid (20 mL). The mixture was concentrated under reduced pressure and extracted with ethyl acetate. The ethyl acetate extract was separated by preparative silica gel-MPLC system equipped with a universal column (Premium 2 L). Subsequently, (*S*)-EtOTP was separated by preparative ODS-MPLC system equipped with a SNAP Ultra C18 column.

2.2. *Preparation of Salted Radish Roots (Takuan-zuke)*

Takuan-zuke was prepared according to a previously reported method with slight modification [8]. The pickling schedule is presented in Figure 2. In this experiment, we used the radish cultivar hoshi-riso daikon (Takii & Co., Kyoto, Japan), which was cultivated in a field at Misato-machi (Gunma, Japan) in August–November 2015. We prepared eight types of salted radish under different concentrations of McIlvaine buffering agent: 0 mmol/kg (only NaCl addition) (SR0/DR0), 10 mmol/kg addition (SR10/DR10), 20 mmol/kg addition (SR20/DR20), and 40 mmol/kg addition (SR40/DR40). The buffering agent was prepared using dipotassium hydrogen phosphate and citrate monohydrate (food additive grade, Kanto Chemical Co., Tokyo, Japan).

Hoshi takuan-zuke (abbreviated DR) was prepared according to the following method. After cutting off the root tips (3–4 cm), the whole daikon (269 kg including green leafy top) was hung to dry in a well-ventilated shady area for two weeks until it became dehydrated and flexible. After chopping the tops (3–4 cm) of root and the leaves, the hoshi daikon (78 kg) was pickled in 8 wt% NaCl (wet weight, the total weight of hoshi daikon and its priming water (33 wt% of daikon)) with pH buffering agents (0–40 mmol/kg of dehydrated daikon) for two months. The inner lid was placed on the daikon, and a stone weight (200 wt% of the dried daikon) was placed on the lid. The salting temperature was maintained at 4 °C.

Shio-oshi takuan-zuke (shio-oshi takuan-zuke, abbreviated SR) was prepared according to the following method. Fresh daikon (234 kg) was pickled with 8 wt% NaCl and 0–40 mmol/kg buffering agent of daikon and pressed under stone weights (200 wt% of daikon). After two days of salt-pressing, 2 wt% NaCl was added to the daikon, and the daikon was dehydrated for 12 days (total: two weeks). Shio-oshi daikon (27–35 kg) were pickled again in buffered saline (9–12 kg; 6 wt% NaCl, and 0–40 mmol/kg buffering agents). The pickling conditions (stone weight and storage temperature) were the same as that for hoshi takuan-zuke.

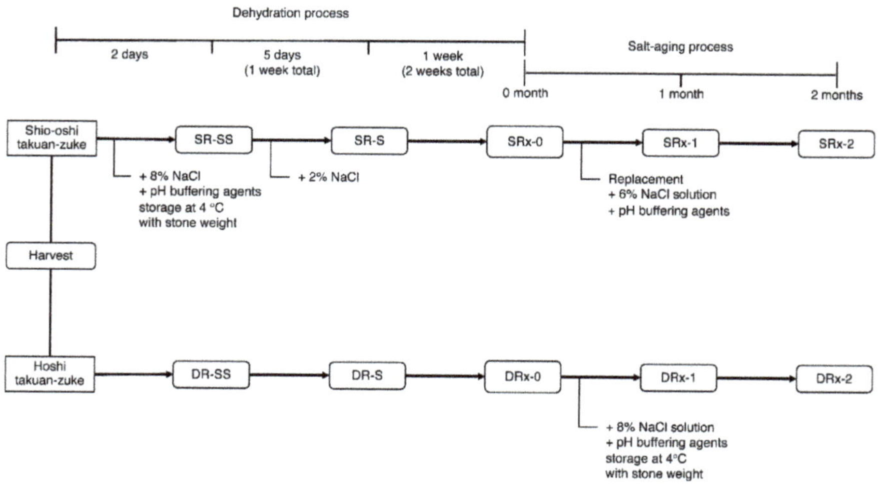

Figure 2. Time schedule for the pickling of daikon. SR indicates dehydration by salting (shio-oshi takuan-zuke). DR indicates dehydration by air-drying (hoshi takuan-zuke); SS indicates two days of dehydration; S indicates one week of dehydration. "X" is the buffer concentration, 0 mmol/kg (only NaCl addition) (SR0/DR0), 10 mmol/kg addition (SR10/DR10), 20 mmol/kg addition (SR20/DR20), and 40 mmol/kg addition (SR40/DR40).

For tryptophan analysis, three types of dehydrated radishes (DR: air-dried daikon with leaves; nDR: air-dried daikon without leaves; SR: salt-pressed daikon without leaves) were produced by dehydrating them for one week using radishes that were collected in 2016. The dehydration process was conducted using the same procedure that was used for the study conducted in 2015.

For the sample, one barrel was prepared for each condition, and two bodies were collected according to the schedule shown in Figure 2. The division and reduced samples were rapidly frozen with liquid N_2 and lyophilized. The lyophilized sample was subjected to freeze grinding and homogenization using a Multi-Beads Shocker (MB901THW(S), Yasui Kikai Co., Osaka, Japan). Lyophilized samples were vacuum packed and then stored at −30 °C.

2.3. *Determination of Moisture, Salt Content, and pH of Takuan-zuke*

The water content was determined by subtracting the lyophilized weight from the wet weight. The salt content was measured using the coulometric titration method (SAT-210, TOA-DKK Co., Tokyo, Japan). The pH of takuan-zuke was measured by first adding pure water to the lyophilized sample to restore the wet weight, and then using a pH meter (HM-30, TOA-DKK Co., Tokyo, Japan).

2.4. *Color Measurement of Takuan-zuke*

Color measurement of takuan-zuke was conducted based on a previous report [8]. The colorimetric change was measured with a Minolta CM-3500d Spectrophotometer, with a D65 illuminant and a 10-degree observer. The Commission Internationale de l'Eclairage (CIE) 1976 L^*, a^*, and b^* color scale values of takuan-zuke were obtained by reflectance color measurement (measurement area: 8 mm diameter). Each value was measured in three places on the skin side of the upper, middle, and lower part of the root.

2.5. *Quantitative Analysis of MTB-GLS in Takuan-zuke*

Determination of glucosinolate in takuan-zuke was performed as previously reported [12]. The lyophilized sample (ca. 100 mg) was extracted by 1.5 mL of 80% MeOH at 75 °C. After the extraction mixture was heated at 75 °C for 10 min, 0.5 μmol sinigrin (internal standard, I.S.) was added. The crude extracts were adsorbed on an anion exchange resin (DEAE Sephadex A-25, GE Healthcare, California, CA, USA) and treated with sulfatase to convert desulfoglucosinolate overnight. The eluate containing the desulfoglucosinolate was quantified by ODS high-performance liquid chromatography (HPLC).

Analytical ODS HPLC was performed with an Agilent 1200-1260 system with a Poroshell 120 EC-C18 (100 × 3.0 mm ø, 2.7 μm; Agilent Technologies, Santa Clara, CA, USA). The flow rate was set at 0.85 mL/min, and the column temperature was set to 35 °C. Elution was achieved using a gradient of two eluents: H_2O as eluent A and acetonitrile as eluent B. The gradient program was set at 0.2% B for 0.25 min, rising to 19.8% B at 6.00 min, and the remaining at 19.8% B to 7.00 min. Finally, the column was equilibrated using 0.2% B from 7.10 to 9.00 min. The results were detected at a wavelength of 229 nm.

2.6. *Quantitative Analysis of TPC in Takuan-zuke*

Quantification of TPC in takuan-zuke was performed by fluorescence derivatization method using 4-(*N*,*N*-dimethylaminosulfonyl)-7-hydrazino-2,1,3-benzoxadiazole (DBD-H) [13]. The lyophilized sample (10–50 mg) was added to 250 μL of 0.1% DBD-H in acetonitrile, 250 μL of 0.2 mM anisaldehyde in acetonitrile, and 500 μL of 0.5% trifluoroacetic acid in 60% (v/v) acetonitrile. The reaction mixture was shaken and incubated at 25 °C for 60 min. To 200 μL of the supernatant after centrifugation, 50 μL of 500 mM McIlvaine buffer (pH 5) and 50 mg of NaCl were added and shaken. The acetonitrile phase was considered the sample for HPLC.

Analytical HPLC was performed with an Agilent 1200–1260 system with a Poroshell HPH-C18 (100 × 3.0 mm ø, 2.7 μm; Agilent Technologies, Santa Clara, CA, USA). The flow rate was set at 0.85 mL/min and the column temperature was set to 40 °C. Elution was achieved using a gradient of two eluents: H_2O as eluent A and acetonitrile as eluent B. The gradient program was: 25% B rising to 73% B at 5.5 min, further increasing to 100% B at 5.6 min, and remaining at 100% B to 5.99 min. Finally, the separation column was equilibrated using 25% B from 5.99 to 8.0 min. Fluorescence was detected with excitation at 450 nm and emission at 565 nm. Anisaldehyde was used as an internal standard.

2.7. *Quantitative Analysis of Yellow Pigment-Related Substances in Takuan-zuke*

The lyophilized sample (ca. 100 mg) was mixed with 250 μL of chloroform, 625 μL of methanol (including 40 nmol/mL EtOTP and 40 nmol/mL ETCA), and 250 μL of H_2O (including 2.25% trifluoroacetic acid and 0.45% semicarbazide). The mixture was shaken and incubated at 37 °C for 30 min. After cooling on ice for 5 min, the mixture was centrifuged at 20,630× *g* for 1 min. To 1 mL of the supernatant, 500 μL chloroform and 500 μL H_2O were added. After cooling on ice for 5 min, the mixture was centrifuged at 20,630× *g* for 1 min. The separated lower layer was obtained as a crude extract for MeSTP analysis, and the upper layer was obtained as a crude extract for L-tryptophan, TPCC, and TPMT analysis. The crude extract for MeSTP was concentrated to dryness using a centrifugal concentrator. The dried sample was dissolved in MeOH (300 μL) for HPLC analysis.

An aliquot of the upper layer extract was diluted to three times the original concentration, with 2% formic acid. A solid phase extraction (SPE) cartridge (Bond Elut Plexa PCX, 30 mg, 1 mL; Agilent Technologies, Santa Clara, CA, USA) was washed with 1 mL each of 1 M NaOH and 1 M HCl and conditioned with 1 mL each of MeOH and 2% formic acid. The diluted samples were loaded onto a PCX cartridge and washed with 1 mL each of 2% formic acid and methanol. The analytes were eluted with 1 mL each of alkaline eluent (30% ammonium hydroxide: 95% methanol = 5:95) in a tube containing 30 μL of concentrated formic acid. The eluate was evaporated to dryness with a centrifugal concentrator. After being dissolved in 200 μL of methanol, the sample solution was irradiated by long-wave ultraviolet (UV) light (UVGL-25, Funakoshi Co., Tokyo, Japan) at 375 nm and analyzed by HPLC.

Analytical HPLC was performed with an Agilent 1200–1260 system with a Poroshell HPH-C18 (100 × 3.0 mm ø, 2.7 µm; Agilent Technologies, Santa Clara, CA, USA). The flow rate was set at 0.7 mL/min, and the column temperature was set to 40 °C. Elution was achieved using a gradient of two eluents: 10 mM phosphate borate buffer (pH 8.2) as eluent A and methanol as eluent B. The gradient program was set at 15% B rising to 25% B at 7.00 min, rising to 100% B at 11.00 min. Finally, the column was equilibrated using 15% B from 11.01 to 13.00 min. The detection wavelengths were as follows: 268 nm for ETCA, EtOTP, and TPCC, 320 nm for MeSTP, and 400 nm for TPMT using a diode array detector. Tryptophan was detected by native fluorescence (excitation wavelength 285 nm, emission wavelength 348 nm). Each isomeric mixture of TPCC, TPMT, and MeSTP was separately quantified, and the results were documented as the sum. Internal standards were EtOTP for MeSTP and ETCA for the others.

2.8. Statistical Analysis

All quantitative data units are expressed in nmol per g of dry weight, and each value is expressed as the mean value ± standard deviation (n = 4). Multiple t-tests were performed using the Holm–Šidák method (α = 0.05). Significant differences between the treatment groups were determined with a two-way Analysis of variance (ANOVA), followed by a Tukey's multiple comparison test using GraphPad Prism ver. 8 for Macintosh (GraphPad Software, Inc., CA, USA).

3. Results

3.1. Basic Data and pH Changes in Takuan-zuke Induced by the pH Buffering Agent

Temporal changes in pH in the two types of dehydrated daikon and takuan-zuke samples are shown in Figure 3. The pH of fresh daikon was 6.4. The pH of takuan-zuke without pH buffering agents gradually decreased to 5.6 in SR0 and 5.8 in DR0 after two months of salting. In contrast, the pH changes of SR- and DR-takuan-zuke with the addition of buffering agents were notably lower than in the case of non-buffered takuan-zuke. The pH values of the acidic buffered takuan-zuke samples decreased within one week of salting for SR groups and one month salting for DR groups. The pH lowering effect on DR samples by buffer addition was concentration-dependent, whereas no notable decrease in pH among SR samples was observed. With the addition of 40 mmol/kg buffer, the pH values of the SR40 and DR40 after two months of salting were 4.2 and 4.5, respectively.

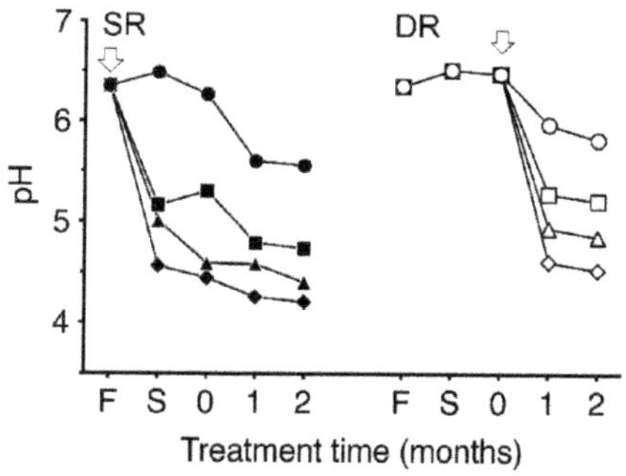

Figure 3. Time-dependent changes in pH during the dehydration and salting process with different concentrations of McIlvaine buffering agent. SR indicates dehydration by salting (shio-oshi takuan-zuke); DR indicates dehydration by sun-drying (hoshi takuan-zuke). The arrows denote the time point of salt addition. F denotes the pH obtained from fresh daikon. S denotes the pH obtained from the dehydrated daikon after one week of salting. "0" denotes the start of salt-aging process. Symbols refer to different concentrations of McIlvaine buffering agent (mmol/kg): ●, SR0; ■, SR10; ▲, SR20; ◆, SR40; ○, DR0; □, DR10; △, DR20; ◇, DR40. Values are mean ± standard deviation (SD) (n = 3). The error bar cannot be displayed because the standard deviation is small.

3.2. Effect of Addition of pH Buffering Agent on the Color of Takuan-Zuke

In the fresh daikon, L^*, a^*, and b^* values were 73.1 ± 4.0, −0.6 ± 0.2, and 10.2 ± 1.1, respectively. Although L^* values for SR groups during salt-aging treatment fluctuated between 65.0 and 74.0, the difference that depended on buffer concentration was negligible. The a^* values during salt-aging treatment changed in the negative direction. Figure 4 shows the b^* values changes during the dehydration and salt-aging processes. Although b^* value in the SR groups with buffered salting increased more than that in SR0, no buffer concentration-dependent changes in color were observed. However, the yellowing reactions in the DR groups were increased in proportion to the buffer strength. The b^* value for DR groups by adding buffer increased significantly compared to that in the case for DR0. The Δb^* value of DR40 after two months salting based on fresh daikon was 2.4-fold that of SR0 and 1.8-fold that of DR0.

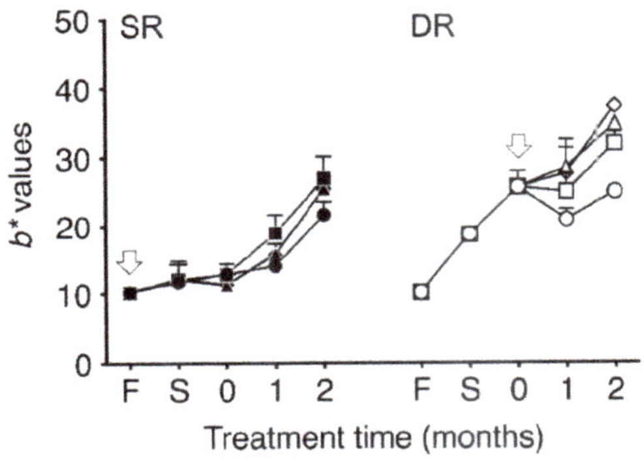

Figure 4. Time-dependent changes in b^* values during the dehydration and salting process with different concentrations of McIlvaine buffering agent. SR indicates dehydration by salting (shio-oshi takuan-zuke); DR indicates dehydration by sun-drying (hoshi takuan-zuke). The arrows denote the time point of salt addition. F denotes the pH obtained from fresh daikon. S denotes the pH obtained from the dehydrated daikon after one week of salting. "0" denotes the start of the salt-aging process. Symbols refer to different concentrations of McIlvaine buffering agent (mmol/kg): ●, SR0; ■, SR10; ▲, SR20; ◆, SR40; ○, DR0; □, DR10; △, DR20; ◇, DR40. Values are mean ± standard deviation (SD) (n = 3).

3.3. Effect of pH on the Yellow Pigment Production Pathway

The temporal changes in the yellowing-related substances in eight takuan-zuke samples are presented in Figure 5. The MTB-GLS level was highest at harvest (51.8 ± 1.0 μmol/g (dry weight; DW)). The levels of MTB-GLS during salt-pressing treatment were decreased and disappeared during the two months of salt-aging treatment. In the buffered SR samples, adding buffering agents slightly suppressed the degradation of MTB-GLS. The degradation rate of MTB-GLS in SR40 was slow compared to that in the other SR groups, and the residue rate in SR40 after two months of salt-aging was 11%. In the DR samples, slight hydrolysis of MTB-GLS was observed during the drying treatment; however, penetration of saline into dried daikon, during the salt-aging process, induced further hydrolysis of MTB-GLS.

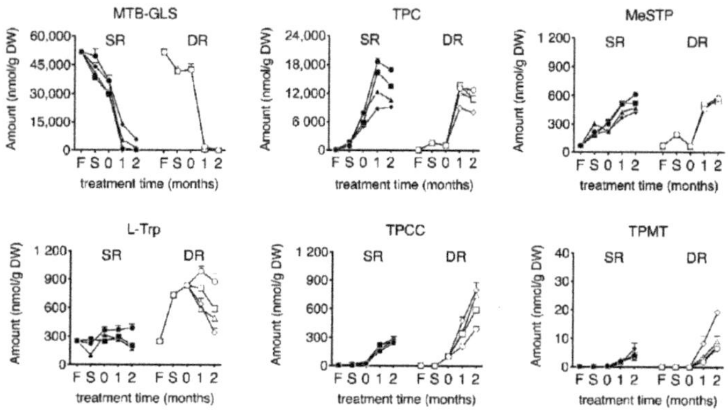

Figure 5. Change in the amount of each yellow pigment and related substances during the dehydrating and salting process. SR indicates dehydration by salting (shio-oshi takuan-zuke); DR indicates dehydration by sun-drying (hoshi takuan-zuke); F indicates the amount of each pigment obtained from fresh daikon; S indicates the amount of each pigment obtained from the dehydrated daikon after one week of salting. "0" time point indicates the start of salt-aging. Symbols refer to different concentrations of McIlvaine buffering agent (mmol/kg): ●, SR0; ■, SR10; ▲, SR20; ◆, SR40; ○, DR0; □, DR10; △, DR20; ◇, DR40. Values are mean ± standard deviation (SD) (n = 4). Data are analyzed using two-way Analysis of variance (ANOVA), followed by Tukey's multiple comparison test. **Abbreviation.** 4-methylthio-3-butenyl glucosinolate (MTB-GLS); 4-methylthio-3-butenyl isothiocyanate (MTB-ITC); 2-thioxo-3-pyrrolidinecarbaldehyde (TPC); 3-(methylthio) methylene-2-thioxopyrolidine (MeSTP); 1-(2-thioxopyrrolidin-3-yl)-1,2,3,4-tetrahydro-β-carboline-3-carboxylic acid (TPCC); 2-[3-(2-Thioxopyrrolidin-3-ylidene)methyl]-tryptophan (TPMT).

MeSTP, as a degradation product of MTB-ITC, was generated from an early stage of salt and/or saline addition. The production levels were nearly equal between SR and DR takuan-zuke. The effect of buffer on MeSTP production was negligible.

TPC, the primary degradation product of MTB-ITC, was generated after salt and/or saline addition. TPC content in the SR groups reached maximum levels after one month of salt-aging (8.8–18.7 μmol/g). In the DR groups, TPC content increased with the addition of saline and reached a maximum after one month of salt-aging (9.0–13.7 μmol/g). TPC level decreased significantly after two months of salt-aging, depending on pH buffering strength.

The content of tryptophan in the fresh daikon was 255 ± 11 nmol/g. The analysis of tryptophan revealed that air-dried dehydration treatment resulted in a significant increase in its levels after harvest (hoshi processing for one week: 2.9-fold, $p < 0.001$; two weeks: 3.4-fold, $p < 0.001$). In contrast, shio-oshi treatment resulted in a slight change in the tryptophan content. Tryptophan in DR0 showed a maximum content value after one month of salt-aging (988 ± 52 nmol/g), and this value was 2.7 times higher than that in SR0 ($p < 0.001$). With subsequent salt-aging, tryptophan content significantly decreased, depending on pH buffering strength.

TPCC was generated immediately after salt and saline addition and increased with the duration of salt-aging after dehydration. The increased TPCC levels during the salt-aging process were significantly different between SR0 and DR0, and the content of TPCC after two months of salt-aging were 244 ± 8 nmol/g and 392 ± 31 nmol/g, respectively. The effect of pH on TPCC formation increased significantly with salt-aging time in the DR group, but no change was observed in the SR group. The TPCC levels in the buffered DR samples after two months of salt-aging were greater than that in DR0 (DR10: 1.5-fold, $p < 0.001$; DR20: 1.9-fold, $p < 0.001$; DR40: 2.0-fold, $p < 0.001$). Table 1 shows the changes in TPCC and tryptophan in the DR group; tryptophan, which is a substrate for TPCC, showedmaximum content after one month of salting (Table 1). The decrease in tryptophan and the production of TPCC were almost equal in the buffered DR sample (DR10: 96%, DR20: 108%, and DR40: 91%).

Table 1. Tryptophan and TPCC content, and their variation in hoshi takuan-zuke.

	Salt-Aging Time (months)	L-tryptophan DR0	L-tryptophan DR10	L-tryptophan DR20	L-tryptophan DR40	TPCC DR0	TPCC DR10	TPCC DR20	TPCC DR40
Amount nmol/g (DW)	1	988 ± 55	804 ± 39	588 ± 32	640 ± 63	210 ± 2	341 ± 24	357 ± 8	481 ± 36
	2	876 ± 82	591 ± 22	494 ± 62	344 ± 46	392 ± 31	590 ± 4	745 ± 18	798 ± 81
Amount of change from DR0-1 nmol/g (DW)	1	-	−184	−400	−348	-	131	147	271
	2	−112	−397	−494	−644	182	380	535	588

The amount of change in tryptophan and 1-(2-thioxopyrrolidin-3-yl)-1,2,3,4-tetrahydro-β-carboline-3-carboxylic acid (TPCC) is shown as an increase or decrease from DR0-1. Concentration of McIlvaine buffering agent: 0 mmol/kg (DR0), 10 mmol/kg (DR10), 20 mmol/kg (DR20), 40 mmol/kg (DR40). Data are expressed as means ± SD.

The conversion of TPCC to TPMT started immediately after salt-aging and continued to increase after two months of salt-aging. The pH buffering strength significantly increased TPMT in the DR groups, with salt-aging time, whereas no effect was observed in the SR groups. The amount of TPMT was 6.7−19.3 nmol/g in the DR samples after two months of salting, which was 1.7–3.0 times greater than that in the SR samples.

3.4. Effect of Dehydration Method on L-tryptophan Metabolism

Temporal change of L-tryptophan in three types of dehydrated daikon using those collected in 2016 (air-dried daikon with leaves; DR, air-dried daikon without leaves; nDR, and salt-pressed daikon without leaves; SR) is shown in Figure 6A. The content of L-tryptophan in DR samples was increased to 1.091 ± 13 nmol/g after seven days of drying, which was 6.7 times higher than that in fresh daikon. In the SR and nDR samples, the extent of increase of L-tryptophan was negligible compared to that in DR samples ($p < 0.0001$).

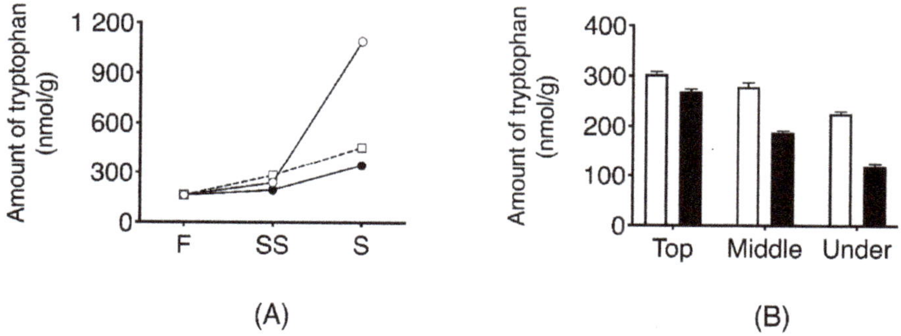

Figure 6. Temporal changes in the quantity of tryptophan obtained from daikon during the dehydrating and salt-aging process (**A**). Effect of dehydration treatment on tryptophan production. F indicates the value derived from fresh daikon. SS indicates the amount of tryptophan obtained from the dehydrated daikon after two days of salting. S indicates the amount of tryptophan collected from the dehydrated daikon after one week of salting. Symbols: ○, dehydrated by sun-drying with leaves (DR); □, dehydrated by sun-drying without leaves (nDR); •, dehydrated by salt-pressing (SR). The error bar cannot be displayed because the standard deviation is small. (**B**). Localization of tryptophan in dehydrated daikon with leaves after two days of salting (DR-SS) white bar, inside radish; black bar, outside radish (on the skin).

The localization of tryptophan in DR-SS is shown in Figure 6B. Tryptophan level was highest inside the upper part of the root at 304 ± 5.3 nmol/g. Tryptophan level inside the root was significantly higher than outside the root, and the levels reduced significantly toward the root tip.

4. Discussion

In this study, we analyzed the effect of acidic buffering agents on the yellowing reaction of short-term aged takuan-zuke. Air-dried radish (hoshi takuan-zuke) and salt-pressed radish (shio-oshi takuan-zuke), prepared under acidic conditions, were the brightest yellow at low temperature and low salinity. In particular, the b^* value of hoshi takuan-zuke was equivalent to that of takuan-zuke, which was prepared by long-term salt-aging under room temperature and high salinity, as discussed in our previous report [8].

As described in Section 2, the pH buffering agent was added based on the wet weight of the daikon; the amount of buffering agent per radish, excluding water and salt, was 0.7 mmol/g (DW) for shio-oshi daikon, and 0.2 mmol/g (DW) for hoshi daikon. Therefore, the amount of buffering agent per dry weight affected the pH of takuan-zuke. There was no difference in the b^* value of shio-oshi takuan-zuke at pH 5 or lower. Therefore, addition of 40 mmol/kg or more buffering agent into hoshi daikon did not affect yellowing.

To clarify the yellowing effects of acidic pH, the dynamics of the known yellowing-related substances were analyzed. In the takuan-zuke manufacturing process, in the absence of pH buffer, the conversion rate of MTB-GLS into TPC via MTB-ITC in the SR and DR samples was 36.1% and 25.5%, respectively. Since hoshi takuan-zuke showed a slight decrease in MTB-GLS during the drying process, the myrosinase reaction was thought to be suppressed under low water activity. During the salt-aging process, degradation of MTB-GLS was suppressed with increasing buffer strength, and the accumulation level of TPC was decreased. Myrosinase activity has been reported to peak at pH 5.7, whereas at pH 3.9, it decreases to 60% relative to the maximum activity level [14]. In addition, enzymatic reaction by myrosinase, under acidic conditions, has been reported to produce not only isothiocyanates but also nitriles [15–17]. Therefore, we suggested that the amount of MTB-ITC and TPC formed decreased due to the inhibition of enzymatic reaction or induction of nitrile formation.

Tryptophan content was significantly different between the two dehydration processes. Tryptophan levels in hoshi takuan-zuke increased from the air-drying dehydration process to early salt-aging. In hoshi takuan-zuke without leaves, no increase in tryptophan was observed, as with shio-oshi takuan-zuke. Tryptophan was seen to accumulate in the upper-inside part of hoshi daikon with leaves. Since tryptophan is reported to be synthesized by chloroplasts in plants [18,19], it was concluded that tryptophan is synthesized in the chloroplast of the daikon leaf and then transferred to the root through the vascular bundle. This result indicated that the increased tryptophan was synthesized in the chloroplasts of leaves during the drying process, rather than in microbial fermentation.

The content of TPCC, which is a yellow pigment precursor, was significantly increased in hoshi takuan-zuke than in shio-oshi takuan-zuke, similar to the results observed for tryptophan. In SR, the tryptophan production rate was the same as the reaction rate of TPCC synthesis, so it was inferred there was no apparent change in the tryptophan content. TPCC synthesis from tryptophan is considered a stoichiometric reaction when a sufficient amount of substrate is used, and tetrahydro-β-carboline synthesis via the Pictet–Spengler reaction has been reported to show high reaction efficiency under low pH conditions [11,20]. TPCC synthesis in takuan-zuke was revealed to be promoted at acidic pH, although the difference in tryptophan biosynthesis across the dehydration methods was a limiting factor for TPCC synthesis during the salt-aging process.

The content of TPMT in hoshi takuan-zuke was increased to a higher level than in shio-oshi takuan-zuke, similar to the results of tryptophan and TPCC. However, in this study, the conversion of TPCC to TPMT under low temperature and low salt conditions was very negligible. Our previous study had shown the optimal pH for TPMT synthesis from TPCC to be either weakly acidic or neutral pH in vitro [7]. In hoshi takuan-zuke at room temperature, the conversion rate to TPMT was 28.7%, and the color of salt-aged takuan-zuke varied to reddish yellow with prolonged salt-aging [8]. In the present study, there was no increase in the a^* value despite the high b^* value during the short-termed salt-aged process. These results suggested the contribution of TPMT to the yellowing of takuan-zuke to be low under acidic condition. In addition, HPLC analysis detected two characteristic peaks in takuan-zuke under acidic conditions. This peak had a maximum absorption wavelength of 392 nm and 404 nm, different from that in TPMT. It was suggested that these compounds are unknown yellow pigments that contribute to the yellowing reaction under acidic conditions.

In this report, we clarified that low pH conditions during takuan-zuke processing promotes an unknown bright-yellowing reaction at low temperature. In addition, we found TPC to be the most abundant among the known yellowing-related substances and an essential intermediate for takuan-zuke coloring. Imai et al. have reported that frozen and grated daikon, adjusted to pH 4 or below using acetic acid, turned yellow upon long-term freezing. This yellow pigment is generated by the condensation of two molecules of TPC [21]. Therefore, we proposed that an unknown yellowing reaction with MTB-ITC and TPC occurs during aging of takuan-zuke at low pH and low temperature. Future research would aim to elucidate the structure and reaction mechanism of the unknown pigment.

5. Conclusions

Temperature and pH conditions during salt-aging are the rate-limiting factors of the yellowing reaction, and we observed that takuan-zuke aged with low salt and at low temperature turns pale yellow. We found that the yellowing reaction was accelerated even at low temperature by the salt-aging of takuan-zuke under acidic conditions. The TPC level, which is one of the important intermediates of the yellow pigment, was highest after one month of salt-aging, regardless of the dehydration treatment. Tryptophan, another important intermediate, was increased only in dried daikon with leaves. The acidified and salt-aged treatment promoted the generation of TPCC, which is a pigment precursor. However, the generation of TPMT, which is a yellow pigment, was marginal compared to that in a previous report. Therefore, it was suggested that the unknown yellow pigment was generated via a pathway different from that described in the previous report regarding the yellow change of takuan-zuke under acidic and low temperature conditions. In future studies, it will be necessary to identify the unknown yellow pigment and the detailed mechanism underlying its generation.

Author Contributions: The experimental design was constructed and supervised by T.K., K.K., A.T., and H.M. The samples were produced with the help of T.K., K.K., A.T., and H.T. Instrumental analyses were performed by T.K. and K.K. The manuscript was drafted and written by T.K. and H.M. All authors have read and agreed to the published version of the manuscript.

Funding: This research received no external funding.

Conflicts of Interest: The authors declare no conflict of interest.

References

1. Ozawa, Y.; Kawakishi, S.; Uda, Y.; Maeda, Y. Isolation and identification of a novel β-carboline derivative in salted radish roots, *Raphanus sativus* L. *Agric. Biol. Chem.* **1990**, *54*, 1241–1245. [CrossRef]
2. Matsuoka, H.; Honzawa, S.; Takahashi, A.; Yoshikawa, H.; Watanabe, E.; Watanabe, T.; Ozawa, Y.; Yamada, Y.; Iizuka, T.; Uda, Y. Photoisomerization of 2-[3-(2-Thioxopyrrolidin-3-ylidene)methyl]-tryptophan, a Yellow Pigment in Salted Radish Roots. *Biosci. Biotechnol. Biochem.* **2008**, *72*, 2262–2268. [CrossRef] [PubMed]
3. Ozawa, Y.; Uda, Y.; Ohshima, T.; Saito, K.; Maeda, Y. Formation of yellow pigment by the reaction of 4-methylthio-3-butenyl isothiocyanate with L-ascorbic acid and some dihydroxyphenolic compounds. *Agric. Biol. Chem.* **1990**, *54*, 605–611. [CrossRef]
4. Uda, Y.; Ozawa, Y.; Ohshima, T.; Kawakishi, S. Identification of enolated 2-thioxo-3-pyrrolidinecarbaldehyde, a new degradation product of 4-methylthio-3-butenyl isothiocyanate. *Agric. Biol. Chem.* **1990**, *54*, 613–617. [CrossRef]
5. Matsuoka, H.; Toda, Y.; Yanagi, K.; Takahashi, A.; Yoneyama, K.; Uda, Y. Formation of Thioxopyrrolidines and Dithiocarbamates from 4-Methylthio-3-butenyl Isothiocyanates, the Pungent Principle of Radish, in Aqueous Media. *Biosci. Biotechnol. Biochem.* **1997**, *61*, 2109–2112. [CrossRef] [PubMed]
6. Ozawa, Y.; Uda, Y.; Kawakishi, S. Generation of β-Carboline Derivative, the Yellowish Precursor of Processed Radish Roots, from 4-Methylthio-3-butenyl Isothiocyanate and l-Tryptophan. *Agric. Biol. Chem.* **1990**, *54*, 1849–1851. [CrossRef]
7. Matsuoka, H.; Takahashi, A.; Ozawa, Y.; Yamada, Y.; Uda, Y.; Kawakishi, S. 2-[3-(2-Thioxopyrrolidin-3-ylidene)methyl]-tryptophan, a Novel Yellow Pigment in Salted Radish Roots. *Biosci. Biotechnol. Biochem.* **2002**, *66*, 1450–1454. [CrossRef] [PubMed]
8. Takahashi, A.; Yamada, T.; Uchiyama, Y.; Hayashi, S.; Kumakura, K.; Takahashi, H.; Kimura, N.; Matsuoka, H. Generation of the antioxidant yellow pigment derived from 4-methylthio-3-butenyl isothiocyanate in salted radish roots (takuan-zuke). *Biosci. Biotechnol. Biochem.* **2015**, *79*, 1512–1517. [CrossRef] [PubMed]
9. Kim, J.-H.; Breidt, F. Development of Preservation Prediction Chart for Long Term Storage of Fermented Cucumber. *J. Life Sci.* **2007**, *17*, 1616–1621. [CrossRef]
10. Ozawa, Y.; Uda, Y.; Matsuoka, H.; Abe, M.; Kawakishi, S.; Osawa, T. Occurrence of Stereoisomers of 1-(2′-Pyrrolidinethione-3′-yl)-1,2,3,4- tetrahydro-β-carboline-3-carboxylic Acid in Fermented Radish Roots and Their Different Mutagenic Properties. *Biosci. Biotechnol. Biochem.* **1999**, *63*, 216–219. [CrossRef] [PubMed]

11. Herraiz, T.; Galisteo, J.; Chamorro, C. l-Tryptophan Reacts with Naturally Occurring and Food-Occurring Phenolic Aldehydes To Give Phenolic Tetrahydro-β-carboline Alkaloids: Activity as Antioxidants and Free Radical Scavengers. *J. Agric. Food Chem.* **2003**, *51*, 2168–2173. [CrossRef] [PubMed]
12. Bjerg, B.; Sørensen, H. Quantitative analysis of glucosinolates in oilseed rape based on HPLC of desulfoglucosinolates and HPLC of intact glucosinolates. In *Glucosinolates in Rapeseeds: Analytical Aspects*; Wathelet, J.-P., Ed.; Springer: Brussels, Belgium, 1987; pp. 125–150.
13. Kobayashi, T.; Kumakura, K.; Kobayashi, W.; Takahashi, A.; Matsuoka, H. Determination of 2-Thioxo-3-pyrrolidinecarbaldehyde in Salted Radish Root (Takuan-zuke) by High-Performance Liquid Chromatography with Fluorescence Detection after Pre-Column Derivatization Using 4-(N,N-dimethylaminosulfonyl)-7-hydrazino-2,1,3-benzoxadiazole. *Separations* **2017**, *4*, 35. [CrossRef]
14. Li, X.; Kushad, M.M. Purification and characterization of myrosinase from horseradish (Armoracia rusticana) roots. *Plant Physiol. Biochem.* **2005**, *43*, 503–511. [CrossRef] [PubMed]
15. Vaughn, S.F.; Berhow, M.A. Glucosinolate hydrolysis products from various plant sources: pH effects, isolation, and purification. *Ind. Crop. Prod.* **2005**, *21*, 193–202. [CrossRef]
16. Gil, V.; MacLeod, A.J. The effects of pH on glucosinolate degradation by a thioglucoside glucohydrolase preparation. *Phytochemistry* **1980**, *19*, 2547–2551. [CrossRef]
17. Rask, L.; Andréasson, E.; Ekbom, B.; Eriksson, S.; Pontoppidan, B.; Meijer, J. Myrosinase: Gene family evolution and herbivore defense in Brassicaceae. *Plant Mol. Biol.* **2000**, *42*, 93–113. [CrossRef] [PubMed]
18. Radwanski, E.R.; Last, R.L. Tryptophan biosynthesis and metabolism: Biochemical and molecular genetics. *Plant Cell* **1995**, *7*, 921–934. [CrossRef]
19. Mano, Y.; Nemoto, K. The pathway of auxin biosynthesis in plants. *J. Exp. Bot.* **2012**, *63*, 2853–2872. [CrossRef]
20. Stöckigt, J.; Antonchick, A.P.; Wu, F.; Waldmann, H. The Pictet-Spengler Reaction in Nature and in Organic Chemistry. *Angew. Chem. Int. Ed.* **2011**, *50*, 8538–8564. [CrossRef] [PubMed]
21. Imai, S.; Tsuge, N.; Kamata, Y. Novel Compound, Production Method and Use Thereof. Available online: https://www.j-platpat.inpit.go.jp/c1800/PU/JP-2001-064285/F674FE2E94611E041BF07688A82C03059BDE9BBAEB88C517978C96CAD1012AA1/11/ja (accessed on 20 April 2020).

Publisher's Note: MDPI stays neutral with regard to jurisdictional claims in published maps and institutional affiliations.

foods

MDPI

Article

A Physiological-Based Model for Simulating the Bioavailability and Kinetics of Sulforaphane from Broccoli Products

Quchat Shekarri and Matthijs Dekker *

Food Quality and Design Group, Wageningen University and Research, P.O. Box 17, 6700 AA Wageningen, The Netherlands; quchat.shekarri@wur.nl
* Correspondence: matthijs.dekker@wur.nl

Abstract: There are no known physiological-based digestion models that depict glucoraphanin (GR) to sulforaphane (SR) conversion and subsequent absorption. The aim of this research was to make a physiological-based digestion model that includes SR formation, both by endogenous myrosinase and gut bacterial enzymes, and to simulate the SR bioavailability. An 18-compartment model (mouth, two stomach, seven small intestine, seven large intestine, and blood compartments) describing transit, reactions and absorption was made. The model, consisting of differential equations, was fit to data from a human intervention study using Mathwork's Simulink and Matlab software. SR urine metabolite data from participants who consumed different broccoli products were used to estimate several model parameters and validate the model. The products had high, medium, low, and zero myrosinase content. The model's predicted values fit the experimental values very well. Parity plots showed that the predicted values closely matched experimental values for the high (r^2 = 0.95), and low (r^2 = 0.93) products, but less so for the medium (r^2 = 0.85) and zero (r^2 = 0.78) myrosinase products. This is the first physiological-based model to depict the unique bioconversion processes of bioactive SR from broccoli. This model represents a preliminary step in creating a predictive model for the biological effect of SR, which can be used in the growing field of personalized nutrition.

Keywords: physiological-based model; sulforaphane; glucoraphanin; compartmental model; broccoli; bioavailability; myrosinase; parameter estimation

Citation: Shekarri, Q.; Dekker, M. A Physiological-Based Model for Simulating the Bioavailability and Kinetics of Sulforaphane from Broccoli Products. *Foods* **2021**, *10*, 2761. https://doi.org/10.3390/foods10112761

Academic Editor: Didier Dupont

Received: 15 October 2021
Accepted: 8 November 2021
Published: 10 November 2021

1. Introduction

To determine the bioavailability of bioactive compounds in foods, it is important to know its composition, structure, how it interacts with other food components, and its fate in the human body after being ingested. Isothiocyanates (ITC) are formed from precursors, glucosinolates (GL), which are found in broccoli and other types of brassica vegetables [1]. Numerous studies investigated the health effects of some ITCs. One such ITC is called sulforaphane (SR), derived from the GL glucoraphanin (GR). Sulforaphane is known to reduce the risk of cancer, and has cardiovascular and central nervous system protection benefits [1,2].

Sulforaphane's health benefits have resulted in studies that investigated the physiological mechanisms involved in digesting plants that contain SR, and SR's absorption, metabolism, and excretion [3,4]. Glucoraphanin is converted to SR by plant endogenous myrosinase (MYR), a β-thioglucosidase hydrolase that catalyzes the removal of glucose to form an O-sulfated thiohydroximate intermediate (Figure 1). GLs and MYR are stored in separate compartments in the broccoli plant cells. Cell structure disruption, from processing (chopping, blanching, powdering etc.), mastication, or plant bruising, is required before MYR can bind GL to facilitate ITC formation [5]. Gut bacteria in the colon also have the capability to facilitate this conversion from GR to SR [6–14]. After ingestion, and transfer through the stomach, SR is absorbed in the intestinal tract into the blood, then distributed to various organs before it is eliminated from the body, mainly via renal excretion [15].

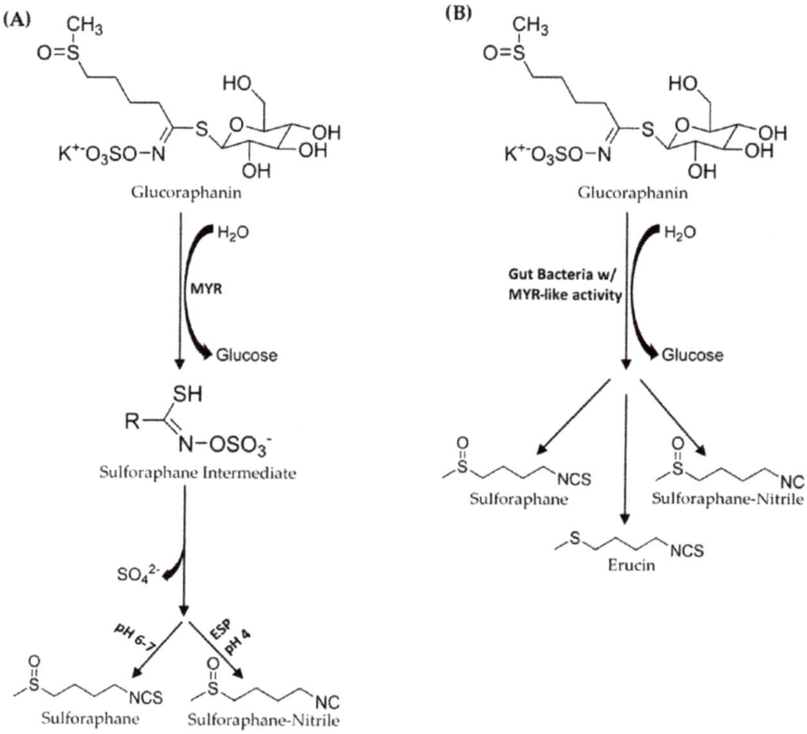

Figure 1. (**A**) In the mouth, myrosinase (MYR) converts glucoraphanin (GR) to an SR-O-sulfated thiohydroximate intermediate. Depending on pH conditions sulforaphane (SR) forms. SR-nitrile formation is preferred at low pH in the presence of epithiospecifier protein (ESP). (**B**) In the gut GR is converted to SR, SR-nitrile, and erucin by gut bacteria.

The bioavailability of SR is described as the fraction of the amount of SR that is ingested and/or formed in the body that reaches systematic circulation [3]. Related to bioavailability is bioaccessibility, which is the fraction of a compound that is released from the food and that reaches the absorption site. In the context of GLs, MYR and ITCs, the bioaccessible ITC is the fraction of ITCs released from the food matrix [3] or the fraction of GLs transformed to ITCs and released in the body. Bioaccessibility of ITCs is affected by the plant's inherent GL content (which varies from 47 to 806 mg/100 g fresh weight of broccoli [1]), processing, the food matrix [16–18], and the digestion [17,19,20].

The bioaccessibility of ITC could increase or decrease depending on the type of processing. Chopping, blending, powdering are particle size reduction methods that rupture the plant tissue and allow for MYR and GLs to diffuse out and bind to each other [5]. Heating affects epithiospecifier proteins (ESP), MYR, and GL content. ESPs are responsible for the conversion of GLs to nitriles [1] and are less heat stable than MYR. Their inactivation allows for the preferential formation of ITCs [21,22]. Any type of prolonged high temperature heating, however, may cause MYR denaturation [5,19,21] and GL thermal degradation [21]. Freeze drying has been shown to retain MYR and GLs [3]. Prior to freeze drying, microwave cooking at adequate power inputs, inactivates ESP while preserving MYR activity which increases bioaccessibility.

SR formation occurs mostly in two organs: the mouth during mastication and in the gut by the microbiota. Research shows that there are differences between individuals in oral processing of foods [23]. Sarvan et al. [20] investigated, in-vivo, the effect of steaming and chewing time on the bioaccessibility of SR and SR-Nitrile, the GR breakdown products after chewing. Results showed that longer chewing times of broccoli with active MYR led to more GR hydrolysis. Compared to raw broccoli, or broccoli steamed for shorter periods, chewing broccoli steamed for 2 min provided the highest amount of SR. Broccoli steamed for 3 min provided the least amount of SR [20]. The effect of chewing on bioavailability was demonstrated by Shapiro et al. [24] who measured the difference in the amount of ITC metabolites excreted from urine when broccoli sprouts were swallowed whole or chewed thoroughly. They found that chewing increased the amount of urine metabolites by 1.5 times.

In broccoli products with inactivated MYR due to prolonged heating, the gut conversion processes of GR to SR and other degradation products become important. The capability of an individual's gut microbiome to convert GR to SR will depend on the types of microbes, their quantities, and how effective their different mechanisms for bioconversion are. Gut bacteria convert GLs to other compounds besides ITCs (Figure 1). Saha et al. [6] used a batch fermentation model with human gut bacteria to demonstrate that gut bacteria is capable of converting GR to SR, SR-nitrile, erucin, and erucin-nitrile. They also showed that the formation of erucin is preceded by the microbial conversion of GR to glucoerucin [6]. Consequently, the bioconversion of GL to non-ITC breakdown products reduces the bioavailability of ITCs.

Capturing the essence of the physiological processes for SR mathematically so that its biological effect can be simulated and predicted, is known as physiological-based modelling. This is an approach that considers the physiological basis of a bioactive compound's interaction with the human body before mathematical concepts are applied. Physiological-based models vary in terms of the number of physiological aspects (i.e., biological mechanisms, organs) considered. Some models only look at the gastrointestinal tract while others consider the whole body [25].

Various types of compartmental model have been described of which the most basic is the compartmental absorption and transit (CAT) model. In the CAT model, the small intestine is divided into a series of compartments and assumes linear transfer kinetics, passive absorption kinetics and well mixed compartments with uniform concentration [26,27]. The transit and absorption of a drug or food component is depicted by the following equation,

$$\frac{dM_n}{dt} = K_t M_{n-1} - K_t M_n - K_a M_n,\ n = 1, 2, \ldots 7 \quad (1)$$

where n is the number of compartments, M is the amount or concentration of the component in the nth compartment, K_t is the transit rate constant between compartments, and K_a is the absorption rate constant of the component into the blood.

Based on the CAT model, the advanced compartmental absorption and transit (ACAT) model was developed to include more details. The ACAT compartmentalizes the stomach and large intestine so that gastric emptying and absorption from the large intestine can be considered. In addition to linear kinetics and passive absorption, the model considers non-linear kinetics due to protein binding, liver metabolism, or active transport and physiochemical factors such as particle size, solubility, density, and permeability [28].

Most physiological-based modeling research available are for pharmaceutical drugs. There are few studies that are related to food components and food products, and even fewer studies for modeling broccoli compounds. Punt et al. [29,30] made whole body eight-compartment models to predict the bioactivation and detoxification of herb estragole in humans and rats. Le Feunteun et al. [31] made a five-compartment model that focused on the digestion of mini-pigs to study the effect of product matrices on the digestion of milk proteins. Strathe et al. [32] also made a model with four main compartments and 38 sub-compartments to study the digestion and absorption of macro-nutrients in growing pigs. Moxon et al. [33] made a two-compartment model to investigate the effect of gastric emptying, luminal viscosity and hydrolysis rate on the rate of glucose absorption.

At the time of writing this article one study was found about the physiological-based modeling of SR from broccoli. Li et al. [34] investigated the kinetics and distribution of sulforaphane in the tissues of mice using a physiological-based model, where the whole body was divided into eight compartments. The mice ingested fresh, steamed, and MYR treated steamed broccoli sprout powders. The difference in kinetics and distribution in the tissues between the three different products were compared. The model did not include SR absorption mechanisms, and it did not include GR to SR conversion processes in the mouth, via myrosinase, and in the gut, via microbes. Also, the study did not extrapolate their results from mice to humans. The conversion of GR in the gut and mouth are important process that affect bioavailability. Therefore, a physiological-based model that considers these processes is needed.

The objective of this study was to make a physiological-based model that describes the kinetics and bioavailability of isothiocyanates from broccoli and to evaluate how the derived parameters are impacted by inter-individual variation. The model is validated against urine excretion sulforaphane data from a previous 2014 Wageningen University in-vivo research study by Oliviero et al. [35]. In this study, the effect of residual myrosinase activity on ITC formation, bioavailability, and excretion kinetics was investigated after 15 test subjects (apparently healthy human volunteers, aged 26–50 years, body mass index 21 ± 2 kg/m^2, six men and nine women, 13 Caucasian, two Asian, and one Latin American), consumed five broccoli products with different levels of myrosinase activity obtained by different levels of microwave heating.

2. Materials and Methods

2.1. Pre-Modeling Data Processing

Participant raw data (measured SR urine conjugate excretion rates) from the Oliviero et al. [35] study was preprocessed for use in Matlab. The data were the time (minutes) and sulforaphane (SR) excretion rate (μmol/min). The technique used to measure SR urine conjugates, solid phase extraction-HPLC-MS/MS [35–37], is associated with experimental error that was quantified by Vermeulen et al. [37]. The relative standard deviation 12, 6, 3% for 1.04, 10.5, and 313 μM SR, respectively, in urine, was used to derive the following exponential equation that helped estimate the experimental error of each data point.

$$y = 0.11505x^{-0.24} \tag{2}$$

The relative error ratio is y, and the concentration of SR is x. The experimental errors were plotted as error bars on the data points for the model fittings.

2.2. Model Description and Assumptions

The model (Figure 2) focuses on the processes involved in the gastro-intestinal transit of glucoraphanin (GR) and sulforaphane (SR). Similar to an advanced compartmental absorption and transit (ACAT) model, it includes the stomach, seven compartments of the small intestine [28], the colon, and a blood compartment for systemic circulation. Unlike an ACAT, the colon, was divided into seven compartments, the stomach into two compartments [31]. A mouth compartment, which is typically not in physiological-based models, was included. As a result of this the full model contains 18 compartments.

The products consumed in the intervention study were portions of 5 g of each broccoli product with 90 mL of water at 40 °C, and with 30 g of raisin bun and water ad libitum. During mastication, myrosinase (MYR) and GRs released from the cell structures react to form an O-sulfated thiohydroximate intermediate, which then immediately converts to SR or SR-nitrile. The amount of the intermediate that is converted to SR versus SR-nitrile is a ratio that is subject to change depending on the individual's chewing pattern and broccoli product. In the mouth, it is assumed that SR-nitrile is the only non-ITC compound formed. Mastication time (30 s) and saliva flow rates (0.033 mL/s) [38] are assumed to be the same for all participants. The volume of the mouth compartment is the product plus the saliva excretions, 0.096 L.

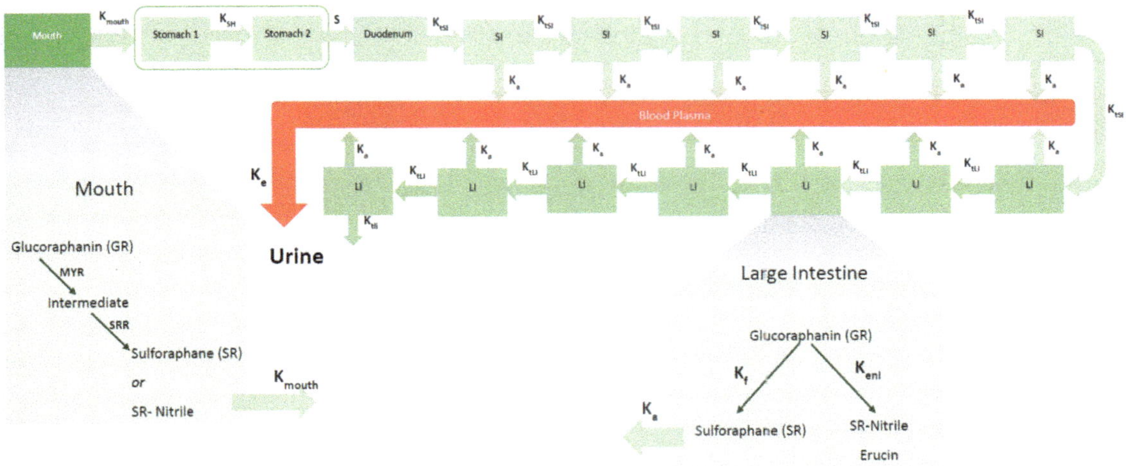

Figure 2. Schematic diagram of model with one mouth compartment, two stomach compartments, seven small intestine (SI) compartments and seven large intestine (LI) compartments, and one blood compartment. The conversion of glucoraphanin (GR) to sulforaphane (SR) and SR-nitrile and erucin in the mouth and large intestine is depicted. Shown are the transit rate constants between the mouth and stomach (K_{mouth}), between the two stomach compartments (K_{SH}), between the stomach and duodenum (S), between the small intestines (K_{tSI}), between the large intestines (K_{tLI}). The absorption of SR from the SI and LI into the blood is represented by the absorption rate constant (K_a). Elimination of SR and SR-conjugates from the blood to urine is represented by the rate constant, K_e. During mastication, GR is converted to an intermediate by myrosinase (MYR) and the subsequently converted to SR or SR-nitrile based on a conversion ratio (SRR). The gut microbial conversion of GR to SR and SR-nitrile/erucin is represented by the rate constants K_f and K_{eni}, respectively.

Swallowing transfers the bolus to the stomach. The first stomach compartment accounts for the disintegration of food particles that are too big to pass through the pyloric sphincter valve leading into the small intestine. Food broken down sufficiently, and mixed with gastric fluids in the first compartment, is moved to the second stomach compartment where it mixes with more gastric juices before emptying into the duodenum. Gastric emptying of solid foods has been described as having a biphasic nature due to the time required for enzymatic and mechanical disintegration before emptying into the intestines [39]. Any MYR is assumed to be deactivated irreversibly in the stomach due to the low pH of the gastric fluids [19,40]. It is also assumed that GLs and ITCs are not absorbed in the stomach. Transit from second stomach compartment to the duodenum depends on the stomach emptying time (30 min) which is assumed to be the same for all participants since the size of the meal is the same. Based on the meal size, the volume of both stomach compartments together is 0.2 L, of which 0.05 L is the volume of the first stomach compartment (Table 1).

The chyme is mixed with duodenal secretions in this first compartment of the small intestine (SI). Due to the differences in the intestinal lining of duodenum (less villi/area for absorption) compared to the rest of the small intestine, less nutrients are absorbed in the duodenum; for simplicity, it is assumed there is no ITC absorption. In the remaining six SI compartments, SR is absorbed into the blood as both GR and SR are transferred from one compartment to the next.

The large intestine is divided into seven compartments where the following processes takes place: formation of SR, nitriles, and erucin by gut bacteria, absorption of SR into the blood, and transit of compounds from one compartment to the next. It is assumed that GRs are not absorbed from both the small and large intestines. Movement of chyme in the intestines are in the forward (towards the rectum) direction. Backward movements are known to occur and are represented by the fact that the compartments are assumed to be well mixed.

ITC is absorbed into the blood plasma, which is represented by one compartment that is presumed to be the same volume between all participants (5 L). Absorption of ITCs across the intestinal wall occurs passively by diffusion and is described according to Fick's Law of Diffusion [41]. ITC metabolites are eliminated from systemic blood circulation via glomerular filtration. No other elimination processes (sweat, defecation, respiration) are considered.

All compartment volumes (Table 1) are assumed to stay constant and the concentration of compounds (GR and SR) per compartment is uniform.

Table 1. Parameters values used in the model for the broccoli products, oral GR conversion processes, gastro-intestinal processes, gut GR conversion processes. Parameter values for the broccoli products MYR, Cgl_0 (initial glucosinolate concentration), and ITC_0 (initial sulforaphane concentration) are separated by product (HighBP, HighBF, MedBF, LowBF, NoBF, each of these products have different MYR content due to different levels of microwave heating). Parameters designated as 'Estimated' were used in model fittings.

		Broccoli Products Composition		
Product	***MYR*** **Myrosinase Content (mg MYR/mg Broccoli)**	***Cgl_0*** **Initial GR Concentration (μM)**	***ITC_0*** **Initial SR Concentration (μM)**	**Reference**
HighBP	3.49×10^{-2}	383.3	354.2	[35,42]
HighBF	3.49×10^{-2}	621.9	115.6	[35,42]
MedBF	6.53×10^{-3}	667.7	69.8 †	[35,42]
LowBF	5.63×10^{-4}	708.3	29.2 †	[35,42]
NoBF	1.13×10^{-5}	734.4	3.1	[35,42]
		Oral GR Conversion		
V_{max}	MMSI V_{max} for glucoraphanin		2070 μmol/min	[42]
K_m	MMSI K_m for glucoraphanin		110.2 μM	[42]
K_i	MMSI K_i for glucoraphanin		893.0 μM	[42]
BR	Amount of broccoli in broccoli product		5000 mg	[35]
SRR	Fraction of sulforaphane converted from GR in mouth		Estimated	
		Gastro-Intestinal		
k_{Mouth}	Mouth to Stomach rate constant		$30\ min^{-1}$ ($60\ min^{-1}$ to $1\ min^{-1}$) *	[43,44]
St	Stomach emptying time		30 min	
S	Gastric rate constant from 2nd stomach to duodenum		$S = -\ln(0.05)/St$	[45]
k_{SH}	Rate constant from 1st to 2nd stomach compartment		Estimated **	
k_{tSI}	Small Intestine transit rate constant		Estimated **	
k_{tLI}	Large Intestine transit rate constant		Estimated **	
k_a	Absorption rate constant		$0.180\ min^{-1}$	[41]
k_e	Elimination rate constant of ITC and ITC conjugates from blood		Estimated **	
N_{SI}	Number of SI compartments (excluding duodenum)		6	[26]
N_{LI}	Number of LI compartments		7	[26]
V_{Mouth}	Product + Saliva		0.096 L (0.095–0.098 L) *	[35,38]
	Stomach: V_{Mouth}+ Raisin bun + gastric secretions		0.2 L (0.167–0.253 L) *	[35,46]
$V_{Stomach\ 1}$	Stomach 1		0.05 L	
$V_{Stomach\ 2}$	Stomach 2		0.15 L	
$V_{duodenum}$	$V_{stomach\ 2}$+ duodenal secretions		0.2 L (0.246–0.332 L) *	[46]
$V_{SI\#}$	SI volume (excluding duodenum)		1.5 L (0.638–1.963 L) *	[47]
	V_{SI}/N_{SI} = SI compartment volume		0.25 L	
$V_{LI\#}$	LI volume		3.4 L (3.347–3.492 L) *	[47]
	V_{LI}/N_{LI}= LI compartment volume		0.5 L	
V_{17}	Blood volume of adult		5 L (4–6 L) *	[48]
		Gut GR Conversion		
k_f	Microbial ITC formation rate constant		Estimated	
k_{eni}	GR to erucin and nitriles		Estimated	

* Ranges for the parameters were determined based on literature. ** These values were estimated for each individual participant based on the model fit of the experimental values. † ITC_0 values used in final fittings for MedBF and LowBF were approximately 3.4 and 9.1%, respectively, of their values in this table due to poor fit results using the original values.

2.3. Compartmental Mathematics

The enzymatic reaction of GL and MYR to form the O-sulfated thiohydroximate intermediate in the mouth is characterized by a Michaelis–Menten equation that accounts for enzyme inhibition. The intermediate instantly reacts to form ITC or ITC-nitrile, therefore the change in GL concentration is negatively proportional to the Michaelis–Menten equation.

$$\frac{dC_{GL}}{dt} = -\left(\frac{V_{max} \times MYR \times BR \times C_{GL\ Mouth}}{K_m + C_{GL\ Mouth} + \frac{C_{GL\ Mouth}^2}{K_i}}\right) \quad (3)$$

MYR is the estimated mg of myrosinase per one mg of dried broccoli. Details on how *MYR* was estimated is found in Appendix A Part I. *BR* is the amount of dried broccoli (5 g) consumed by the participants. The maximum rate, V_{max} (µM/min*mg *MYR*), the Michaelis constant, K_m (µM), and the inhibition constant, K_i (µM) were derived using glucoraphanin data from Roman et al.'s [42] MYR kinetic study. Details on the derivation of these variables can be seen in Appendix A Part II.

The amounts of ITC and nitriles formed is expressed as a fraction of the amount of intermediate (Equations (4) and (5)).

$$\frac{dC_{ITC\ Mouth}}{dt} = SRR \times \frac{dC_{GL}}{dt} \quad (4)$$

and

$$\frac{dC_{Nitrile\ Mouth}}{dt} = (1 - SRR) \times \frac{dC_{GL}}{dt} \quad (5)$$

SRR is the fraction of hydrolyzed *GR* converted to *SR* in the mouth. The remaining, 1 − *SRR*, is converted to nitriles.

Transfer of ITCs and GLs out of the mouth, as well as into and out of the stomach and intestinal compartments are first order rate reactions (Equations (6) and (7)).

$$\frac{dC_{ITC\ i}}{dt} = k_{transfer\ (i-1\ to\ i)} \times \frac{V_{i-1}}{V_i} \times C_{ITC\ i-1} \quad (6)$$

and

$$\frac{dC_{GL\ i}}{dt} = k_{transfer\ (i-1\ to\ i)} \times \frac{V_{i-1}}{V_i} \times C_{GL\ i-1} \quad (7)$$

The rate constants for transfer from the mouth, k_{mouth}, for the stomach compartments, k_{SH} and *S*, and intestinal compartments, k_{tsi} and k_{tli}, are the inverses of the residence time of each compartment (Table 1). Due to differences in some compartment volumes (V), volume ratios are considered.

The gut formation of ITCs, nitriles, and erucin are also represented by the following first order rate reactions,

$$\frac{dC_{ITC}}{dt} = k_f \times C_{GL} \quad (8)$$

$$\frac{dC_{Nitrile\ \&\ Erucin}}{dt} = k_{eni} \times C_{GL} \quad (9)$$

where k_f is the rate constant of formation for ITC, and k_{eni} is the rate constant of formation for erucin and nitrile. The change in GL concentration in the gut is proportional to the formation of ITC, nitrile, and erucin.

$$\frac{dC_{GL}}{dt} = -(k_f \times C_{GL}) - (k_{eni} \times C_{GL}) \quad (10)$$

Absorption into the blood is defined by the following equation,

$$\frac{dC_{ITC}}{dt} = k_a \times C_{ITC} \quad (11)$$

where k_a, is the rate of absorption (passive diffusion) from the intestines and it is proportional to the effective permeability (P_{eff}) of sulforaphane and inversely proportional to the radius, R, of the intestines [49].

$$k_a = \frac{2P_{eff}}{R} \tag{12}$$

Each compartment is defined by mass balanced differential equations that contain the processes just described. For example, the full ITC and GL differential equations for the seventh small intestine compartment and first large intestine compartment are shown, Equations (13)–(16),

Seventh small intestine compartment:

$$\frac{dC_{ITC\ SI7}}{dt} = \frac{V_{SI6}}{V_{SI7}} \underbrace{\times (k_{tSI} \times C_{ITC\ SI6})}_{\text{ITC In}} \underbrace{-(k_{tSI} \times C_{ITC\ SI7})}_{\text{ITC Out}} \underbrace{-(k_a \times C_{ITC\ SI7})}_{\text{ITC Absorption}} \tag{13}$$

$$\frac{dC_{GL\ SI7}}{dt} = \frac{V_{SI6}}{V_{SI7}} \underbrace{\times (k_{tSI} \times C_{GL\ SI6})}_{\text{GL In}} \underbrace{-(k_{tSI} \times C_{GL\ SI7})}_{\text{GL Out}} \tag{14}$$

First large intestine compartment:

$$\frac{dC_{ITC\ LI1}}{dt} = \frac{V_{SI7}}{V_{LI1}} \underbrace{\times (k_{tSI} \times C_{ITC\ SI7})}_{\text{ITC In}} \underbrace{+(k_f \times C_{GL\ LI1})}_{\text{ITC Formation}} \underbrace{-(k_{tLI} \times C_{ITC\ LI1})}_{\text{ITC Out}} \underbrace{-(k_a \times C_{ITC\ LI1})}_{\text{ITC Out}} \tag{15}$$

$$\frac{dC_{GL\ LI1}}{dt} = \frac{V_{SI7}}{V_{LI1}} \underbrace{\times (k_{tSI} \times C_{GL\ SI7})}_{\text{GL In}} \underbrace{-(k_f \times C_{GL\ LI1})}_{\text{ITC Formation}} \underbrace{-(k_{tLI} \times C_{GL\ LI1})}_{\text{GL Out}} \underbrace{-(k_{eni} \times C_{GL\ LI1})}_{\text{Erucin/Nitrile Formation}} \tag{16}$$

Full list of differential equations for all compartments are in the Supplementary Files.

2.4. *Simulink Model*

The mathematical equations used to represent the different compartmental processes were translated to a block diagram on Matlab's Simulink application (Matlab R2020a). Code scripts written on the MATLAB (MathWorks) interface integrated with the Simulink block diagram model to run sensitivity analyses and to perform fitting on the five different broccoli products.

2.5. *Matlab Coding and Fittings*

Sensitivity Analysis. Sensitivity analysis was performed to see the effect of changing parameter values on the model output. A parameter was changed within literature determined ranges while other parameters were kept constant. The analysis was performed for the different broccoli products, NoBF, LowBF, MedBF, HighBF, and HighBP.

Model Fittings. After the sensitivity analysis, parameters with the most influence on the simulation output, were used for fitting the model to each participant's data set. The fitting was done using a Trust-Region-Reflective Least Squares algorithm from the least squares data fitting solver of Matlab's Optimization Toolbox. The simulation period was 1600 min (26 h). Values of parameters used for fitting are in Table 1. This fitting procedure yielded parameter estimates that gave the best fit between the model simulations and the experimental data of the intervention study.

2.6. *Statistical and Data Analysis*

Confidence Intervals for each parameter estimate were determined at 90% confidence using Matlab's non-linear parameter confidence interval function. This function also provided covariance matrices that were used to calculate the correlation coefficients.

Using Matlab's trapezoidal numerical integration function, the cumulative SR excreted per participant for the experimental and predicted data sets were calculated.

Full codes are provided in the supplementary files.

3. Results

3.1. Sensitivity Analysis and Parameters Selection

The model's sensitivity towards twelve parameters was tested. Based on the analysis, three parameters were used in model fittings for HighBP and HighBF, seven for Med and Low BF, and five for NoBF (Table 2).

Table 2. Summary of parameters that influenced the simulation outputs for each broccoli product during the sensitivity analysis. Parameters with a check mark (✓) were used for model fittings; parameters with an O, influenced simulation output but were not used in model fittings; parameters with an x did not affect output during the sensitivity analysis and were excluded from fitting. Three parameters were used to fit HighBP and HighBF, seven for MedBF and LowBF, and five for NoBF.

	k_{SH}	SRR	k_f	k_e	k_{tSI}	k_{tLI}	k_{eni}	Cgl_0	ITC_0	MYR	k_a	St
HighBP	✓	✓	x	o	✓	x	x	o	o	o	x	x
HighBF	✓	✓	x	o	✓	x	x	o	o	o	x	x
MedBF	✓	✓	✓	✓	✓	✓	✓	o	o	o	x	x
LowBF	✓	✓	✓	✓	✓	✓	✓	o	o	o	x	x
NoBF	✓	x	✓	x	✓	✓	✓	o	o	o	x	x

Parameters that did not affect model output were excluded. Gastric emptying time (St), and the rate constant of absorption (k_a), were excluded from all model fittings because they had insignificant effects on the output (example in Figure 3A,B). As expected, myrosinase content (MYR), initial GR concentration (Cgl_0), initial SR concentration (ITC_0), and the ratio of GR converted to SR in the mouth (SRR), caused a direct and proportional upward shift to the HighBP, HighBF, MedBF, and LowBF simulation outputs (example in Figure 3C,D). The MYR and ITC_0 contents were low for NoBF, therefore, of the broccoli product related parameters, only Cgl_0 affected the output. Furthermore, the conversion ratio of GR to SR in the mouth (SRR), which depends on myrosinase content, did not affect the output of NoBF during the sensitivity analysis since myrosinase was inactive (Figure 3J). The first stomach rate constant (K_{SH}), small intestine transit rate constant (k_{tSI}) and SR elimination from the blood (k_e), also caused proportional upward shifts but a narrowing of the curves was observed (example Figure 3L). The gut parameters, ITC formation (k_f) and erucin and nitrile formation (k_{eni}) rate constants, had opposite effects on outputs. Increases in k_{eni} resulted in downward shifts and narrowing of the output curves (Figure 3H), while the curves shifted upwards for k_f (Figure 3G). Increasing gut transit rate constant (k_{tLI}), decreased the size of the second peak for MedBF and LowBF, and the single peak for NoBF (example in Figure 3F). Changes in k_{tLI} did not affect outputs of HighBP and HighBF products (Figure 3E).

3.2. Model Fittings

The model was successfully fit to the data of each participant and for each product. Figure 4 shows the model fits for five of the participants and for all five broccoli products. 75 fittings were possible (15 participants times five products) but only 72 fitting were performed. Three data sets were excluded due to lack of data. While four data sets had to be preprocessed before fitting.

Few data sets had poor fits. Participant *m*, for the MedBF product, was underfitted while participants *q*, MedBF, and *g*, LowBF, were overfitted (Figures 4 and S1: Model Fittings Results for All Participants in the Supplementary Files).

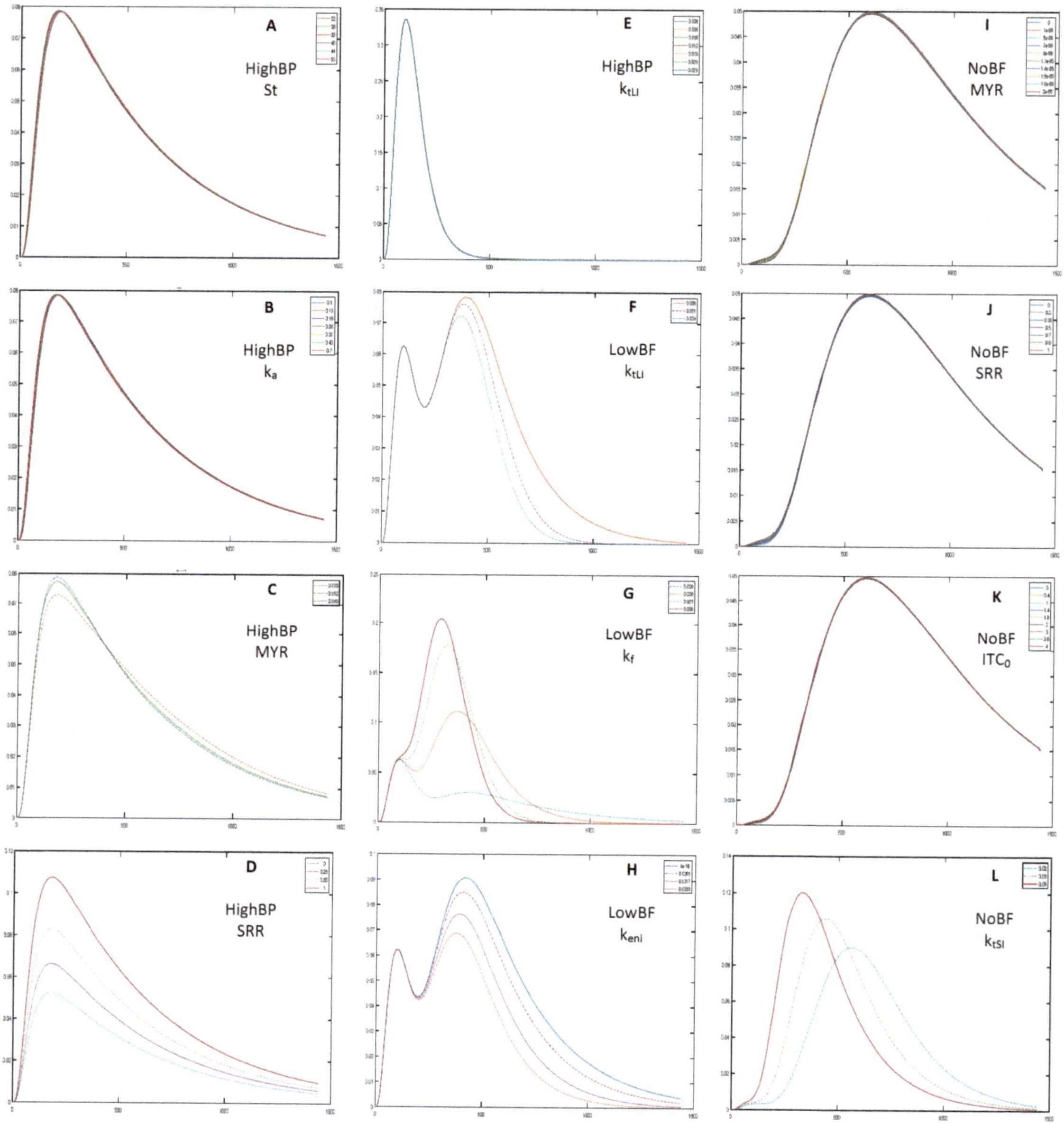

Figure 3. HighBP sensitivity analysis for (**A**) gastric emptying time, St; (**B**) the rate constant of absorption, k_a; (**C**) myrosinase content, MYR; (**D**) the ratio of GR converted to SR in the mouth, SRR; (**E**) large intestine transit rate constant, k_{tLI}. LowBF sensitivity analysis for (**F**) large intestine transit rate constant, k_{tLI}; (**G**) ITC formation rate constant in the gut, k_f; and (**H**) erucin and nitrile formation rate constant in the gut, k_{eni}. NoBF sensitivity analysis for (**I**) myrosinase content, MYR; (**J**) the ratio of GR converted to SR in the mouth, SRR; (**K**) and initial ITC concentration (ITC_0) and (**L**) small intestine transit rate constant, k_{tSI}. The sensitivity analysis shows how the simulation output changes when all parameters are kept constant while one changes.

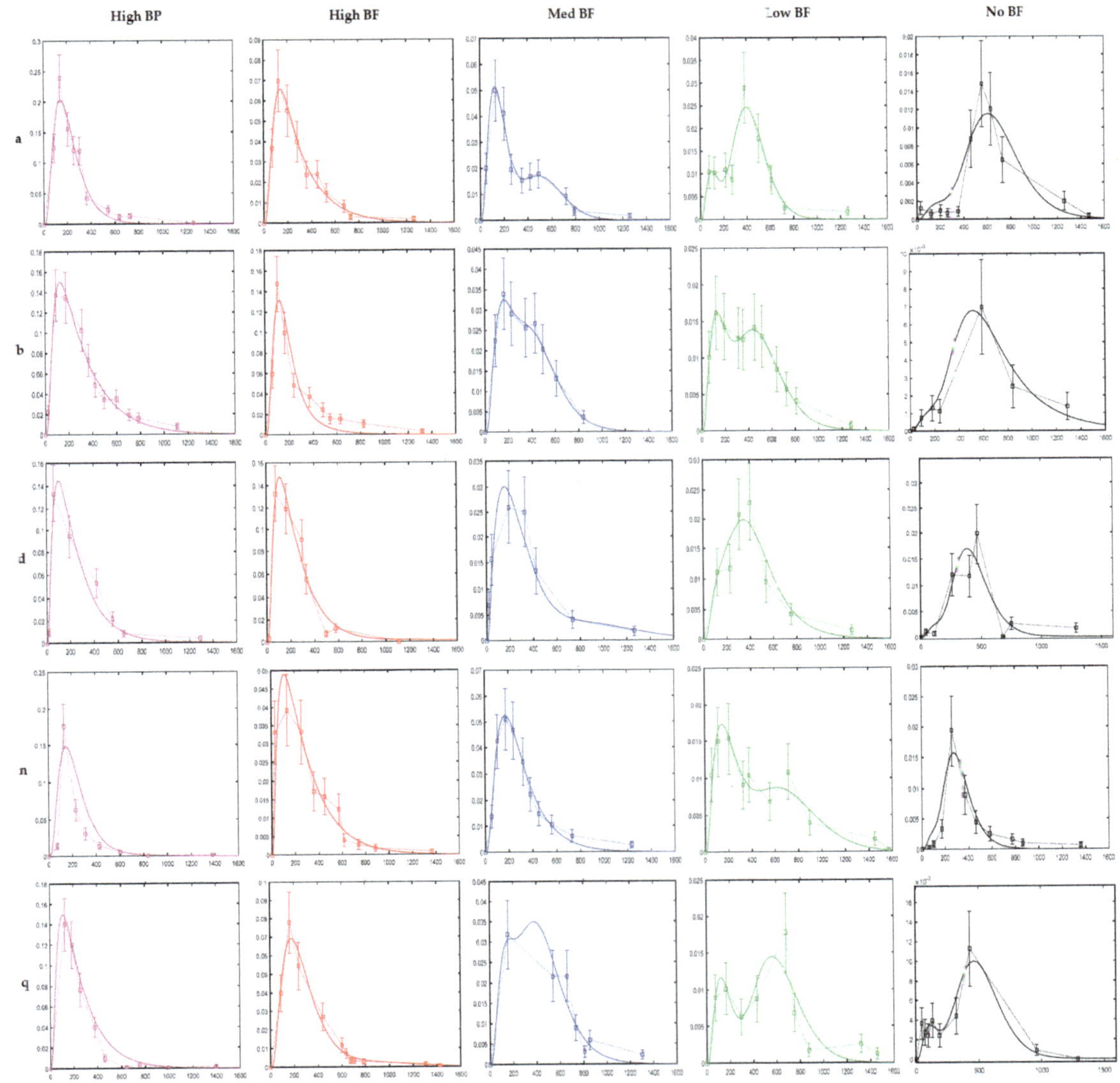

Figure 4. Model fittings for participants a, b, d, n, q. X and Y axis for each graph is, time (min) and sulforaphane conjugate excretion rate (µmol/min), respectively. Experimental data are the square bullets, and the solid lines are the model fits. Error bars represent potential experimental error from the analytical techniques used by Oliviero et al. to measure the amounts of ITC conjugates in urine. All participant model fittings are in Figure S1: Model Fittings Results for All Participants in the Supplementary Files.

3.3. *Bioavailability of Sulforaphane*

Bioavailability was calculated by dividing the cumulative amounts of SR by the amount of GR in the broccoli products. Cumulative amounts of SR for the experimental data and model data were determined using Matlab's trapezoid function to integrate. There were small differences between the predicted bioavailability and experimental bioavailability (Table 3). Average HighBP predictions were 2% larger than the calculated experimental bioavailability. The difference was 1% for MedBF, 0.9% for LowBF, and 0.1% for HighBF and NoBF.

Table 3. Average sulforaphane (SR) bioavailability values ±SD for the experimental data and model. There were minor differences between the experimental bioavailability and model bioavailability for each broccoli product type.

	Average SR Bioavailability (%)		
	Experimental Data	**Model**	**Difference**
HighBP	63 ± 0.2	65 ± 0.1	2%
HighBF	33 ± 0.1	33 ± 0.1	0.1%
MedBF	25 ± 0.1	24 ± 0.1	1%
LowBF	19 ± 0.1	18 ± 0.1	0.9%
NoBF	10 ± 0.04	10 ± 0.04	0.1%

3.4. Mouth and Gut Parameter Estimations

Tables 4–6 shows the distribution (boxplots) of mouth and gut parameter estimates for each product type. Interquartile ranges (IQR) for the parameter estimates varied widely. The distributions of most parameter estimates are positively skewed indicating that 50% of participants are less variable within the first two quartiles than the 50% of participants in quartile three and four.

Table 4. SRR estimation results (horizontal red bar: median, blue box second and third quartile, whiskers first and fourth quartiles).

High BP (*n* = 15)	**High BF (*n* = 15)**	**Med BF (*n* = 14)**	**Low BF (*n* = 14)**	**No BF**
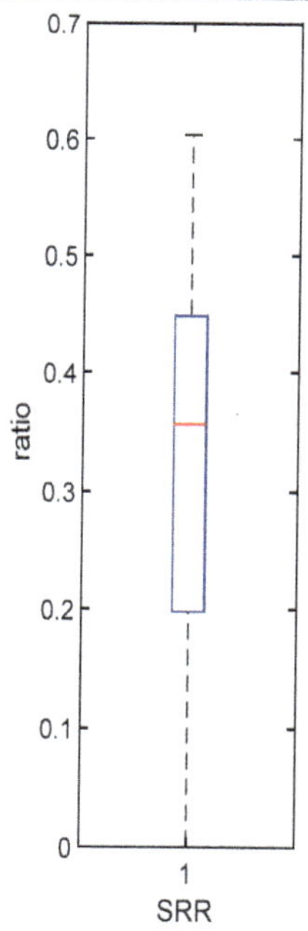	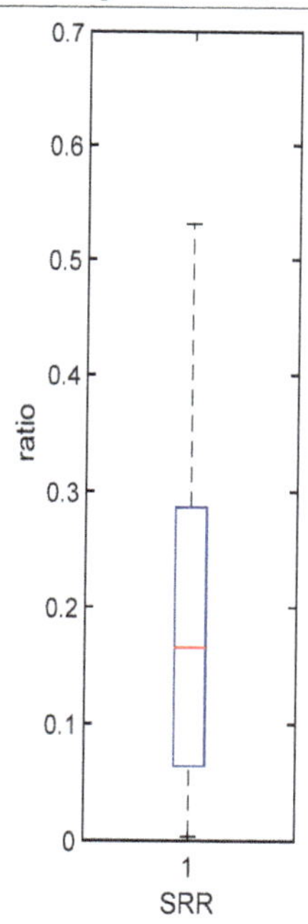	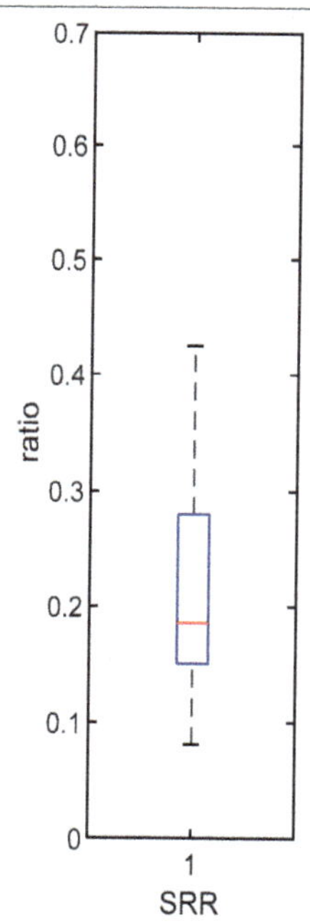	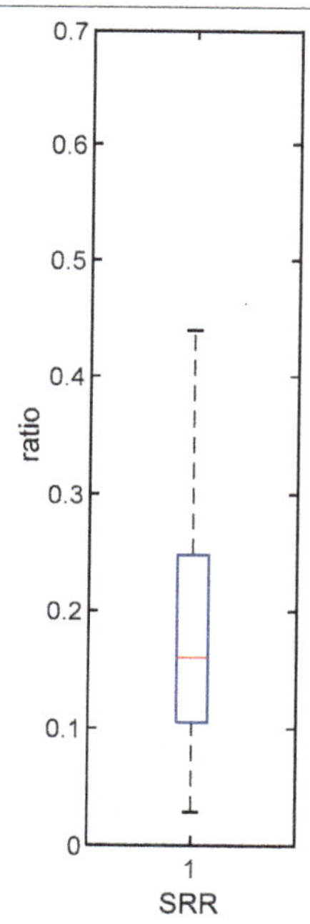	—
Median: 0.357 IQR: 0.251 IQR/Med: 0.7 Outliers: None	Median: 0.165 IQR: 0.223 IQR/Med: 1.4 Outliers: None	Median: 0.186 IQR: 0.13 IQR/Med: 0.7 Outliers: None	Median: 0.160 IQR: 0.143 IQR/Med: 0.9 Outliers: None	SRR parameter not fitted. Fixed at 0.047.

Table 5. k_f estimation results (horizontal red bar: median, blue box second and third quartile, whiskers first and fourth quartiles, +: outliers).

High BP(n = 15)	High BF(n = 15)	Med BF(n = 14) *	Low BF(n = 14) *	No BF(n = 14) *
—	—	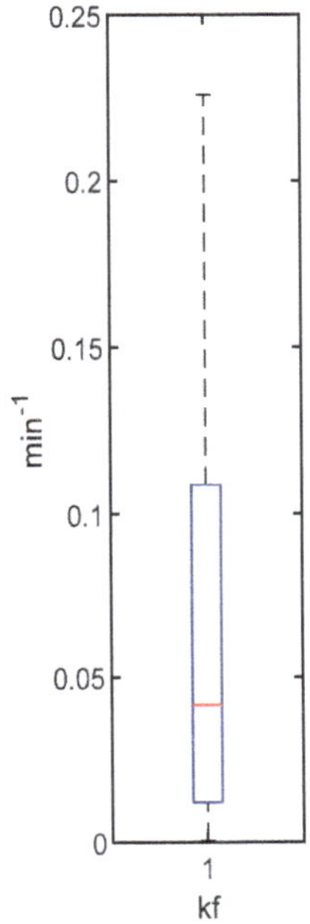	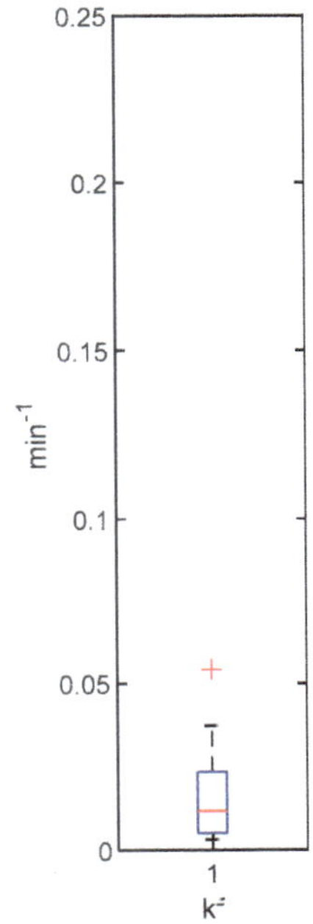	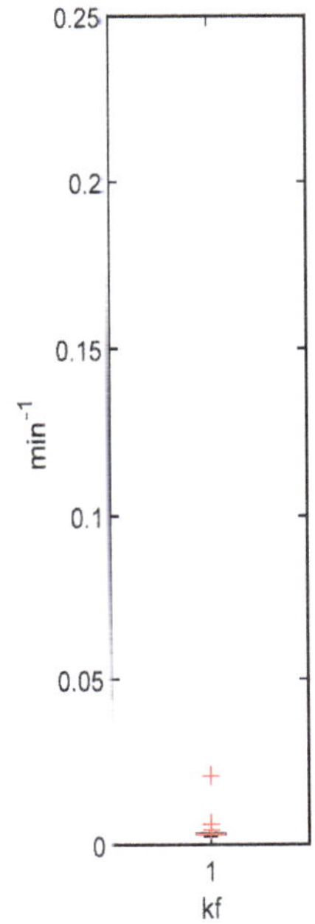
Not fitted. k_f fixed at 0.0033 min^{-1}.	Not fitted. k_f fixed at 0.0033 min^{-1}.	Median: 0.042 min^{-1} IQR: 0.097 min^{-1} IQR/Med: 2.3 Outliers: c (0.555 min^{-1})	Median: 0.012 min^{-1} IQR: 0.018 min^{-1} IQR/Med: 1.5 Outliers n (0.054 min^{-1})	Median: 0.003 min^{-1} IQR: 0.001 min^{-1} IQR/Med: 0.3 Outliers: g (0.006 min^{-1}) h (0.005 min^{-1}) q (0.021 min^{-1})

* Some or all outliers excluded for better visual presentation and comparison of boxplots. Outliers are values more than 1.5 times the IQR. See Supplementary Files for plotted outliers and for expanded view of the NoBF k_f boxplot.

SRR, the ratio of GR that gets converted to SR in the mouth, was estimated for each participant who consumed the HighBP, HighBF, Med and Low BF products (Table 4). SRR was not fitted for the NoBF product since the amount of MYR is very low. Therefore, any GR converted to SR in the mouth is insignificant for the NoBF product. The medians for HighBF, MedBF, and LowBF mean that half of the participants converted less than 19% GR to SR in the mouth. HighBP distribution, on the other hand, is negatively skewed with half the participants converting 35 to 60% of GR (median = 0.357) to SR.

K_f and k_{eni} represent the rate of formation of sulforaphane and erucin and nitriles, respectively, by gut bacteria in the large intestine. These parameters were estimated for the MedBF, LowBF and NoBF products but not for the High myrosinase products. K_f was fixed at 0.0033 min^{-1} and k_{eni} was fixed at 0.0015 min^{-1} for both products. All distributions are positively skewed indicating that at 50% of the participant metabolize GR to SR slower than 0.042 min^{-1} for MedBF, 0.012 min^{-1} for LowBF, and 0.003 min^{-1} for NoBF (Table 5); and they metabolize GR to erucin and nitrile slower than 0.017 min^{-1} for MedBF, 0.033 min^{-1} for LowBF, and 0.003 min^{-1} for NoBF (Table 6).

The results of other parameters are discussed in Appendix B.

Table 6. k_{eni} estimation results (horizontal red bar: median, blue box second and third quartile, whiskers first and fourth quartiles, +: outliers).

High BP(n = 15)	High BF(n = 15)	Med BF(n = 14) *	Low BF(n = 14)	No BF(n = 14)
—	—	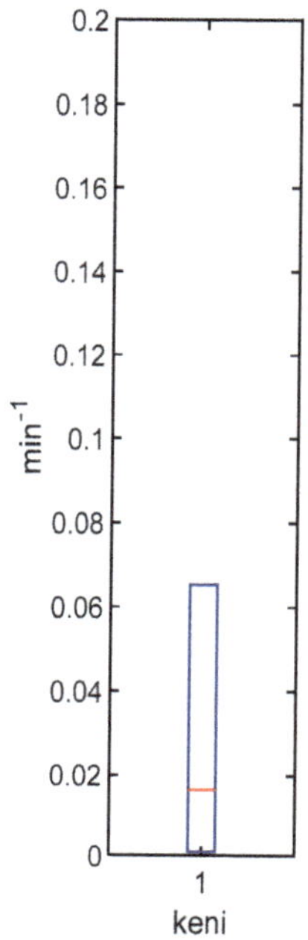	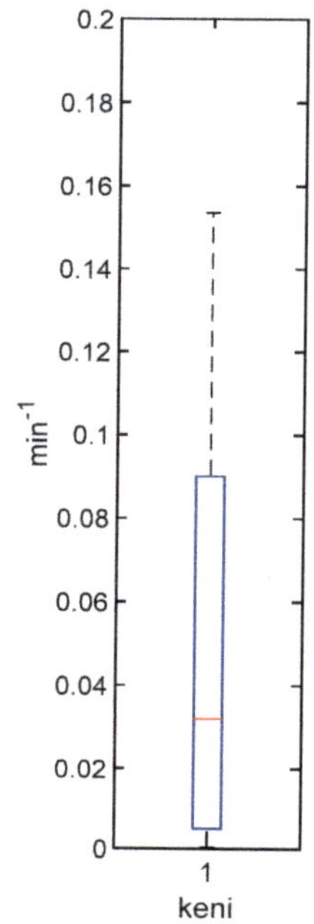	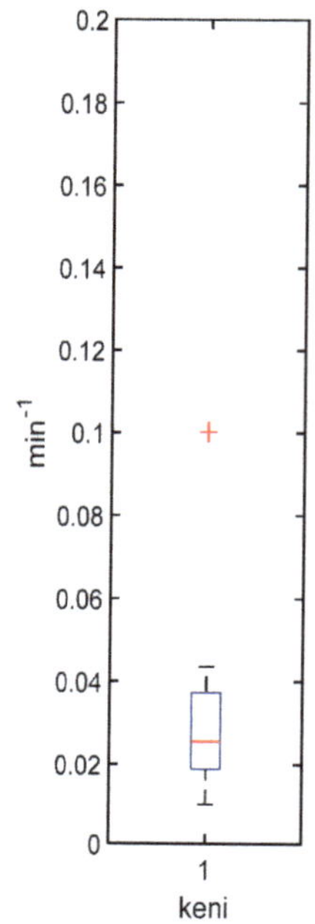
Not fitted. k_{eni} fixed at 0.0015 min^{-1}.	Not fitted. k_{eni} fixed at 0.0015 min^{-1}.	Median: 0.017 min^{-1} IQR: 0.064 min^{-1} IQR/Med: 3.8 Outliers: c (1.95 min^{-1}) h (0.576 min^{-1}) n (0.26995 min^{-1})	Median: 0.033 min^{-1} IQR: 0.085 min^{-1} IQR/Med: 2.6 Outliers: None	Median: 0.026 min^{-1} IQR: 0.018 min^{-1} IQR/Med: 0.7 Outliers: g (0.100 min^{-1})

* Some or all outliers excluded for better visual presentation and comparison of boxplots. Outliers are values more than 1.5 times the IQR. See Supplementary Files for plotted outliers.

3.5. Certainty of Parameter Estimates

For each participant and broccoli product, 90% confidence intervals were calculated to determine the range of parameters that are likely to include the parameter estimates. The confidence intervals were very large and are therefore not a good measure of certainty in the parameter estimates. The percentage of participants that had intervals larger than 30% on either side of the estimated parameter, was 62% of participants for HighBP, 53% for HighBF, 96% for MedBF, 98% for LowBF, and 100% for NoBF. The large confidence intervals were due to the limited data of each participant.

3.6. Goodness of Fit

A comparison between the experimental and the predicted values are shown in Figure 5.

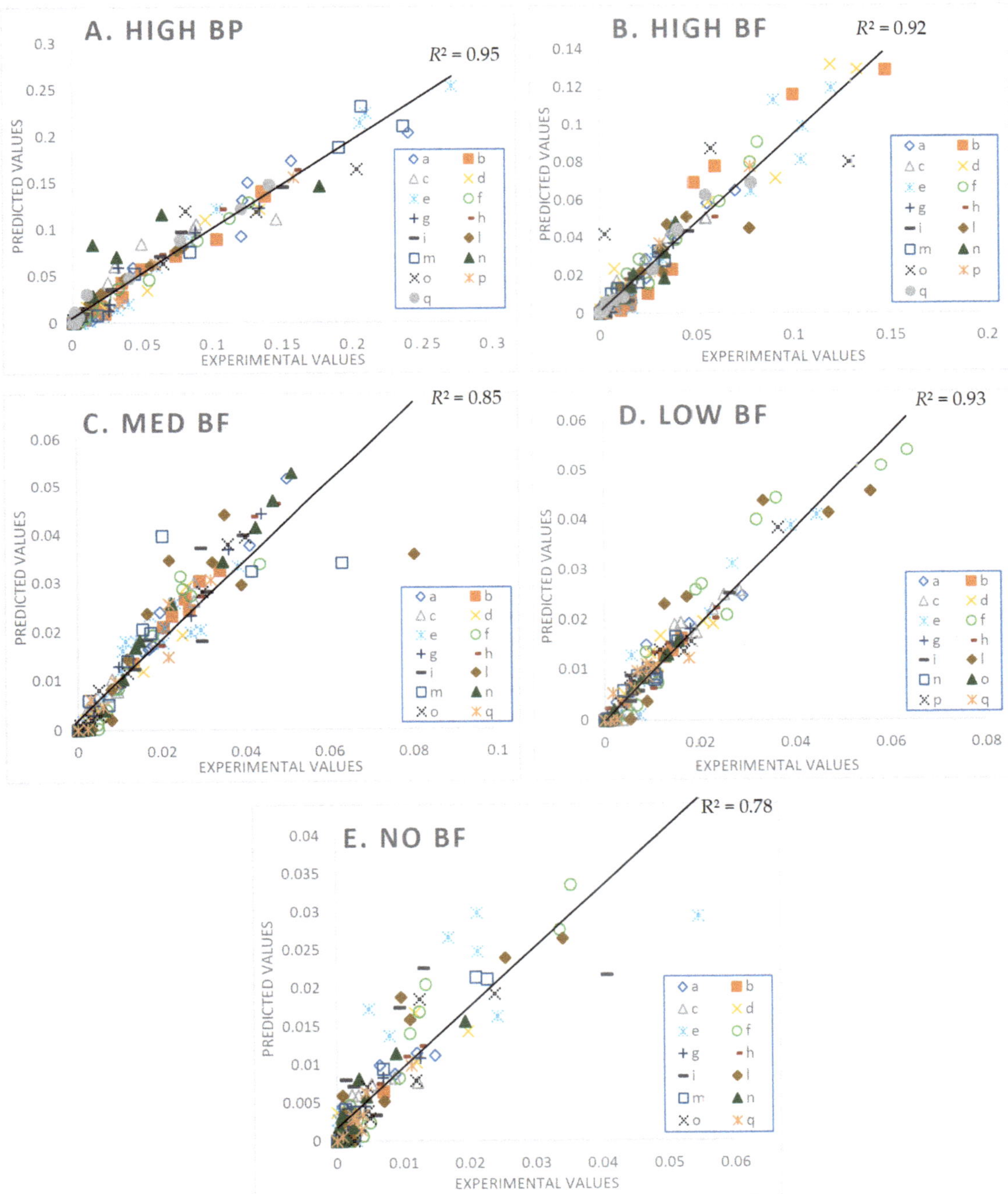

Figure 5. Parity plot of High MYR Broccoli Powder (**A**), High MYR Broccoli Florets (**B**), Medium MYR Broccoli Florets (**C**), Low MYR Broccoli Florets (**D**), and No MYR Broccoli Florets (**E**) with lumped participant data. Participants are designated by letters.

The best fits are for the HighBP product (R^2 = 0.95), followed by LowBF (R^2 = 0.93), HighBF (R^2 = 0.92), MedBF (R^2 = 0.85), and lastly NoBF (R^2 = 0.78). Most predicted values for NoBF are higher than their corresponding experimental values (Figure 5E). Of the five broccoli products NoBF was the most difficult to fit due to the narrow peaks and wider base for most of the excretion curves compared to the other products (see NoBF simulation results in Figure S1: Model Fittings Results for All Participants in the supplementary files.). A noticeable feature of the model for the HighBF, MedBF, LowBF, and NoBF products (Figures 5B–E and S1), is that the model solutions for some participants (e.g., participants e, l, n) approached excretion rates of 0 faster than the data points. The fit on the tail end of the data points, as well as the general fits for NoBF, may be improved by updating the model to include future physiological information on GR gut conversions, ITC absorption, distribution, and elimination.

4. Discussion

4.1. Sensitivity Analysis and Selection of Parameters

Sensitivity analysis for each product type was conducted to understand how changing one parameter and keeping the other constant would affect the model simulation output. Parameters to be estimated in model fittings were selected based on the results of the sensitivity analysis and preliminary model fittings. Given the low amount of data (6–15 data points) per participant, this procedure helped reduce the number of parameters fitted to the most necessary, thereby increasing the degrees of freedom.

Parameters that affected model output but were excluded from final fittings were k_e, MYR, Cgl_0, and ITC_0. The model was sensitive to k_e for the HighBP, HighBF, MedBF, and LowBF products. However, k_e was not included in the fittings for HighBP and BF due to the poor model convergence. For all products, MYR and Cgl_0 were fixed at the concentrations calculated from the Oliviero study (Table 1 and Matlab codes in the Supplementary Files). The initial concentration of SR (ITC_0) calculated for HighBP, HighBF, and NoBF were kept at the values derived from the Oliviero study, while for MedBF and LowBF, ITC_0 was reduced in order to get a good fit (Table 1). The reduction in ITC_0 may be explained by the high variability in the measured amounts of SR in the broccoli products. Oliviero et al. [35] measured SR in triplicates and the standard deviation for MedBF was 28.4% of the average, while for LowBF, it was 64.3%; the standard deviations were the two highest out of the five products. ITC_0 values used in the model were averages based on a sample size of three. Therefore, it is possible that most participants were consuming less SR initially. With the lower ITC_0 values, the fittings for MedBF and LowBF had lower means squared error (MSE) values and better fits visually. As with k_e, the model function appeared to be stuck in a local minimum when ITC_0 was used at its original values. A better approach for future modeling might be to fit MYR, Cgl_0, and ITC_0 for all participants simultaneously. With the estimated MYR, Cgl_0, and ITC_0 kept constant, the remaining parameters sensitive to the model would be fitted.

More data are needed per participant, >15 data points, to get higher degrees of freedom. The number of data points per participant becomes important the greater the number of parameters being estimated. Obtaining a specified number of urine excretion data points from an intervention study is understandably not easy to achieve. Collecting blood samples during the intervention period in addition to urine samples would provide additional data for the model fittings that may improve parameter estimations as well as increase the number of parameters being estimated.

4.2. Model Fittings

The model succeeded in fitting the data of the various broccoli products well within the experimental error for most of the participants (Figures 4 and S1). It also described the myrosinase mediated conversion of GR in the mouth and the microbial conversion of GR in the gut well. The quick appearance (within 2–3 h after consumption) of the single excretion peaks for HighBP and HighBF products, represents excretion rates of SR that were initially consumed or formed in the mouth, and were absorbed in the small intestine. The NoBF product, which had an insignificant amount of myrosinase, had single peaks that appeared much later (7–8 h after consumption). The NoBF peaks represent excretion rates of SR that was formed by gut bacteria and absorbed only in the large intestine. MedBF and LowBF curves tend to have two peaks. The first peak representing absorption from the small intestine and the second peak from the large intestine. This excretion pattern due to differences in myrosinase content have been observed by other authors [35,36,50,51].

4.3. Bioavailability of Sulforaphane

This compartment model is a good predictor of bioavailability. Other researchers observed that using a compartmental absorption and transit (CAT) model was better at predicting bioavailability than a single compartment model [52].

4.4. Mouth and Gut Parameter Estimations

The parameter estimates clearly show the variability between individuals in the amount of GR converted to SR during mastication and in their gut bacteria activity. The distributions (Tables 4–6) also vary across product categories. The median GR conversion ratio, SRR, is higher for HighBP compared to the HighBF, MedBF, and LowBF broccoli products. 50% of the HighBP participants have conversion ratios similar to what Sarvan et al. [20] found in their study for 0.5 min and 1 min cooked broccoli, where after chewing, 41% and 60% GR was converted to SR. The fact that the overall distribution for HighBP is shifted to higher values compared to HighBF demonstrates the importance of the product matrix. Although, both had the same myrosinase content, most of the participants could convert more GR to SR after consuming the powder. General differences in chewing patterns between individuals have been documented [23]. However, more specific studies on the effect of chewing patterns correlated to conversion GR ratios may provide insights into the variations observed between individuals in the parameter distributions.

The variation between products is significant for the gut parameters, k_f and k_{eni}. The IQR for LowBF and NoBF were expected to be similar since most of the myrosianse was inactivated in both products. However, the NoBF IQR was significantly smaller, which implied that the participants in the Oliviero study had very similar gut bacteria or that their overall bacterial activity was similar. Unfortunately, data on the gut microbial population of the participants was not available to correlate to the parameter estimation results. It is well known that gut bacterial populations differ between individuals and populations [7,53–55]. Forty-seven bacterial species having been identified as having GL metabolizing activities in-vitro [9], but only a few have been investigated for their GL bioconversion mechanisms: *Enterobacter cloacae* [56], *Lactobacillus agilis R16* and *Escherichia coli VL8* [57], *Bacteroides thetaiotaomicorn* [58].

5. Conclusions

A physiological-based multicompartment digestion and absorption model, was developed to describe the kinetics and bioavailability of sulforaphane (SR) from broccoli, and to evaluate how the derived parameters are impacted by inter-individual variation. The model included reactions during digestion in the mouth and gut. It successfully fit participant data and was able to describe bioavailability of SR very well as there were minimal differences between the predicted and experimentally bioavailability. The parameters estimated during the model fitting represented physiological aspects of the digestion process, which were also sources of inter-individual variability. For the digestion of broccoli, the parameters that represented sources for variation between individuals were SRR, the ratio of GR converted to SR during mastication, and k_f and k_{eni}, the conversion rate constants of GR to SR or other break down products. The inter-individual variability between participants was captured in the variability of some these estimates. However, it was not possible to correlate the variability between participants to specific physical attributes, such as chewing patterns or predominant gut microbes, as that information for the participants was not available.

The model's predicted values fit the experimental values very well, especially for the high and low myrosinase products. The lower quality of the fit for the no myrosinase product, indicates the need to improve the model's representation of microbial gut conversions. The work completed in this study is a preliminary step in creating a validated model, which, in the future, could be a useful tool in being able to predict the biological effects of SR and possibly other bioactive compounds. A future predictive model has the potential to positively influence the growing field of personalized nutrition.

Supplementary Materials: The following are available online at https://www.mdpi.com/article/10.3390/foods10112761/s1. Table S1: Compartment Equations, Tables S2–S8: Parameter Box Plots, Figure S1: Model Fittings Results for All Participants, Matlab Codes and Simulink Model

Author Contributions: Q.S.: Conceptualization, methodology, software, investigation, writing—original draft preparation, visualization. M.D.: Conceptualization, methodology, writing—review and editing, supervision. All authors have read and agreed to the published version of the manuscript.

Funding: This research received no external funding.

Data Availability Statement: Not applicable.

Acknowledgments: Frank Sommerhage for his assistance with Matlab.

Conflicts of Interest: The authors declare no conflict of interest.

Appendix A. Myrosinase Calculations

The amount of myrosinase (mg MYR/mg Broccoli) was calculated based on the methods of Oliviero et al. [35] and the research of Roman et al. [42] Oliviero et al. used a spectrophotometric method to determine the myrosinase activity in the different broccoli products. Roman et al. investigated two mechanisms for substrate inhibition during the conversion of GLs to ITCs by modeling their experimental results (enzyme reaction rate vs. sinigrin concentration) using the modified Michaelis–Menten K_i = netics with Substrate Inhibition (MMSI) model (Equation (A1)).

$$v = \frac{V_{max} \cdot [S]}{K_m + [S] + \frac{[S]^2}{K_i}} \tag{A1}$$

V_{max} (μM/min*mg MYR) is the maximum rate of the system, K_m (μM) is the Michaelis-Menten constant, K_i (μM) is the inhibition constant, S is the substrate concentration, and v is the reaction rate. This physiological-based model assumes substrate inhibition occurs in the catalytic site.

Part I. Steps to determining myrosinase content in each broccoli product.

1. Experimental data points were extracted from the reaction rate vs. sinigrin concentration graph [42].

Sinigrin Concentration (μM)	Initial Reaction Rate (μmol/min)
5	0.03
10	0.05
25	0.16
40	0.24
50	0.29
75	0.42
100	0.66
149	0.58
199	0.59
248	0.57
298	0.52

2. The MMSI equation (Equation (A1)) was used to model the data. The parameters, V_{max}, K_m, and K_i, were solved by minimizing the sum of squares difference using Excel's SUMXMY2 function and solver.

Parameter Estimates	
V_{max}	0.96 µM/min
K_m	86.51 µM
K_i	780.05 µM

3. The specific enzyme activity of Myrosinase was calculated using information from the materials Oliviero et al. used to determine activity and the parameter estimates from step 2.

Concentration of sinigrin in reaction mixture used by Oliviero et al. to determine myrosinase activity was calculated as follows:

$$\frac{30\text{ mg}}{\text{mL}} \times \frac{\text{mol}}{397.5\text{ g}} \times \frac{\text{g}}{1000\text{ mg}} \times 10^6 = 75.5\,\frac{\text{umol}}{\text{mL}}$$

$$75.5\,\frac{\mu\text{mol}}{\text{mL}} \times \frac{0.05\text{ mL sinigrin solution}}{1.105\text{ mL}} \times 1000\,\frac{\text{mL}}{\text{L}} = 3415\text{ uM Sinigrin in reaction mixture}$$

Equation (A1) was used to calculate specific enzyme activity

$$\frac{\left(0.96\frac{\mu\text{M}}{\text{min}*\text{mg MYR}}\right) \times 3415\ \mu\text{M}}{86.51\ \mu\text{M} + 3415\ \mu\text{M} + \frac{(3415\ \mu\text{M})^2}{780.05\ \mu\text{M}}} = 0.178\frac{\text{umol}}{\text{mg MYR} * \text{min}}$$

4. The Oliviero *MYR* activity (column A below) for each product type was divided by 5000 mg to determine the µmol MYR/mg broccoli * min (column B). Column B was divided by the specific enzyme activity (0.178 µmol/mg MYR*min) to obtain the mg MYR/mg Broccoli (column C).

	A	B	C
	MYR Activity (Units/5 g dry wt Broccoli)	**µmol/mg Broccoli * min**	**mg MYR/mg Broccoli**
High MYR BP	31	0.0062	0.035
High MYR BF	31	0.0062	0.035
Medium MYR BF	5.8	0.00116	0.007
Low MYR BF	0.5	0.0001	0.000563
No MYR BF (<0.01)	0.01	0.000002	0.0000113

Part II. Estimating V_{max}, K_m, and K_i, for the myrosinase conversion of GR in the mouth.

1. Experimental data points were extracted from the reaction rate vs. glucoraphanin concentration graph [42].

Glucoraphanin Concentration (µM)	Initial Reaction Rate (µmol/min)
5	0.02
8	0.09
10	0.11
25	0.25
50	0.45
75	0.7
90	1.14
100	1.33
150	1.25
200	1.19
250	1.18
300	1.05

2. The MMSI equation (Equation (A1)) was used to model the data. The parameters, V_{max}, K_m, and K_i, were solved by minimizing the sum of squares difference using Excel's SUMXMY2 function and solver.

Parameter Estimates	
V_{max}	2070 µmol/min
K_m	110.16 µM
K_i	893.02 µM

Appendix B. Parameter Estimates Results and Discussion for K_{SH}, K_{tSI}, K_{tLI}, and K_e

Stomach emptying was described as biphasic by having two stomach compartments. The first stomach compartment was for the delay caused by the disintegration of large food particles before it is emptied into the duodenum. Delayed emptying has been observed for wheat products [59] and other solid foods [31,39]. Since participants ate a 30-g raisin bun, right after ingesting 5 g of broccoli florets, a compartment was included to represent disintegration of both food products. K_{SH}, the rate constant from the first stomach compartment to the second, is the inverse of the time required for food disintegration. Parameter estimates for K_{SH} are shown in Table A1. Excluding the outliers, HighBP (IQR = 0.008 min^{-1}) and HighBF (IQR = 0.004 min^{-1}), have normal distributions compared to Med, Low and NoBF products. Comparatively, the median values of HighBP (0.009 min^{-1}) and High BF (0.007 min^{-1}) are smaller than the medians of the MedBF (0.017 min^{-1}), LowBF (0.016 min^{-1}), and NoBF (0.014 min^{-1}) products. These values imply that most participants who consumed MedBF, LowBF and NoBF, required less time for stomach disintegration of the broccoli and the raisin bun than HighBF and HighBP. Since all participants consumed the same amounts of florets and raisin bun, differences in food disintegration were only expected between individuals due to physiological differences. However, the median and distribution differences between the High MYR products and the remaining three were not expected. Kong et al. [60] observed longer disintegration times in their stomach model for raw and 2 min cooked carrots versus their 6 min cooked carrots. Since the high MYR broccoli florets, HighBF, were less heat processed, a harder product texture may be the cause of its longer disintegration time. However, this does not explain why the powdered product, HighBP does not have shorter disintegration times compared to the MedBF, LowBF, and NoBF. It is possible that the raisin bun consumed at the time participants consumed HighBP was harder in texture than when the other broccoli products were consumed.

The transit rate constants for each compartment of the small and large intestine are k_{tSI} and k_{tLI}, respectively. The rate constants were determined based on the measured time it takes for food contents to transit through the small and large intestines. Excluding outliers, the distribution of k_{tSI} parameters for MedBF (IQR = 0.014 min^{-1}), LowBF (IQR = 0.013 min^{-1}), and NoBF (IQR = 0.009 min^{-1}) are less variable compared to the High BP (IQR = 0.098 min^{-1}) and High BF (IQR = 0.057 min^{-1}) broccoli products (Table A2). Except for MedBF (outliers excluded), the parameter distributions for the four products are positively skewed. The median values for all broccoli products are similar, around 0.02 min^{-1}, which is within the range of literature cited values—0.01228 min^{-1}–0.2333 min^{-1} [26]. Across all products transit through the small intestine compartments takes 50 min (1/0.02 min^{-1}) or longer for 50% of the participants. For each product, the percentage of k_{tSI} parameter estimates that fall within the literature cited range are: 100% NoBF, 93% LowBF, 71% MedBF, 67% HighBF, and 73% HighBP. K_{tLI} was not fitted for the HighBP and HighBF; it was fixed at 0.003 min^{-1} for both products (Table A3) because sensitivity analysis showed this parameter did not have an influence on the model's output. The medians for MedBF (0.280 min^{-1}), LowBF (0.220 min^{-1}), and NoBF (0.031 min^{-1}) are larger than literature ranges (0.002 min^{-1}–0.003 min^{-1}) for gut intestinal transit [61]. The parameter distributions are variable especially NoBF in which the interquartile range is 560% of the median. Furthermore, the distribution for NoBF is positively skewed, indicating that 50% of the participants had colon transit times longer than 32 min (1/0.031 min^{-1})

per compartment or 3.8 h for the colon. Based on the skewness, the participants with the long transit times are not as variable as the 50% of participants with transit times shorter than 32 min per compartment. Nine out of the 15 participants in the Oliviero study were women and it is known that women have longer intestinal transit times compared to men [62]. The gender of the participants with their corresponding data were not provided, so it is impossible to correlate transit times to gender.

Table A1. K_{SH} estimation results.

High BP(n = 15) *	High BF(n = 15) *	Med BF(n = 14) *	Low BF(n = 14)	No BF(n = 14)
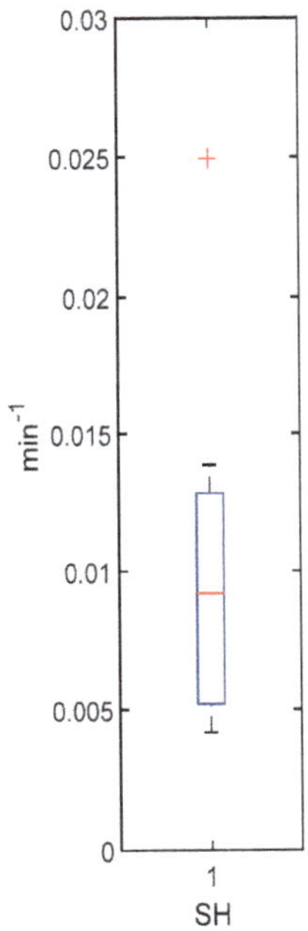	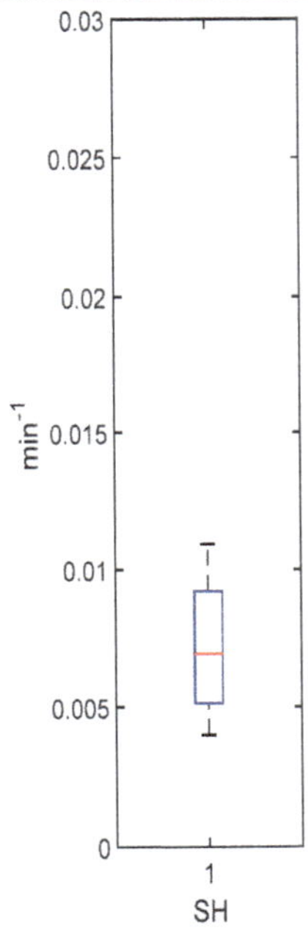	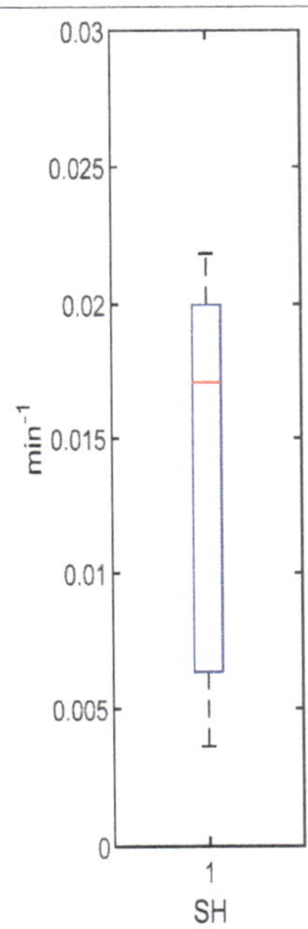	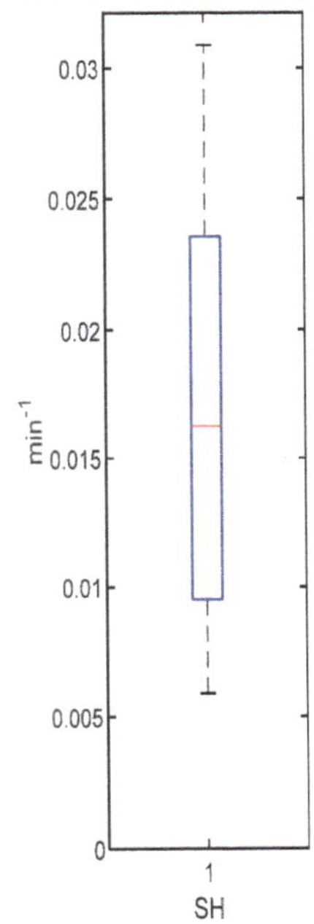	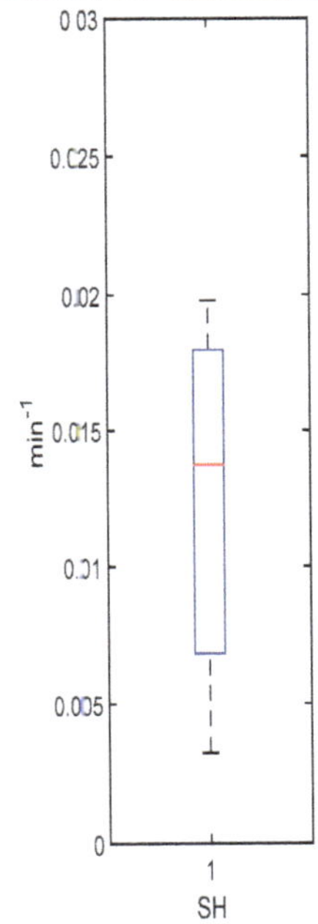
Median: 0.009 min^{-1} IQR: 0.008 min^{-1} IQR/Med: 0.9 Outliers: d (9.408 min^{-1}) l (0.752 min^{-1}) p (0.025min^{-1})	Median: 0.007 min^{-1} IQR: 0.004 min^{-1} IQR/Med: 0.6 Outliers: d (0.105 min^{-1})	Median: 0.017 min^{-1} IQR: 0.014 min^{-1} IQR/Med: 0.8 Outliers: d (0.181 min^{-1})	Median: 0.016 min^{-1} IQR: 0.014 min^{-1} IQR/Med: 0.9 Outliers: None	Median: 0.014 min^{-1} IQR: 0.011 min^{-1} IQR/Med: 0.8 Outliers: None

* Some or all outliers excluded for better visual presentation and comparison of boxplots. Outliers (+) are values more than 1.5 times the IQR. See supplementary tables for plotted outliers.

The rate constant for sulforaphane elimination from the blood, k_e, was not fitted for HighBP and HighBF products, but rather fixed at 0.024 min^{-1} (Table A4). The median rates of elimination for MedBF, LowBF and NoBF are 0.020 min^{-1}, 0.025 min^{-1}, and 0.017 min^{-1}, respectively. The IQR is largest for MedBF (0.043 min^{-1}), followed by LowBF (0.028 min^{-1}) and NoBF (0.01 min^{-1}). Vermeulen et al. [36] fitted the plasma concentration data to one compartmental model and determined the elimination half-lives ($t_{0.5}$) of SR for cooked and raw broccoli to be 4.6 h and 3.8 h, respectively. From half-lives ($t_{0.5}$ = 0.693/ke), the elimination rate constants were calculated to be 0.002511 min^{-1} (cooked) and 0.003039 min^{-1} (raw) [36]. Vermeulen's values are lower than all estimated parameters of MedBF, LowBF, and NoBF products. The difference is most likely due to the method used in determining k_e, because Vermeulen et al. uses a more empirical model than the one described in this thesis.

Table A2. k_{tSI} estimation results.

High BP(*n* = 15) *	**High BF(*n* = 15) ***	**Med BF(*n* = 14) ***	**Low BF(*n* = 14)**	**No BF(*n* = 14)**
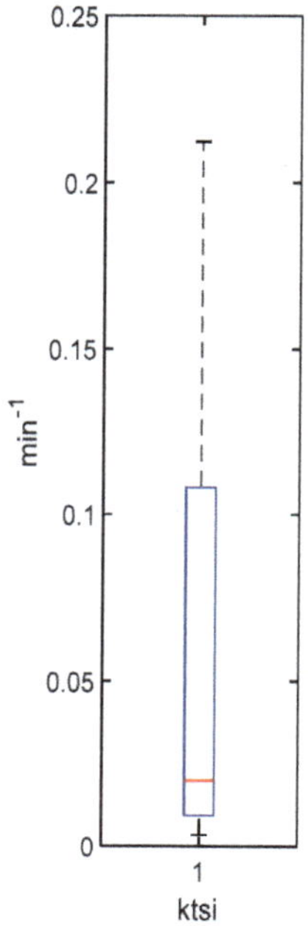	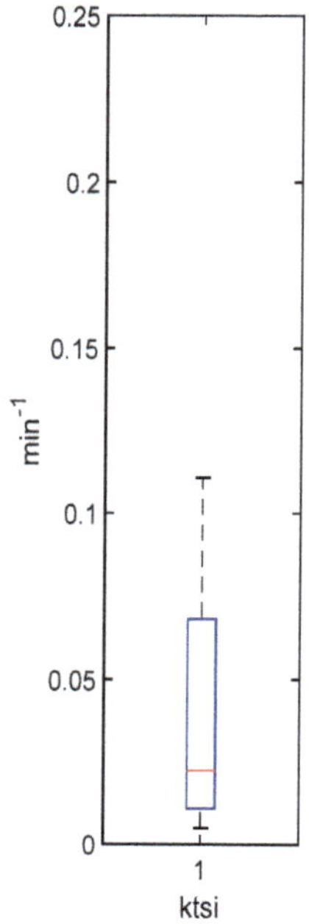	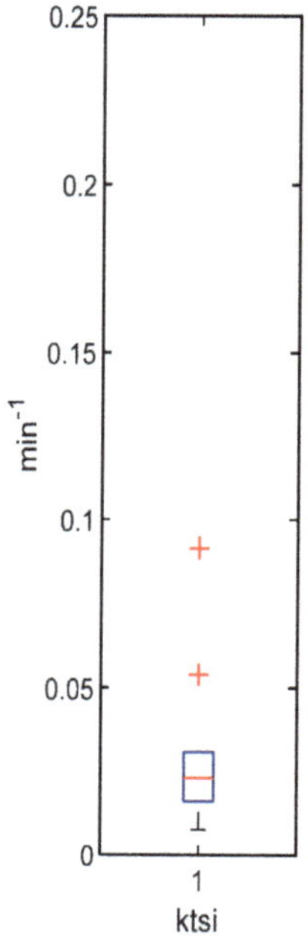	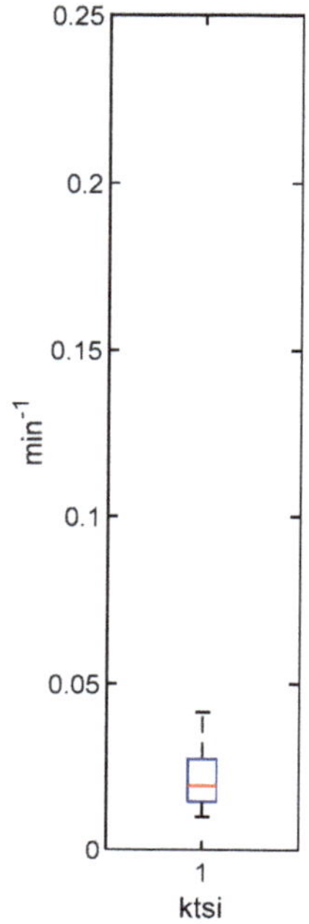	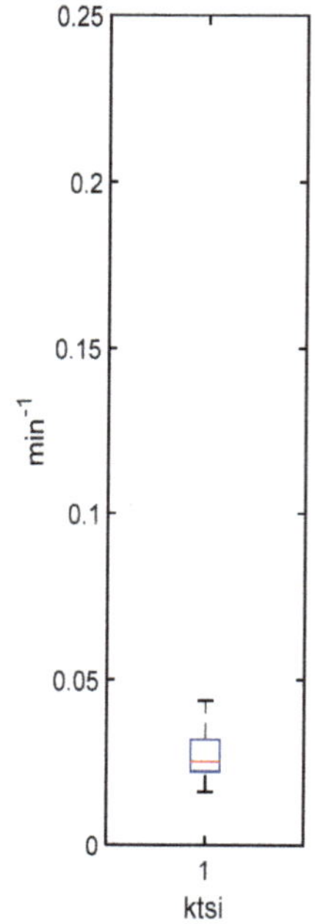
Median: 0.020 min^{-1} IQR: 0.098 min^{-1} IQR/Med: 4.9 Outliers: c (5.192 min^{-1})	Median: 0.023 min^{-1} IQR: 0.057 min^{-1} IQR/Med: 2.5 Outliers: n (1.369 min^{-1})	Median: 0.023 min^{-1} IQR: 0.014 min^{-1} IQR/Med: 0.6 Outliers: h (0.366 min^{-1}) i (0.092 min^{-1}) o (0.054 min^{-1})	Median: 0.020 min^{-1} IQR: 0.013 min^{-1} IQR/Med: 0.7 Outliers: none	Median: 0.026 min^{-1} IQR: 0.009 min^{-1} IQR/Med: 0.3 Outliers: none

* Some or all outliers excluded for better visual presentation and comparison of boxplots. Outliers (+) are values more than 1.5 times the IQR. See supplementary tables for plotted outliers.

Table A3. k_{tLI} estimation results.

High BP(n = 15)	High BF(n = 15)	Med BF(n = 14) *	Low BF(n = 14) *	No BF(n = 14) *
—	—	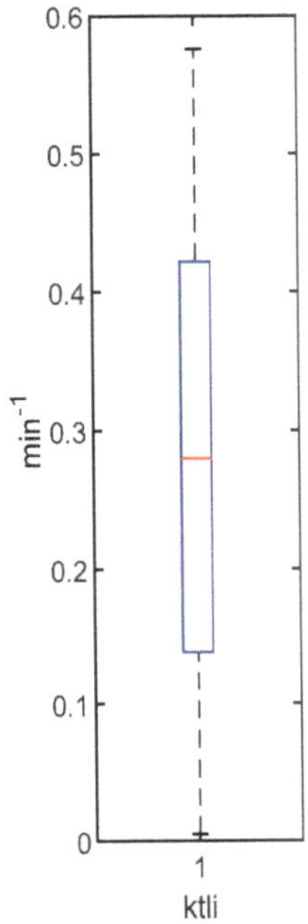	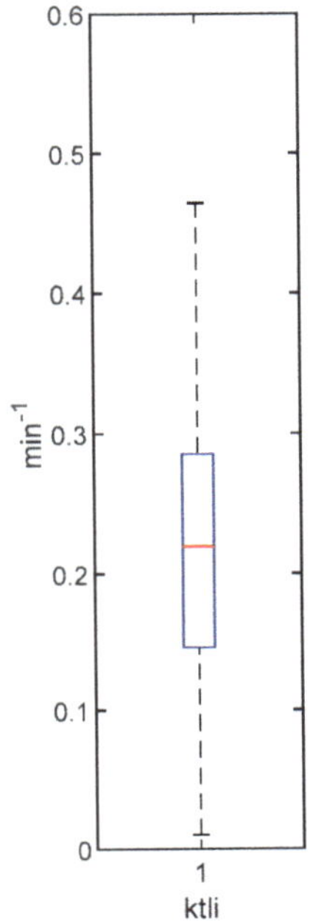	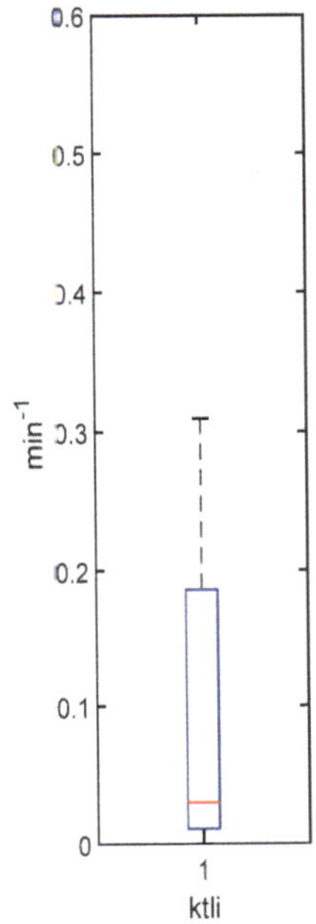
Not fitted. K_{tLI} fixed at 0.003 min^{-1}.	Not fitted. K_{tLI} fixed at 0.003 min^{-1}.	Median: 0.280 min^{-1} IQR: 0.284 min^{-1} IQR/Med: 1.0 Outliers: h (0.96575 min^{-1}) i (0.92453 min^{-1})	Median: 0.220 min^{-1} IQR: 0.139 min^{-1} IQR/Med: 0.6 Outliers: n (0.9998 min^{-1})	Median: 0.031 min^{-1} IQR: 0.175 min^{-1} IQR/Med: 5.6 Outliers: q (1.0385 min^{-1})

* Some or all outliers excluded for better visual presentation and comparison of boxplots. Outliers (+) are values more than 1.5 times the IQR. See supplementary tables for plotted outliers.

Table A4. k_e estimation results.

High BP(n = 15)	High BF(n = 15)	Med BF(n = 14) *	Low BF(n = 14)	No B(n = 14) *
—	—	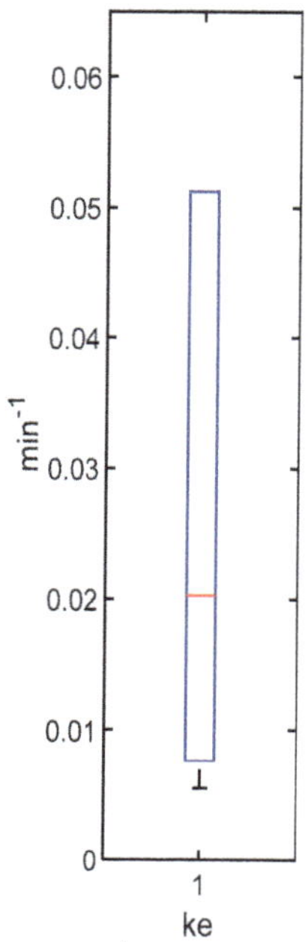	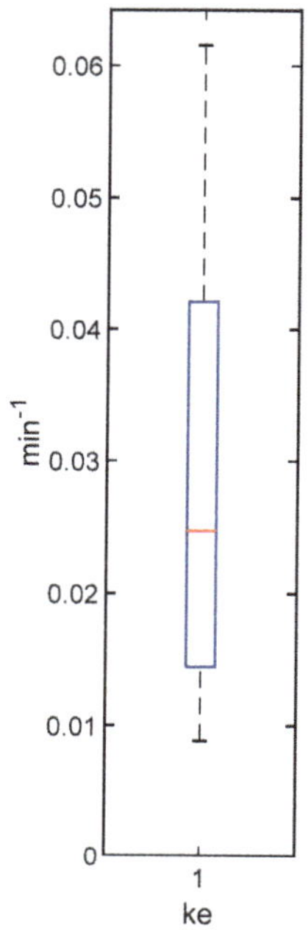	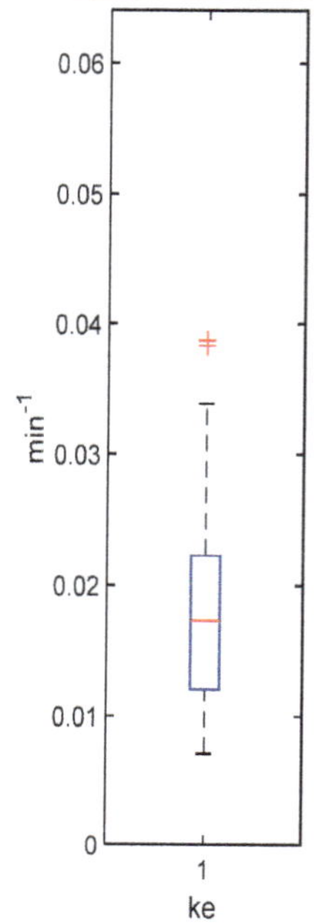
Not fitted. k_e fixed at 0.024 min^{-1}.	Not fitted. k_e fixed at 0.024 min^{-1}.	Median: 0.020 min^{-1} IQR: 0.043 min^{-1} IQR/Med: 2.2 Outliers: c (0.277 min^{-1}) g (0.319 min^{-1}) o (0.295 min^{-1})	Median: 0.025 min^{-1} IQR: 0.028 min^{-1} IQR/Med: 1.1 Outliers: None	Median: 0.017 min^{-1} IQR: 0.01 min^{-1} IQR/Med: 0.6 Outliers: f (0.038 min^{-1}) g (0.039 min^{-1})

* Some or all outliers excluded for better visual presentation and comparison of boxplots. Outliers (+) are values more than 1.5 times the IQR. See supplementary tables for plotted outliers.

References

1. Mérillon, J.M.; Ramawat, K.G. *Glucosinolates;* Springer: Cham, Switzerland, 2017.
2. Dinkova-Kostova, A.T.; Kostov, R.V. Glucosinolates and isothiocyanates in health and disease. *Trends Mol. Med.* **2012**, *18*, 337–347. [CrossRef] [PubMed]
3. Oliviero, T.; Verkerk, R.; Dekker, M. Isothiocyanates from brassica Vegetables-Effects of processing, cooking, mastication, and digestion. *Mol. Nutr. Food Res.* **2018**, *62*, e1701069. [CrossRef]
4. Verkerk, R.; Schreiner, M.; Krumbein, A.; Ciska, E.; Holst, B.; Rowland, I.; De Schrijver, R.; Hansen, M.; Gerhäuser, C.; Mithen, R.; et al. Glucosinolates in brassica vegetables: The influence of the food supply chain on intake, bioavailability and human health. *Mol. Nutr. Food Res.* **2009**, *53*, S219. [CrossRef]
5. Nugrahedi, P.Y.; Verkerk, R.; Widianarko, B.; Dekker, M. A mechanistic perspective on Process-Induced changes in glucosinolate content in brassica vegetables: A review. *Crit. Rev. Food Sci. Nutr.* **2015**, *55*, 823–838. [CrossRef]
6. Saha, S.; Hollands, W.; Teucher, B.; Needs, P.W.; Narbad, A.; Ortori, C.A.; Barrett, D.A.; Rossiter, J.T.; Mithen, R.F.; Kroon, P.A. Isothiocyanate concentrations and interconversion of sulforaphane to erucin in human subjects after consumption of commercial frozen broccoli compared to fresh broccoli. *Mol. Nutr. Food Res.* **2012**, *56*, 1906–1916. [CrossRef]
7. Li, F.; Hullar, M.A.J.; Beresford, S.A.A.; Lampe, J.W. Variation of glucoraphanin metabolism in vivo and ex vivo by human gut bacteria. *Br. J. Nutr.* **2011**, *106*, 408–416. [CrossRef] [PubMed]
8. Mullaney, J.A.; Kelly, W.J.; McGhie, T.K.; Ansell, J.; Heyes, J.A. lactic acid bacteria convert glucosinolates to nitriles efficiently yet differently from enterobacteriaceae. *J. Agric. Food Chem.* **2013**, *61*, 3039–3046. [CrossRef] [PubMed]
9. Narbad, A.; Rossiter, J.T. Gut glucosinolate metabolism and isothiocyanate production. *Mol. Nutr. Food Res.* **2018**, *62*, e1700991. [CrossRef]

10. Palop, M.L.; Smiths, J.P.; ten Brink, B. Degradation of sinigrin by Lactobacillus agilis strain R16. *Int. J. Food Microbiol.* **1995**, *26*, 219–229. [CrossRef]
11. Rabot, S.; Guerin, C.; Nugon-Baudon, L.; Szylit, O. Glucosinolate degradation by bacterial strains isolated from a human intestinal microflora. In *9th International Rapeseed Congress*; GCIRC: Cambridge, UK; Paris, France, 1995.
12. Elfoul, L.; Rabot, S.; Khelifa, N.; Quinsac, A.; Duguay, A.; Rimbault, A. Formation of allyl isothiocyanate from sinigrin in the digestive tract of rats monoassociated with a human colonic strain of Bacteroides thetaiotaomicron. *FEMS Microbiol. Lett.* **2001**, *197*, 99–103. [CrossRef]
13. Luang-In, V.; Narbad, A.; Nueno-Palop, C.; Mithen, R.; Bennett, M.; Rossiter, J.T. The metabolism of methylsulfinylalkyl- and methylthioalkyl-glucosinolates by a selection of human gut bacteria. *Mol. Nutr. Food Res.* **2014**, *58*, 875–883. [CrossRef] [PubMed]
14. Cheng, D.L.; Hashimoto, K.; Uda, Y. In vitro digestion of sinigrin and glucotropaeolin by single strains of Bifidobacterium and identification of the digestive products. *Food Chem. Toxicol.* **2004**, *42*, 351–357. [CrossRef] [PubMed]
15. Kühn, C.; Kupke, F.; Baldermann, S.; Klopsch, R.; Lamy, E.; Hornemann, S.; Pfeiffer, A.F.H.; Schreiner, M.; Hanschen, F.S.; Rohn, S. Diverse excretion pathways of benzyl glucosinolate in humans after consumption of nasturtium (*Tropaeolum majus* L.)—A pilot study. *Mol. Nutr. Food Res.* **2018**, *62*, 1800588. [CrossRef] [PubMed]
16. Rein, M.J.; da Silva Pinto, M. Improvement of bioaccessibility and bioavailability: From molecular interactions to delivery systems. In *Engineering Foods for Bioactives Stability and Delivery*; Roos, Y.H., Livney, Y.D., Eds.; Springer: New York, NY, USA, 2017; pp. 401–416.
17. Capuano, E.; Pellegrini, N. An integrated look at the effect of structure on nutrient bioavailability in plant foods. *J. Sci. Food Agric.* **2019**, *99*, 493–498. [CrossRef]
18. Aguilera, J.M. The food matrix: Implications in processing, nutrition and health. *Crit. Rev. Food Sci. Nutr.* **2019**, *59*, 3612–3629. [CrossRef]
19. Sarvan-Kruse, I.; Kramer, E.; Bouwmeester, H.; Dekker, M.; Verkerk, R. Sulforaphane formation and bioaccessibility are more affected by steaming time than meal composition during in vitro digestion of broccoli. *Food Chem.* **2017**, *214*, 580–586. [CrossRef]
20. Sarvan, I.; van der Klauw, M.; Oliviero, T.; Dekker, M.; Verkerk, R. The effect of chewing on oral glucoraphanin hydrolysis in raw and steamed broccoli. *J. Funct. Foods* **2018**, *45*, 306–312. [CrossRef]
21. Hanschen, F.S.; Kühn, C.; Nickel, M.; Rohn, S.; Dekker, M. Leaching and degradation kinetics of glucosinolates during boiling of Brassica oleracea vegetables and the formation of their breakdown products. *Food Chem.* **2018**, *263*, 240–250. [CrossRef]
22. Matusheski, N.V.; Juvik, J.A.; Jeffery, E.H. Heating decreases epithiospecifier protein activity and increases sulforaphane formation in broccoli. *Phytochemistry* **2004**, *65*, 1273–1281. [CrossRef]
23. Hutchings, S.C.; Foster, K.D.; Bronlund, J.E.; Lentle, R.G.; Jones, J.R.; Morgenstern, M.P. Mastication of heterogeneous foods: Peanuts inside two different food matrices. *Food Qual. Prefer.* **2011**, *22*, 332–339. [CrossRef]
24. Shapiro, T.A.; Fahey, J.W.; Wade, K.L.; Stephenson, K.K.; Talalay, P. Chemoprotective glucosinolates and isothiocyanates of broccoli sprouts. *Cancer Epidemiol. Biomark.* **2001**, *10*, 501.
25. Feunteun, S.L.; Al-Razaz, A.; Dekker, M.; George, E.; Laroche, B.; Van Aken, G. Physiologically based modeling of food digestion and intestinal microbiota: State of the art and future challenges. An INFOGEST review. *Annu. Rev. Food Sci. Technol.* **2021**, *12*, 149–167. [CrossRef] [PubMed]
26. Yu, L.X.; Crison, J.R.; Amidon, G.L. Compartmental transit and dispersion model analysis of small intestinal transit flow in humans. *Int. J. Pharm.* **1996**, *140*, 111–118. [CrossRef]
27. Huang, W.; Lee, S.L.; Yu, L.X. Mechanistic approaches to predicting oral drug absorption. *AAPS J.* **2009**, *11*. 217–224. [CrossRef]
28. Yu, L.X. An integrated model for determining causes of poor oral drug absorption. *Pharm. Res.* **1999**, *16*, 1883–1887. [CrossRef] [PubMed]
29. Punt, A.; Paini, A.; Boersma, M.G.; van Bladeren, P.J.; Rietjens, I.M.C.M.; Delatour, T.; Scholz, G.; Schilter, B.; Freidig, A.P. Use of physiologically based biokinetic (PBBK) modeling to study estragole bioactivation and detoxification in humans as compared with male rats. *Toxicol. Sci.* **2009**, *110*, 255–269. [CrossRef]
30. Punt, A.; Freidig, A.P.; Delatour, T.; Scholz, G.; Boersma, M.G.; Schilter, B.; van Bladeren, P.J.; Rietjens, I.M.C.M. A physiologically based biokinetic (PBBK) model for estragole bioactivation and detoxification in rat. *Toxicol. Appl. Pharmacol.* **2008**, *231*, 248–259. [CrossRef]
31. Le Feunteun, S.; Barbé, F.; Rémond, D.; Ménard, O.; Le Gouar, Y.; Dupont, D.; Laroche, B. Impact of the dairy matrix structure on milk protein digestion kinetics: Mechanistic modelling based on mini-pig in vivo data. *Food Bioprocess. Technol.* **2014**, *7*, 1099–1113. [CrossRef]
32. Strathe, A.B.; Danfær, A.; Chwalibog, A. A dynamic model of digestion and absorption in pigs. *Anim. Feed. Sci. Technol.* **2008**, *143*, 328–371. [CrossRef]
33. Moxon, T.E.; Gouseti, O.; Bakalis, S. In silico modelling of mass transfer & absorption in the human gut. *J. Food Eng.* **2016**, *176*, 110–120.
34. Li, Y.; Zhang, T.; Li, X.; Zou, P.; Schwartz, S.J.; Sun, D. Kinetics of sulforaphane in mice after consumption of sulforaphane-enriched broccoli sprout preparation. *Mol. Nutr. Food Res.* **2013**, *57*, 2128–2136. [CrossRef]
35. Oliviero, T.; Verkerk, R.; Vermeulen, M.; Dekker, M. In vivo formation and bioavailability of isothiocyanates from glucosinolates in broccoli as affected by processing conditions. *Mol. Nutr. Food Res.* **2014**, *58*, 1447–1456. [CrossRef]

36. Vermeulen, M.; Klöpping-Ketelaars, I.W.; van den Berg, R.; Vaes, W.H. Bioavailability and kinetics of sulforaphane in humans after consumption of cooked versus raw broccoli. *J. Agric. Food Chem.* **2008**, *56*, 10505–10509. [CrossRef] [PubMed]
37. Vermeulen, M.; van Rooijen, H.J.; Vaes, W.H. Analysis of isothiocyanate mercapturic acids in urine: A biomarker for cruciferous vegetable intake. *J. Agric. Food Chem.* **2003**, *51*, 3554–3559. [CrossRef] [PubMed]
38. Sreebny, L.M. Saliva in health and disease: An appraisal and update. *Int. Dent. J.* **2000**, *50*, 140–161. [CrossRef] [PubMed]
39. Siegel, J.A.; Urbain, J.L.; Adler, L.P.; Charkes, N.D.; Maurer, A.H.; Krevsky, B.; Knight, L.C.; Fisher, R.S.; Malmud, L.S. Biphasic nature of gastric emptying. *Gut* **1988**, *29*, 85–89. [CrossRef] [PubMed]
40. Mahn, A.; Angulo, A.; Cabañas, F. Purification and characterization of broccoli (*Brassica oleracea* var. *italica) Myrosinase (β-Thioglucosidase Glucohydrolase). J. Agric. Food Chem.* **2014**, *62*, 11666–11671. [CrossRef] [PubMed]
41. Lennernäs, H. Intestinal permeability and its relevance for absorption and elimination. *Xenobiotica* **2007**, *37*, 1015–1051. [CrossRef] [PubMed]
42. Román, J.; Castillo, A.; Cottet, L.; Mahn, A. Kinetic and structural study of broccoli myrosinase and its interaction with different glucosinolates. *Food Chem.* **2018**, *254*, 87–94. [CrossRef] [PubMed]
43. Watanabe, S.; Dawes, C. A comparison of the effects of tasting and chewing foods on the flow rate of whole saliva in man. *Arch. Oral Biol.* **1988**, *33*, 761–764. [CrossRef]
44. Bornhorst, G.M.; Singh, R.P. Bolus formation and disintegration during digestion of food carbohydrates. *Compr. Rev. Food Sci. Food Saf.* **2012**, *11*, 101–118. [CrossRef]
45. Van Liere, E.J.; Sleeth, C.K.; Northup, D. The Relation of the size of the meal to the emptying time of the human stomach. *Am. J. Physiol.-Leg. Content* **1937**, *119*, 480–482. [CrossRef]
46. Yu, A.; Jackson, T.; Tsume, Y.; Koenigsknecht, M.; Wysocki, J.; Marciani, L.; Amidon, G.L.; Frances, A.; Baker, J.R.; Hasler, W.; et al. Mechanistic fluid transport model to estimate gastrointestinal fluid volume and its dynamic change over time. *AAPS J.* **2017**, *19*, 1682–1690. [CrossRef]
47. Helander, H.F.; Fändriks, L. Surface area of the digestive tract–revisited. *Scand. J. Gastroenterol.* **2014**, *49*, 681–689. [CrossRef]
48. Weiss, C.; Jelkmann, W. Functions of the blood. In *Human Physiology*; Schmidt, R.F., Thews, G., Eds.; Springer: Berlin/Heidelberg, Germany, 1989; pp. 402–438.
49. Louis, L.; Harvey, W. Predicting oral drug absorption: Mini review on physiologically-based pharmacokinetic models. *Pharmaceutics* **2017**, *9*, 41.
50. Oliviero, T.; Lamers, S.; Capuano, E.; Dekker, M.; Verkerk, R. Bioavailability of isothiocyanates from broccoli sprouts in protein, lipid, and fiber gels. *Mol. Nutr. Food Res.* **2018**, *62*, 1447–1456. [CrossRef]
51. Conaway, C.C.; Getahun, S.M.; Liebes, L.L.; Pusateri, D.J.; Topham, D.K.; Botero-Omary, M.; Chung, F.L. Disposition of glucosinolates and sulforaphane in humans after ingestion of steamed and fresh broccoli. *Nutr. Cancer* **2000**, *38*, 168–178. [CrossRef] [PubMed]
52. Yu, L.X.; Amidon, G.L. A compartmental absorption and transit model for estimating oral drug absorption. *Int. J. Pharm.* **1999**, *186*, 119–125. [CrossRef]
53. Charron, C.S.; Vinyard, B.T.; Ross, S.A.; Seifried, H.E.; Jeffery, E.H.; Novotny, J.A. Absorption and metabolism of isothiocyanates formed from broccoli glucosinolates: Effects of BMI and daily consumption in a randomised clinical trial. *Br. J. Nutr.* **2018**, *120*, 1370–1379. [CrossRef]
54. Dominianni, C.; Sinha, R.; Goedert, J.J.; Pei, Z.; Yang, L.; Hayes, R.B.; Ahn, J. Sex, body mass index, and dietary fiber intake influence the human gut microbiome. *PLoS ONE* **2015**, *10*, e0124599. [CrossRef] [PubMed]
55. Haro, C.; Rangel-Zúñiga, O.A.; Alcalá-Díaz, J.F.; Gómez-Delgado, F.; Pérez-Martínez, P.; Delgado-Lista, J.; Quintana-Navarro, G.M.; Landa, B.B.; Navas-Cortés, J.A.; Tena-Sempere, M.; et al. Intestinal Microbiota Is Influenced by Gender and Body Mass Index. *PLoS ONE* **2016**, e0154090. [CrossRef]
56. Tani, N.; Ohtsuru, M.; Hata, T. Isolation of myrosinase producing microorganism. *Agric. Biol. Chem.* **1974**, *38*, 1617–1622. [CrossRef]
57. Luang-In, V.; Narbad, A.; Cebeci, F.; Bennett, M.; Rossiter, J.T. Identification of proteins possibly involved in glucosinolate metabolism in *L. agilis* R16 and *E. coli* VL8. *Protein J.* **2015**, *34*, 135–146. [CrossRef] [PubMed]
58. Liou, C.S.; Sirk, S.J.; Diaz, C.A.C.; Klein, A.P.; Fischer, C.R.; Higginbottom, S.K.; Erez, A.; Donia, M.S.; Sonnenburg, J.L.; Sattely, E.S. A Metabolic pathway for activation of dietary glucosinolates by a human gut symbiont. *Cell* **2020**, *180*, 717–728. [CrossRef] [PubMed]
59. Bornhorst, G.; Singh, R.; Heldman, D. Rate kinetics of bread bolus disintegration during in vitro digestion. In *International Congress of Engineering and Food*; Elsevier Procedia: Amsterdam, The Netherlands, 2011.
60. Kong, F.; Singh, R.P. A model stomach system to investigate disintegration kinetics of solid foods during gastric digestion. *J. Food Sci.* **2008**, *73*, 202–210. [CrossRef]
61. Basile, M.; Neri, M.; Carriero, A.; Casciardi, S.; Comani, S.; Del Gratta, C.; Di Donato, L.; Di Luzio, S.; Macri, M.A.; Pasquarelli, A.; et al. Measurement of segmental transit through the gut in man. A novel approach by the biomagnetic method. *Dig. Dis. Sci.* **1992**, *37*, 1537–1543. [CrossRef]
62. Degen, L.P.; Phillips, S.F. Variability of gastrointestinal transit in healthy women and men. *Gut* **1996**, *39*, 299. [CrossRef]

foods

MDPI

Article

Immunomodulating Effect of the Consumption of Watercress (*Nasturtium officinale*) on Exercise-Induced Inflammation in Humans

Hendrik Schulze [1], Johann Hornbacher [2], Paulina Wasserfurth [1], Thomas Reichel [3], Thorben Günther [4], Ulrich Krings [4], Karsten Krüger [3], Andreas Hahn [1], Jutta Papenbrock [2] and Jan P. Schuchardt [1,*]

1 Institute of Food Science and Human Nutrition, Leibniz University Hannover, D-30167 Hannover, Germany; schulze@nutrition.uni-hannover.de (H.S.); wasserfurth@nutrition.uni-hannover.de (P.W.); hahn@nutrition.uni-hannover.de (A.H.)
2 Institute of Botany, Leibniz University Hannover, D-30419 Hannover, Germany; j.hornbacher@botanik.uni-hannover.de (J.H.); jutta.papenbrock@botanik.uni-hannover.de (J.P.)
3 Department of Exercise Physiology and Sports Therapy, Institute of Sport Science, Justus Liebig University Giessen, D-35394 Giessen, Germany; thomas.reichel@sport.uni-giessen.de (T.R.); karsten.krueger@sport.uni-giessen.de (K.K.)
4 Institute of Food Chemistry, Leibniz University Hannover, D-30167 Hannover, Germany; thorben.detering@lci.uni-hannover.de (T.G.); krings@lci.uni-hannover.de (U.K.)
* Correspondence: schuchardt@nutrition.uni-hannover.de; Tel.: +49-511-7622987

Citation: Schulze, H.; Hornbacher, J.; Wasserfurth, P.; Reichel, T.; Günther, T.; Krings, U.; Krüger, K.; Hahn, A.; Papenbrock, J.; Schuchardt, J.P. Immunomodulating Effect of the Consumption of Watercress (*Nasturtium officinale*) on Exercise-Induced Inflammation in Humans. *Foods* **2021**, *10*, 1774. https://doi.org/10.3390/foods10081774

Academic Editors: Franziska S. Hanschen, Sascha Rohn and Laura Jaime

Received: 21 May 2021
Accepted: 28 July 2021
Published: 30 July 2021

Abstract: The vegetable watercress (*Nasturtium officinale* R.Br.) is, besides being a generally nutritious food, a rich source of glucosinolates. Gluconasturtiin, the predominant glucosinolate in watercress, has been shown to have several health beneficial properties through its bioactive breakdown product phenethyl isothiocyanate. Little is known about the immunoregulatory effects of watercress. Moreover, anti-inflammatory effects have mostly been shown in in vitro or in animal models. Hence, we conducted a proof-of-concept study to investigate the effects of watercress on the human immune system. In a cross-over intervention study, 19 healthy subjects (26.5 ± 4.3 years; 14 males, 5 females) were given a single dose (85 g) of fresh self-grown watercress or a control meal. Two hours later, a 30 min high-intensity workout was conducted to promote exercise-induced inflammation. Blood samples were drawn before, 5 min after, and 3 h after the exercise unit. Inflammatory blood markers (IL-1β, IL-6, IL-10, TNF-α, MCP-1, MMP-9) were analyzed in whole blood cultures after ex vivo immune cell stimulation via lipopolysaccharides. A mild pro-inflammatory reaction was observed after watercress consumption indicated by an increase in IL-1β, IL-6, and TNF-α, whereas the immune response was more pronounced for both pro-inflammatory and anti-inflammatory markers (IL-1β, IL-6, IL-10, TNF-α) after the exercise unit compared to the control meal. During the recovery phase, watercress consumption led to a stronger anti-inflammatory downregulation of the pro-inflammatory cytokines IL-6 and TNF-α. In conclusion, we propose that watercress causes a stronger pro-inflammatory response and anti-inflammatory counter-regulation during and after exercise. The clinical relevance of these changes should be verified in future studies.

Keywords: watercress; cruciferous vegetables; glucosinolates; gluconasturtiin; anti-inflammatory; pro-inflammatory

1. Introduction

With the revival of domestic greens watercress (*Nasturtium officinale* R.Br.), a member of the Brassicaceae family, gains a growing interest in science. The semi-aquatic plant species native to Europe and Asia is often consumed as a salad or garnish, or as part of a soup, especially in the Mediterranean kitchen. It is valued for its high nutrient density caused by a low energy content and high amounts of vitamins (B1, B2, B3, B6, C, E), minerals (calcium, iron), and phytochemicals (polyphenols, terpenes) [1–3]. Like all members of the

Brassicaceae family, watercress contains mustard oil glycosides or glucosinolates (GLS), of which gluconasturtiin is the predominant GLS in watercress. As a precursor, it is converted into the bioactive compound phenethyl isothiocyanate (PEITC) upon tissue disruption due to the action of the thioglucosidase myrosinase.

Several studies investigated the health beneficial effects of watercress and PEITC including antioxidative, anti-inflammatory, antidiabetic, anti-allergic, antibacterial, hypolipemic, cardioprotective, and anticancer effects (reviewed in [2]). While most of these effects have been observed in vitro or in animal studies, only a few human intervention studies with watercress have been carried out. Moreover, human intervention studies that administered watercress have mainly focused on antioxidative [4,5] and anticancer effects [6]. The influence of watercress on the immune system, in particular anti-inflammatory activity, has barely been investigated in human studies thus far. There has yet been no confirmation that watercress and its ingredients gluconasturtiin/PEITC act in a similar way in humans compared to effects observed in vitro, namely, by inhibiting the pro-inflammatory nuclear factor kappa B (NfκB) pathway [7]. Because the NfκB pathway can be stimulated directly by reactive oxygen species (ROS) [8] or indirectly by the ROS-dependent heat shock response [9,10], antioxidants might attenuate the exercise-induced inflammation [11]. As a consequence, it is necessary to determine the levels of antioxidants and their capacity in watercress. Moreover, it remains unknown as to whether other gluconasturtiin metabolites are formed in vivo and how they contribute to an antioxidative effect such as that indicated for benzenepropanenitrile [12].

We performed a pilot study with four subjects to examine the effect of a single dose of fresh watercress on various biomarkers of exercise-induced inflammation [13]. On the basis of the results of that previous study, where we observed indications for anti-inflammatory effects, we conducted this follow-up study with a greater number of subjects to further characterize the inflammatory response of watercress consumption. After consuming 85 g of fresh watercress, untrained subjects had to complete a high-intensity workout to induce a pro-inflammatory condition. Inflammatory blood markers (IL-1β, IL-6, IL-10, TNF-α, MCP-1, MMP-9) were analyzed in whole blood cultures after ex vivo immune cell stimulation via lipopolysaccharides (LPS).

2. Materials and Methods

2.1. Plant Material

The administered watercress *(Nasturtium officinale)* was obtained from the Institute of Botany, Leibniz University Hannover. Cuttings were taken for propagation and cultivated in a hydroponic greenhouse system using a Hoagland solution. After 8 weeks, the plant material was harvested freshly on a daily basis. Following a 45 min wet transport, it was cut as little as necessary for consumption.

2.2. Analysis of Glucosinolates by HPLC/LC–MS

GLS were analyzed by HPLC–UV according to Hornbacher et al. [14]. The GLS content of the watercress samples were measured in triplicate. All standard substances were checked for identity. For the identification of the GSL in *N. officinale*, samples were analyzed by liquid chromatography–mass spectrometry (LC–MS). A volume of 10 μL was injected into the HPLC system (Shimadzu, Darmstadt, Germany) and separated on a Knauer Vertex Plus column (250 × 4 mm, 5 μm particle size, packing material ProntoSIL 120-5 C18-H) equipped with a pre-column (Knauer, Berlin, Germany). A water (solvent A)–methanol (solvent B), both containing 2 mM ammonium acetate, gradient was used with a flow rate of 0.8 mL/min at 30 °C. For measuring the samples, the following gradient was used: 10–90% B for 35 min, 90% for 2 min, 90–10% B for 1 min, and 10% B for 2 min. Detection of the spectra in the range 190–800 nm was performed with a diode array detector (SPD-M20A, Shimadzu, Darmstadt, Germany). The HPLC system was coupled to an AB Sciex TripleTOF mass spectrometer (AB Sciex TripleTOF 4600, Canby, OR, USA). At a temperature of 600 °C and an ion spray voltage floating of −4500 V, the negative electrospray ionization (ESI) was performed. For the ion source gas one and two 50 psi were used and for the curtain gas 35 psi. In the range of 100–1500 Da in the TOF range, the mass spectra as well as the MS/MS spectra from 150–1500 Da at a collision energy of −10 eV were recorded. Peaks were identified by analyzing the characteristic mass fragments of ds-4-methoxyglucobrassicin (195, 398, 433, 795) and ds-glucoarabishirsutain (195, 382, 417, 763). Due to lack of standards of the GSLs fractions of the measured samples were collected in a fraction collector (FRC-10A Shimadzu, Darmstadt, Germany), dried in a vacuum centrifuge, and dissolved in 300 µL ultrapure water. The retention time for every GSL was determined by measuring either the collected fraction or the authentic standard (Phytolab, Vestenbergsgreuth, Germany) with the HPLC system, as described above.

2.3. Measurement of Antioxidant Contents and Antioxidant Capacity

The measurements of carotenoid, total phenol and total flavonoid contents, as well as the measurement of the oxygen radical absorbance capacity (ORAC), were performed according to Boestfleisch et al. [15].

Tocopherol contents were analyzed according to Cruz et al. [16] with modifications. Small portions of finely ground fresh sample (1.0 g) were weighed accurately into amber glass vials containing ascorbic acid (50 mg), butylated hydroxytoluene (1 mg), and internal standard (1 µg δ-tocopherol). Samples were homogenized with methanol (2 mL) by vortex mixing for 1 min. Then, dichloromethane (4 mL) was added and vortex-mixed for 1 min. Subsequently, 0.9% (w/v) NaCl (1 mL) was added, the mixture was homogenized (1 min) and centrifuged (3 min, 14,000× g), and the clear lower layer was transferred to an amber flask. Extraction was repeated twice with dichloromethane. The extracts were combined and vacuum-dried in a vacuum centrifuge (Eppendorf, Hamburg, Germany) at 25 °C. The extract was recovered with 1 mL of n-hexane and anhydrous sodium sulfate was added (around 100 mg). After an additional centrifugation (5 min, 14,000× g), the supernatant was analyzed immediately. Analysis was performed with an HPLC system equipped with a Nucleodur C18 column (250 mm × 4.6 mm; Macherey-Nagel, Düren, Germany). Tocopherols were separated with an isocratic gradient consisting of 90% n-hexane and 10% diethyl ether at room temperature and a flow rate of 1 mL/min. Analytes were monitored with a fluorescence detector (Shimadzu, Duisburg, Germany). Excitation was performed at 289 nm and fluorescence of analytes was analyzed at 331 nm.

2.4. Human Study Design and Subjects

An overview of the timeline of the study and the interventions in particular is shown in Figure 1. The inclusion criteria of the cross-over study were age between 18 and 35 years, BMI between 18 and 30 kg/m^2, and less than two hours of moderate exercise per week, classifying these participants as untrained. For the questionnaire-based assessment of the training status, we factored in leisure time physical activities such as jogging or weight-lifting, as well as daily non-athletic exertions such as movement by foot or bike. The exclusion criteria were cardiovascular or metabolic disease, smoking, pregnancy, drug or alcohol dependency, concurrent participation in another clinical trial or in another study within the last 30 days, and intake of antioxidative or antiphlogistic medicine or dietary supplements.

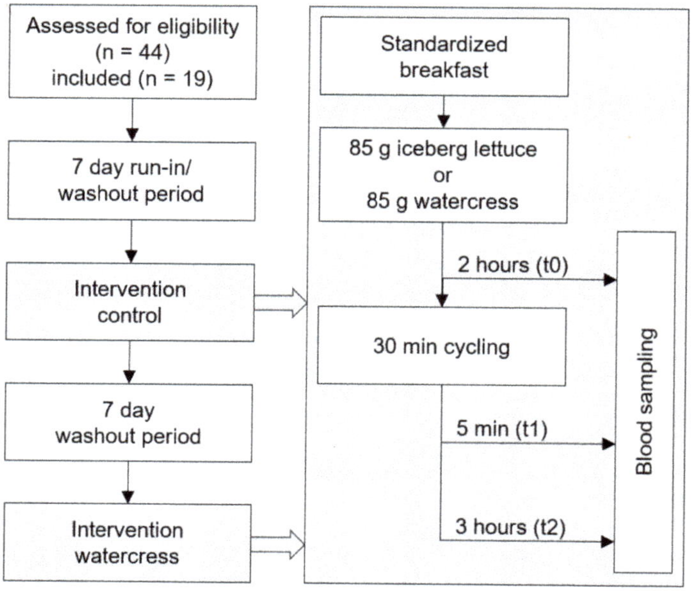

Figure 1. Flow diagram showing the timeline of the study.

As part of a run-in/washout phase, subjects refrained from consuming foods rich in polyphenols, vitamin C and E (mainly berries, nuts, and vegetables of the Brassicaceae family) seven days before each examination. To ensure compliance, participants received written instructions and a list of foods that should not be consumed. On the basis of the consumed amount in similar studies [4,5,13], subjects ate a single dose of 85 g of fresh watercress accompanied by a standard breakfast (two buns, cream cheese or oat spread, yoghurt or balsamic dressing). In the control group, 85 g of iceberg lettuce was administered instead. Two hours after consumption, the first blood sample (t0) was taken. Immediately after blood draw, the subjects completed a 30 min high-intensity endurance workout on echo bikes within a parameter range of 80–92% HR_{max}, 120–145 W, and a final rating of perceived exertion of 17.8. Throughout the workout, heart rates were recorded using a heart rate monitor watch with a Bluetooth heart rate sensor chest strap (RC 14.11, Sigma-Elektro, Neustadt, Germany). Additional blood samples were taken 5 min (t1) and 3 h (t2) after the end of the exercise. The subjects were served a lunch in the meantime consisting of a potato soup. All blood samples were obtained by venipuncture of an arm vein using Multifly needles (Sarstedt, Nürnbrecht, Germany) into heparin plasma monovettes (Sarstedt, Nürnbrecht, Germany). To assure that subjects did not enter the examination with elevated inflammatory markers due to an infection, we determined C-reactive protein in serum using serum monovettes (Sarstedt, Nürnbrecht, Germany). Both interventions (control and watercress) were conducted identically with an intermediary washout phase of 7 days.

The study was carried out following the rules of the Declaration of Helsinki and was approved by the Ethics Committee at the Medical Chamber of Lower Saxony (30/37/2020, Hannover, Germany, 10/2020).

2.5. Measurement of Inflammatory Markers by Bio-Plex Multiplex Immunoassay

The freshly drawn blood samples were stimulated ex vivo in whole blood cultures via lipopolysaccharides (LPS). Therefore, samples were immediately diluted 1:5 with the cell culture medium RPMI 1640 including 20 mmol HEPES and L-glutamine (Sigma-Aldrich, Hamburg, Germany) and added antibiotics (100 U/mL penicillin and 100 µg/mL streptomycin; Sigma-Aldrich, Hamburg, Germany). The samples were seeded into 12-well microtiter plates and mixed with 10 ng/mL (final concentration) LPS from *Escherichia coli* (Sigma-Aldrich, Hamburg, Germany). The plates were incubated for 24 h at 37 °C without a CO_2 application. The used HEPES buffer is able to stabilize the pH over 24 h. The supernatants were frozen at −80 °C until analysis.

The levels of the inflammatory markers (IL-1β, IL-6, IL-10, TNF-α, MCP-1, MMP-9) in whole blood culture supernatant were simultaneously determined using a human Magnetic Luminex Assay (Bio-Techne, Abingdon, Oxon, UK) and a Magpix Luminex instrument (Luminex Corp, Austin, TX, USA).

2.6. Data Analysis and Statistical Methods

Data are presented as the means ± standard deviation. All variables were tested for normal distribution by Shapiro–Wilk test. In the case of not normally distributed data, a suitable transformation was applied, and parametric tests were used. Differences among inflammatory markers were analyzed using ANOVA with repeated measures. In addition, groups were compared using a *t*-test for dependent means. To calculate correlations, we utilized Pearson correlation (parametric data) and Spearman's rho correlation (non-parametric data). Statistical significance was regarded as values of $p \leq 0.05$. Analyses were conducted using Infostat (version 2012; University of Córdoba, Córdoba, Argentine) and SPSS (version 27; SPSS Inc., Chicago, IL, USA).

3. Results

Of the 21 recruited subjects, 19 completed the study (Table 1). Incomplete study data were excluded from statistical analyses. One subject failed to participate because of illness. Another subject showed increased serum levels of C-reactive protein (>0.5 mg/L), indicating an elevated systemic inflammation. No health- or workout-related incidents occurred during the study, with the exception of one subject taking a two-minute break from the exercise due to total exhaustion.

Table 1. Characterization of the study population.

Parameters	
Sex (*n*, females/males)	5/14
Age (years)	26.5 ± 4.3
Weight (kg)	72.9 ± 12.5
Height (m)	1.77 ± 0.09
BMI (kg/m^2)	23.3 ± 3.5
WHR (females/males)	0.74 ± 0.04/0.84 ± 0.05

BMI = body mass index, WHR = waist-to-hip ratio.

3.1. Plant Material

3.1.1. Levels of Glucosinolates

Gluconasturtiin was recognized as the predominant GLS in watercress with a fraction of 90.5 ± 1.1% of the total GLS content. In addition, minor amounts of glucoarabishirsutain (5.3 ± 0.8%), glucobrassicin (1.4 ± 0.4%), neoglucobrassicin (1.4 ± 0.3%), and 4-methoxyglucobrassicin (1.4 ± 0.4%) were found (Figure 2, Table A1). In the course of the study, GLS contents varied notably with the most impactful difference in gluconasturtiin of 36% at day 3 compared to the previous day. This resulted in a difference of 34% of the total GLS content while maintaining the partial composition of the GLS profile.

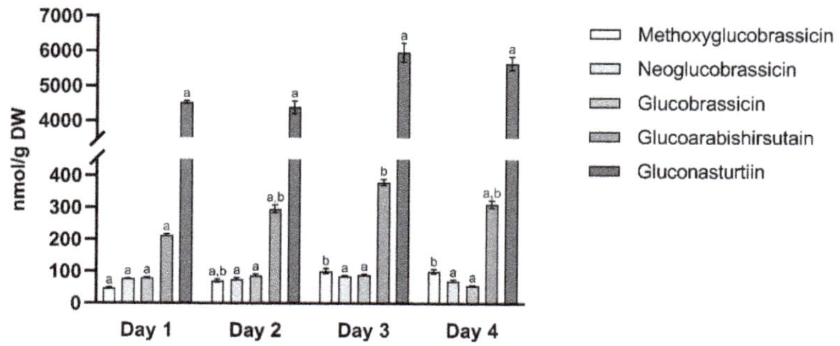

Figure 2. Mean concentration of different glucosinolates (GLS) (nmol/g DW) in watercress at different sampling times. The standard deviation represents the values for three technical replicates. Analysis of variance (ANOVA) was performed with Infostat. Means with a common letter are not significantly different.

3.1.2. Levels of Antioxidants and Antioxidant Capacity

Contents of flavonoids as well as ascorbic acid in analyzed watercress were similar at all sampling days (Table A2). Contents of total phenols were slightly higher in samples taken at day 3 and day 4, whereas carotenoid contents were slightly higher at day 2 and day 4. Tocopherol contents as well as ORAC were similar at all sampling days.

3.2. *Levels of Inflammatory Blood Markers*

Statistical analysis using the ANOVA with repeated measures showed significant differences between the sampling times for all analyzed parameters ($p \leq 0.001$). Hence, the acute exercise affected concentrations of all measured inflammatory markers. In addition, the same analysis pointed out that the consumption of watercress led to significant differences for the levels of IL-1β, IL-6, and IL-10 across all sampling times (IL-1β ($p = 0.006$), IL-6 ($p = 0.006$), IL-10 ($p \leq 0.001$)). The remaining parameters showed no significant differences (TNF-α ($p = 0.382$), MCP-1 ($p = 0.192$), MMP-9 ($p = 0.118$)). With regards of the varying GLS levels of the plant material, no correlations with the inflammatory markers were found.

Significantly higher concentrations of the pro-inflammatory cytokines IL-1β (16%), IL-6 (33%), and TNF-α (30%), as well as the enzyme MMP-9 (22%), were observed in the watercress group compared to the control group two hours after the watercress consumption (t0) (Figure 3, Table A3). Upon exercise-stimulation (t1), the control group showed a significant rise in all inflammatory markers, specifically IL-6 (33%), IL-10 (47%), MCP-1 (53%), and MMP-9 (53%) with IL-1β and TNF-α showing no reaction. Compared to the control breakfast, levels of IL-1β, IL-6, IL-10, and TNF-α were significantly higher after the watercress breakfast (31%, 32%, 51%, and 23%, respectively). To determine if the consumption of watercress resulted in a stronger increase of the cytokines regardless of the pre-exercise levels, we compared the differences (t1 − t0). Thereby, a significant stronger upregulation of the anti-inflammatory cytokine IL-10 was found. After the recovery phase (t2), all inflammatory markers except MMP-9 decreased. A comparison of the differences (t2 − t1) between the watercress and the control group revealed a significantly stronger downregulation of the pro-inflammatory cytokines IL-6 and TNF-α and a trend in IL-1β ($p = 0.062$). In the case of TNF-α, the level at t2 was even lower compared to the pre-exercise state of the control group. This post-exercise downregulation can also be observed in the ratio of the anti-inflammatory IL-10 and the pro-inflammatory IL-1β (Figure 4). The watercress consumption influenced the ratio 3 h after exercise towards the anti-inflammatory reaction.

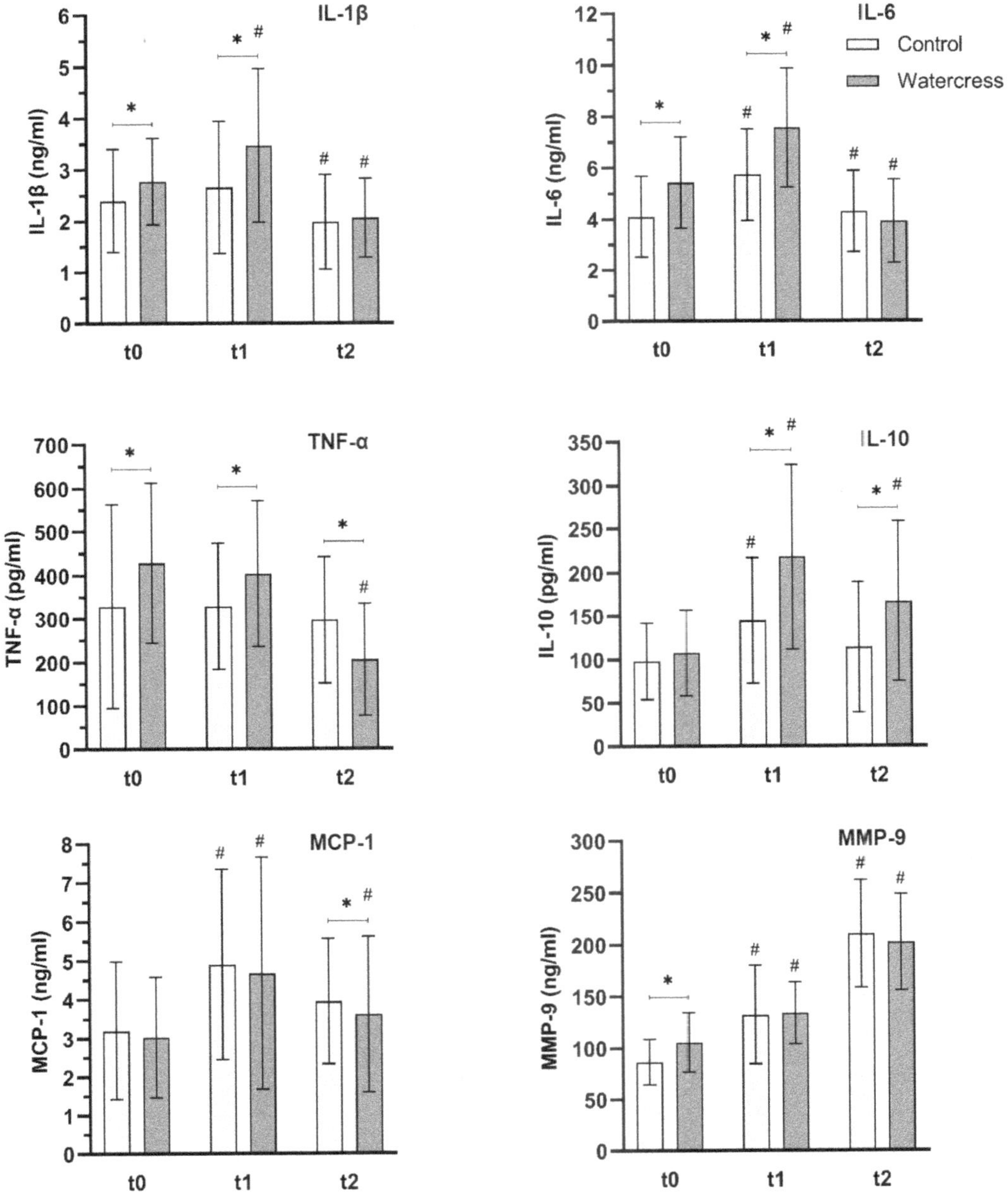

Figure 3. Effect of acute watercress consumption on blood markers of inflammation (IL-1β, IL-6, IL-10, TNF-α, MCP-1, MMP-9) in ex vivo LPS-stimulated whole blood cultures after high-intensity workout in untrained subjects. t0, pre-exercise; t1, 5 min post-exercise; t2, 3 h post-exercise. Analysis of variance (ANOVA) was performed with SPSS. * Significant difference between control and watercress. # Significant difference to previous sampling point.

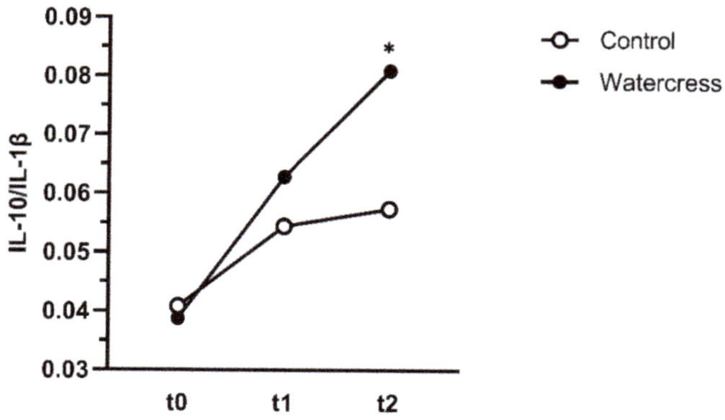

Figure 4. Effect of acute watercress consumption on the ratio of IL-10 (anti-inflammatory) and IL-1β (pro-inflammatory). t0, pre-exercise; t1, 5 min post-exercise; t1, 3 h post-exercise. Analysis of variance (ANOVA) was performed with SPSS. * Significant difference between control and watercress.

4. Discussion

Many food compounds are known to assure the maintenance of the immune system or improve its performance [17]. By interacting with ROS or affecting cytokine biology, their intake can modulate immune function. A wide variety of anti-inflammatory nutrients has been investigated thus far. While many act as antioxidants and, hence, indirectly modulate the ROS-dependent NfκB activation [18] such as ascorbic acid, glutathione, or carotenoids, others operate as pro-resolving mediators such as omega-3 fatty acids [19].

A common misconception is that pro-inflammatory processes at all times need to be annihilated to prevent the body from harm. Similar to oxidative stress, pro-inflammatory processes can have both detrimental and beneficial effects to the human body. When inflammatory processes become chronic, they often have negative impacts. Chronic low-grade inflammation is associated with a wide range of chronic conditions, such as the metabolic syndrome, cardiovascular disease, type 2 diabetes, and non-alcoholic fatty liver disease. However, if the body is able to resolve inflammatory processes, they can provide signals for adaptation in physiological contexts such as sport [20]. In its acute form, it is generally a beneficial procedure, which removes stimuli and initiates the repair system. As the first line of defense, pro-inflammatory cytokines such as IL-1β, IL-6, and TNF-α promote the activation and secretion of more cytokines and acute-phase proteins as well as the proliferation and differentiation of T- and B-cells. The time-delayed increase of the cytokine IL-10 has been widely recognized as a suppression of the inflammatory response by its downregulating effects on TNF-α and IL-1 [21–23]. The immune system acts through a fluctuation between a pro-inflammatory response and an anti-inflammatory or inflammation resolving counter-regulation. An increase in the magnitude of the fluctuation could be interpreted as positive for the capacity of the system. Thus, a mild activation of the immune system through dietary components might aid in resolving inflammation by a preceding mobilization. We assume that this immunomodulating effect also applies to watercress and its ingredients.

Upon exercise- and LPS-stimulation, a solid, non-excessive immune reaction was observed in IL-6 and IL-10, with a greater response after the watercress consumption. Interestingly, the subsequent inflammatory counter-regulation of IL-6, TNF-α, and IL-1β was more pronounced in the watercress group, which likely was a result of the higher exercise-induced levels of IL-10. This shift towards an anti-inflammatory response after the watercress consumption was supported by the increased ratio of IL-10 and IL-1β. Our results suggest that watercress intake stimulates the immune system, which might enhance its metabolic capacity. The finding that the consumption of watercress causes an initial pro-inflammatory reaction followed by a greater exercise-induced response in both pro- and anti-inflammatory markers is remarkable and has not been observed thus far. We assume that gluconasturtiin, in particular PEITC, is responsible for the observed immunomodulating effect, where PEITC might operate as an activator in T-lymphocytes or macrophages or their receptors and thereby stimulates or sensitizes the cytokine production. It is also conceivable that PEITC activates the heat shock response, which leads to a stimulation of the pro-inflammatory NfκB pathway [24]. Another explanation is that specific components of watercress might interfere with the immunometabolism and thereby stimulate its function. Although the average contents of secondary metabolites besides gluconasturtiin and the ORAC are in the lower range of vegetables, it is likely that the observed effects are mainly caused by the GLS. The extraction procedure for the evaluation of the ORAC uses methanol as organic solvent, which inhibits the hydrolysis of gluconasturtiin to PEITC. However, PEITC would contribute only very little to the overall ORAC, since its capacity to scavenge ROS was reported to be 1.9 μg TE/mg PEITC [25]. The levels of gluconasturtiin in the obtained watercress resemble the results of the pilot study [13] and were in the same range of reported contents for raw watercress [26]

In contrast to our results, PEITC and watercress extracts have thus far been shown to possess only anti-inflammatory and no pro-inflammatory properties [27–31]. The pilot study, on which this investigation is based on, aligns with these observations [13]. In vitro studies presume the mechanism behind this effect of PEITC in the inhibition of the pro-inflammatory NfκB pathway in macrophages, possibly through the modulation of toll-like receptors [31–33]. Due to the redox-sensitivity of NfκB, the activation of the antioxidative nuclear factor erythroid 2-related factor 2 (Nrf2) pathway by PEITC plays a considerable role in its anti-inflammatory effect [34]. Previous human intervention studies have focused mainly on those antioxidative effects of watercress. They showed that a regular consumption as well as a single dose of watercress reduces the oxidative stress in various biomarkers [4,5]. On the basis of the antioxidative and thereby assumed anti-inflammatory effects, the outcome of this study does not align with the previous literature. Although PEITC is able to promote oxidative stress at very high concentrations, presumably acting as a scavenger for glutathione [35], it remains debatable as to whether a single dose of 85 g of watercress provides the body with the necessary range of concentration. In consequence of the redox-sensitivity of NfκB, an initial pro-oxidative effect after the consumption of watercress could explain the simultaneously increased cytokine levels of IL-1β, IL-6, and TNF-α. The pro-oxidative state would be further enhanced by the exercise, which subsequently results in an even more pronounced pro-inflammatory response in the watercress group and a stronger induction of the counter-regulatory Nrf2-mediated antioxidant response that is observed in the recovery phase.

5. Conclusions

The course of inflammatory markers after the watercress consumption with initially increasing pro-inflammatory markers and a higher release of an anti-inflammatory marker in the recovery phase has not yet been described in the literature. We interpret this observation with a mild activation of the immune system resulting in a stronger pro-inflammatory reaction, which is more effectively resolved by a powerful anti-inflammatory counter-regulation. This thesis and the clinical relevance have to be investigated in future studies. Moreover, cell culture studies are necessary to explore the underlying effects of watercress components on leukocytes. Likewise, it must not be ignored that the immune cells were stimulated ex vivo by LPS. The results and the possible effects on the immune system can therefore not be directly transferred to the situation in vivo. Because watercress is a complex food with considerable amounts of other potentially immunomodulating substances besides PEITC, a causal relationship between the observed effects and gluconasturtiin/PEITC cannot be stated. A clinical trial with isolated gluconasturtiin/PEITC should be conducted in order to confirm the effect and whether its magnitude is influenced by other components. The variable GLS levels of watercress show the need for developing standardized extracts or supplements. Thereby, future clinical trials are provided with a standardized GLS dosage and can overcome logistic and sensory barriers of administering raw watercress.

Author Contributions: Conceptualization, K.K., A.H. and J.P.S.; methodology, J.H., K.K. and J.P.S.; formal analysis, H.S., J.H. and T.R.; investigation, H.S., J.H., P.W., T.R., U.K. and T.G.; resources, J.H., K.K., A.H. and J.P.S.; writing—original draft preparation, H.S. and J.P.S.; writing—review and editing, H.S., J.H., J.P., K.K. and J.P.S.; visualization, H.S.; supervision, J.P. and J.P.S.; project administration, J.P.S. All authors have read and agreed to the published version of the manuscript.

Funding: This research received no external funding.

Institutional Review Board Statement: The study was conducted according to the guidelines of the Declaration of Helsinki, and approved by the Ethics Committee at the Medical Chamber of Lower Saxony (30/37/2020, Hannover, Germany, 10/2020).

Informed Consent Statement: Informed consent was obtained from all subjects involved in the study.

Acknowledgments: We would like to thank Julia Volker for help in the laboratory. The publication of this article was funded by the Open Access Fund of Leibniz University Hannover.

Conflicts of Interest: The authors declare no conflict of interest.

Appendix A

Table A1. Mean concentration of different glucosinolates (GLS) (nmol/g DW) in watercress at different sampling times. The standard deviation represents the values for three technical replicates. Analysis of variance (ANOVA) was performed with Infostat. Means with a common letter are not significantly different.

GLS	Day 1	Day 2	Day 3	Day 4
4-Methoxyglucobrassicin	46.8 ± 2.1 [a]	70.6 ± 5.0 [a,b]	101.3 ± 8.3 [b]	101.5 ± 6.3 [b]
Glucobrassicin	79.3 ± 2.2 [a]	87.8 ± 3.2 [a]	90.5 ± 1.6 [a]	56.6 ± 1.3 [a]
Neoglucobrassicin	76.6 ± 1.4 [a]	76.4 ± 3.7 [a]	85.5 ± 1.7 [a]	71.3 ± 3.3 [a]
Glucoarabishirsutain	213.5 ± 2.6 [a]	296.4 ± 12.4 [a,b]	379.6 ± 10.1 [b]	311.8 ± 11.8 [a,b]
Gluconasturtiin	4525 ± 45 [a]	4393 ± 180 [a]	5966 ± 266 [a]	5655 ± 196 [a]
Total	4942 ± 46 [a]	4924 ± 201 [a]	6623 ± 285 [a]	6196 ± 215 [a]

Table A2. Mean contents of antioxidative substances (total flavonoids and phenols, ascorbic acid, α- and γ-tocopherol) and antioxidative capacity (ORAC) in watercress at different sampling times. The standard deviation represents the values for three technical replicates. Analysis of variance (ANOVA) was performed with Infostat. Means with a common letter are not significantly different.

Parameters	Day 1	Day 2	Day 3	Day 4
Total flavonoids (ng CE/g FW)	415 ± 20 [a]	470 ± 43 [a,b]	509 ± 49 [a,b]	529 ± 46 [b]
Total phenols (ng GAE/g FW)	832 ± 41 [a]	980 ± 90 [a,b]	1145 ± 42 [b]	1178 ± 128 [b]
Ascorbic acid (μg/g FW)	578 ± 31 [a]	638 ± 55 [a]	750 ± 85 [a]	687 ± 31 [a]
Carotenoids (μg/g FW)	84.9 ± 8.8 [a]	108.9 ± 7.7 [a]	90.3 ± 22.4 [a]	118.9 ± 16.3 [a]
α-Tocopherol (μg/g FW)	3.63 ± 0.02 [a]	5.11 ± 0.11 [b]	5.58 ± 0.17 [b]	5.27 ± 0.03 [b]
γ-Tocopherol (ng/g FW)	70.8 ± 5.5 [a]	86.9 ± 0.4 [a]	72.4 ± 3.6 [a]	60.6 ± 5.1 [a]
ORAC (μmol TE/g FW)	19.6 ± 2.7 [a]	20.3 ± 1.1 [a]	21.4 ± 1.5 [a]	20.7 ± 2.4 [a]

CE = catechin equivalents (standard curve was performed with catechin), GAE = gallic acid equivalents (standard curve was performed with gallic acid), ORAC = oxygen radical absorbance capacity, TE = trolox equivalents (standard curve was performed with trolox).

Table A3. Effect of acute watercress consumption on blood markers of inflammation (IL-1β, IL-6, IL-10, TNF-α, MCP-1, MMP-9) in ex vivo LPS-stimulated whole blood cultures after high-intensity workout in untrained subjects.

Parameter	Control			Watercress		
	t0	t1	t2	t0	t1	t2
IL-1β (ng/mL)	2.40 ± 0.98 *	2.66 ± 1.26 *	1.98 ± 0.90 #	2.77 ± 0.82 *	3.47 ± 1.45 *,#	2.06 ± 0.75 #
IL-6 (ng/mL)	4.09 ± 1.54 *	5.73 ± 1.73 *,#	4.29 ± 1.55 #	5.43 ± 1.73 *	7.56 ± 2.25 *,#	3.91 ± 1.59 #
IL-10 (pg/mL)	98.0 ± 42.6	144.5 ± 70.3 *,#	113.7 ± 73.2 *	107.6 ± 48.1	217.8 ± 103.3 *,#	166.6 ± 89.2 *,#
TNF-α (pg/mL)	328.3 ± 228.1 *	328.0 ± 140.5 *	296.6 ± 141.1 *	428.0 ± 179.3 *	403.4 ± 163.2 *	205.3 ± 126.1 *#
MCP-1 (ng/mL)	3.20 ± 1.73	4.89 ± 2.39 #	3.94 ± 1.57 *,#	3.02 ± 1.52	4.67 ± 2.91	3.60 ± 1.96 *,#
MMP-9 (ng mL)	86.6 ± 21.5 *	132.2 ± 46.5 #	210.2 ± 50.3 #	105.5 ± 28.3 *	133.9 ± 29.2 #	202.0 ± 45.2 #

t0, pre-exercise; t1, 5 min post-exercise; t2, 3 h post-exercise. * Significant difference between control and watercress. # Significant difference to previous sampling point.

References

1. Di Noia, J. Defining Powerhouse Fruits and Vegetables: A Nutrient Density Approach. *Prev. Chronic Dis.* **2014**, *11*, 130390. [CrossRef]
2. Klimek-Szczykutowicz, M.; Szopa, A.; Ekiert, H. Chemical Composition, Traditional and Professional Use in Medicine, Application in Environmental Protection, Position in Food and Cosmetics Industries, and Biotechnological Studies of Nasturtium Officinale (Watercress)—A Review. *Fitoterapia* **2018**, *129*, 283–292. [CrossRef] [PubMed]
3. de Souza, H.C.; dos Santos, A.M.P.; Fortunato, D.M.N.; Lima, D.C.; Fragoso, W.D.; Ferreira, S.L.C. Determination of the Mineral Composition of Watercress and Data Evaluation Using Multivariate Analysis. *Anal. Lett.* **2011**, *44*, 1758–1768. [CrossRef]
4. Fogarty, M.C.; Hughes, C.M.; Burke, G.; Brown, J.C.; Davison, G.W. Acute and Chronic Watercress Supplementation Attenuates Exercise-Induced Peripheral Mononuclear Cell DNA Damage and Lipid Peroxidation. *Br. J. Nutr.* **2013**, *109*, 293–301. [CrossRef]
5. Gill, C.I.; Haldar, S.; Boyd, L.A.; Bennett, R.; Whiteford, J.; Butler, M.; Pearson, J.R.; Bradbury, I.; Rowland, I.R. Watercress Supplementation in Diet Reduces Lymphocyte DNA Damage and Alters Blood Antioxidant Status in Healthy Adults. *Am. J. Clin. Nutr.* **2007**, *85*, 504–510. [CrossRef]
6. Hofmann, T.; Kuhnert, A.; Schubert, A.; Gill, C.; Rowland, I.R.; Pool-Zobel, B.L.; Glei, M. Modulation of Detoxification Enzymes by Watercress: In Vitro and in Vivo Investigations in Human Peripheral Blood Cells. *Eur. J. Nutr.* **2009**, *48*, 483–491. [CrossRef]
7. Cheung, K.L.; Kong, A.-N. Molecular Targets of Dietary Phenethyl Isothiocyanate and Sulforaphane for Cancer Chemoprevention. *AAPS J.* **2010**, *12*, 87–97. [CrossRef]
8. Schoonbroodt, S.; Ferreira, V.; Best-Belpomme, M.; Boelaert, J.R.; Legrand-Poels, S.; Korner, M.; Piette, J. Crucial Role of the Amino-Terminal Tyrosine Residue 42 and the Carboxyl-Terminal PEST Domain of I KBα in NF-KB Activation by an Oxidative Stress. *J. Immunol.* **2000**, *164*, 4292–4300. [CrossRef] [PubMed]
9. Fehrenbach, E.; Niess, A.M. Role of Heat Shock Proteins in the Exercise Response. *Exerc. Immunol. Rev.* **1999**, *5*, 57–77. [PubMed]
10. Asea, A.; Rehli, M.; Kabingu, E.; Boch, J.A.; Bare, O.; Auron, P.E.; Stevenson, M.A.; Calderwood, S.K. Novel Signal Transduction Pathway Utilized by Extracellular HSP70: Role of Toll-like Receptor (TLR) 2 and TLR4. *J. Biol. Chem.* **2002**, *277*, 15028–15034. [CrossRef] [PubMed]
11. Vassilakopoulos, T.; Karatza, M.-H.; Katsaounou, P.; Kollintza, A.; Zakynthinos, S.; Roussos, C. Antioxidants Attenuate the Plasma Cytokine Response to Exercise in Humans. *J. Appl. Physiol.* **2003**, *94*, 1025–1032. [CrossRef]
12. Azarmehr, N.; Afshar, P.; Moradi, M.; Sadeghi, H.; Sadeghi, H.; Alipoor, B.; Khalvati, B.; Barmoudeh, Z.; Abbaszadeh-Goudarzi, K.; Doustimotlagh, A.H. Hepatoprotective and Antioxidant Activity of Watercress Extract on Acetaminophen-Induced Hepatotoxicity in Rats. *Heliyon* **2019**, *5*, e02072. [CrossRef] [PubMed]
13. Schuchardt, J.P.; Hahn, A.; Greupner, T.; Wasserfurth, P.; Rosales-López, M.; Hornbacher, J.; Papenbrock, J. Watercress—Cultivation Methods and Health Effects. *J. Appl. Bot. Food Qual.* **2019**, *92*, 232–239. [CrossRef]
14. Hornbacher, J.; Rumlow, A.; Pallmann, P.; Turcios, A.E.; Riemenschneider, A.; Papenbrock, J. The Levels of Sulfur-Containing Metabolites in Brassica Napus Are Not Influenced by the Circadian Clock but Diurnally. *J. Plant Biol.* **2019**, *62*, 359–373. [CrossRef]
15. Boestfleisch, C.; Wagenseil, N.B.; Buhmann, A.K.; Seal, C.E.; Wade, E.M.; Muscolo, A.; Papenbrock, J. Manipulating the Antioxidant Capacity of Halophytes to Increase Their Cultural and Economic Value through Saline Cultivation. *AoB Plants* **2014**, *6*, plu046. [CrossRef] [PubMed]
16. Cruz, R.; Casal, S. Validation of a Fast and Accurate Chromatographic Method for Detailed Quantification of Vitamin E in Green Leafy Vegetables. *Food Chem.* **2013**, *141*, 1175–1180. [CrossRef] [PubMed]
17. Wu, D.; Lewis, E.D.; Pae, M.; Meydani, S.N. Nutritional Modulation of Immune Function: Analysis of Evidence, Mechanisms, and Clinical Relevance. *Front. Immunol.* **2019**, *9*, 1–19. [CrossRef]
18. Schreck, R.; Rieber, P.; Baeuerle, P.A. Reactive Oxygen Intermediates as Apparently Widely Used Messengers in the Activation of the NF-Kappa B Transcription Factor and HIV-1. *EMBO J.* **1991**, *10*, 2247–2258. [CrossRef]
19. Norling, L.V.; Ly, L.; Dalli, J. Resolving Inflammation by Using Nutrition Therapy: Roles for Specialized Pro-Resolving Mediators. *Curr. Opin. Clin. Nutr. Metab. Care* **2017**, *20*, 145–152. [CrossRef]

20. Gomez-Cabrera, M.C.; Viña, J.; Ji, L.L. Role of Redox Signaling and Inflammation in Skeletal Muscle Adaptations to Training. *Antioxidants* **2016**, *5*, 48. [CrossRef]
21. Cassatella, M.A.; Meda, L.; Bonora, S.; Ceska, M.; Constantin, G. Interleukin 10 (IL-10) Inhibits the Release of Proinflammatory Cytokines from Human Polymorphonuclear Leukocytes. Evidence for an Autocrine Role of Tumor Necrosis Factor and IL-1 Beta in Mediating the Production of IL-8 Triggered by Lipopolysaccharide. *J. Exp. Med.* **1993**, *178*, 2207–2211. [CrossRef]
22. Gérard, C.; Bruyns, C.; Marchant, A.; Abramowicz, D.; Vandenabeele, P.; Delvaux, A.; Fiers, W.; Goldman, M.; Velu, T. Interleukin 10 Reduces the Release of Tumor Necrosis Factor and Prevents Lethality in Experimental Endotoxemia. *J. Exp. Med.* **1993**, *177*, 547–550. [CrossRef]
23. O'Farrell, A.-M.; Liu, Y.; Moore, K.W.; Mui, A.L.-F. IL-10 Inhibits Macrophage Activation and Proliferation by Distinct Signaling Mechanisms: Evidence for Stat3-Dependent and -Independent Pathways. *EMBO J.* **1998**, *17*, 1006–1018. [CrossRef]
24. Naidu, S.D.; Suzuki, T.; Yamamoto, M.; Fahey, J.W.; Dinkova-Kostova, A.T. Phenethyl Isothiocyanate, a Dual Activator of Transcription Factors NRF2 and HSF1. *Mol. Nutr. Food Res.* **2018**, *62*, 1700908. [CrossRef] [PubMed]
25. Coscueta, E.R.; Reis, C.A.; Pintado, M. Phenylethyl Isothiocyanate Extracted from Watercress By-Products with Aqueous Micellar Systems: Development and Optimisation. *Antioxidants* **2020**, *9*, 698. [CrossRef]
26. Giallourou, N.; Oruna-Concha, M.J.; Harbourne, N. Effects of Domestic Processing Methods on the Phytochemical Content of Watercress (Nasturtium Officinale). *Food Chem.* **2016**, *212*, 411–419. [CrossRef] [PubMed]
27. Shahani, S.; Behzadfar, F.; Jahani, D.; Ghasemi, M.; Shaki, F. Antioxidant and Anti-Inflammatory Effects of Nasturtium Officinale Involved in Attenuation of Gentamicin-Induced Nephrotoxicity. *Toxicol. Mech. Methods* **2017**, *27*, 107–114. [CrossRef] [PubMed]
28. Tsai, J.-T.; Liu, H.-C.; Chen, Y.-H. Suppression of Inflammatory Mediators by Cruciferous Vegetable-Derived Indole-3-Carbinol and Phenylethyl Isothiocyanate in Lipopolysaccharide-Activated Macrophages. *Mediat. Inflamm.* **2010**, *2010*, 1–5. [CrossRef]
29. Sadeghi, H.; Mostafazadeh, M.; Sadeghi, H.; Naderian, M.; Barmak, M.J.; Talebianpoor, M.S.; Mehraban, F. In Vivo Anti-Inflammatory Properties of Aerial Parts of Nasturtium Officinale. *Pharm. Biol.* **2014**, *52*, 169–174. [CrossRef] [PubMed]
30. Rose, P.; Won, Y.K.; Ong, C.N.; Whiteman, M. β-Phenylethyl and 8-Methylsulphinyloctyl Isothiocyanates, Constituents of Watercress, Suppress LPS Induced Production of Nitric Oxide and Prostaglandin E2 in RAW 264.7 Macrophages. *Nitric Oxide* **2005**, *12*, 237–243. [CrossRef]
31. Park, H.-J.; Kim, S.-J.; Park, S.-J.; Eom, S.-H.; Gu, G.-J.; Kim, S.H.; Youn, H.-S. Phenethyl Isothiocyanate Regulates Inflammation through Suppression of the TRIF-Dependent Signaling Pathway of Toll-like Receptors. *Life Sci.* **2013**, *92*, 793–798. [CrossRef]
32. Xu, C.; Shen, G.; Chen, C.; Gélinas, C.; Kong, A.-N.T. Suppression of NF-KB and NF-KB-Regulated Gene Expression by Sulforaphane and PEITC through IκBα, IKK Pathway in Human Prostate Cancer PC-3 Cells. *Oncogene* **2005**, *24*, 4486–4495. [CrossRef] [PubMed]
33. Prawan, A.; Saw, C.L.L.; Khor, T.O.; Keum, Y.-S.; Yu, S.; Hu, L.; Kong, A.-N. Anti-NF-KB and Anti-Inflammatory Activities of Synthetic Isothiocyanates: Effect of Chemical Structures and Cellular Signaling. *Chem. Biol. Interact.* **2009**, *179*, 202–211. [CrossRef] [PubMed]
34. Boyanapalli, S.S.S.; Paredes-Gonzalez, X.; Fuentes, F.; Zhang, C.; Guo, Y.; Pung, D.; Saw, C.L.L.; Kong, A.-N.T. Nrf2 Knockout Attenuates the Anti-Inflammatory Effects of Phenethyl Isothiocyanate and Curcumin. *Chem. Res. Toxicol.* **2014**, *27*, 2036–2043. [CrossRef]
35. Kassie, F.; Knasmüller, S. Genotoxic Effects of Allyl Isothiocyanate (AITC) and Phenethyl Isothiocyanate (PEITC). *Chem. Biol. Interact.* **2000**, *127*, 163–180. [CrossRef]

MDPI
St. Alban-Anlage 66
4052 Basel
Switzerland
Tel. +41 61 683 77 34
Fax +41 61 302 89 18
www.mdpi.com

Foods Editorial Office
E-mail: foods@mdpi.com
www.mdpi.com/journal/foods

www.ingramcontent.com/pod-product-compliance
Lightning Source LLC
Chambersburg PA
CBHW040827250926
51850CB00007BA/1283
9783036529776